Computational Analysis of Communication

Computational Analysis of Communication

A Practical Introduction to the Analysis of Texts, Networks, and Images with Code Examples in Python and R

Wouter van Atteveldt
Vrije Universiteit Amsterdam
Amsterdam, Netherlands

Damian Trilling
University of Amsterdam
Amsterdam, Netherlands

Carlos Arcila Calderón
University of Salamanca
Salamanca, Spain

WILEY Blackwell

This edition first published 2022

Registered Office
John Wiley & Sons, Inc., 111 River Street, Hoboken, NJ 07030, USA

Editorial Office
The Atrium, Southern Gate, Chichester, West Sussex, PO19 8SQ, UK

For details of our global editorial offices, customer services, and more information about Wiley products visit us at www.wiley.com.

Wiley also publishes its books in a variety of electronic formats and by print-on-demand. Some content that appears in standard print versions of this book may not be available in other formats.

For general information on our other products and services or for technical support, please contact our Customer Care Department within the United States at (800) 762-2974, outside the United States at (317) 572-3993 or fax (317) 572-4002.

Library of Congress Cataloging-in-Publication Data
Names: van Atteveldt, Wouter, author. | Trilling, Damian, 1983- author. | Calderón, Carlos Arcila, author.
Title: Computational analysis of communication : a practical introduction to the analysis of texts, networks, and images with code examples in Python and R / Wouter van Atteveldt, Vrije Universiteit Amsterdam, Damian Trilling, University of Amersterdam, Carlos Arcila Calderón, University of Salamanca.
Description: Hoboken, NJ : John Wiley & Sons, [2022] | Includes bibliographical references and index.
Identifiers: LCCN 2021058779 (print) | LCCN 2021058780 (ebook) | ISBN 9781119680239 (paperback) | ISBN 9781119680277 (pdf) | ISBN 9781119680284 (epub)
Subjects: LCSH: Social sciences--Network analysis. | Communication--Network analysis. | Computational linguistics--Network analysis. | Communication--Data processing.
Classification: LCC HM741 .A88 2021 (print) | LCC HM741 (ebook) | DDC 302.3072--dc23
LC record available at https://lccn.loc.gov/2021058779
LC ebook record available at https://lccn.loc.gov/2021058780

Cover image: © 4X image/Getty Images
Cover design by Wiley

Set in 9.5/12.5pt STIXTwoText by Integra Software Services Pvt. Ltd, Pondicherry, India
Printed and bound by CPI Group (UK) Ltd, Croydon, CR0 4YY

C106161_250122

To our patient spouses

Contents

Preface *xi*
Acknowledgments *xiii*

1 Introduction *1*
1.1 The Role of Computational Analysis in the Social Sciences *1*
1.2 Why Python and/or R? *3*
1.3 How to Use This Book *4*
1.4 Installing R and Python *5*
1.4.1 Installing R and RStudio *7*
1.4.2 Installing Python and Jupyter Notebook *9*
1.5 Installing Third-Party Packages *12*

2 Getting Started: Fun with Data and Visualizations *13*
2.1 Fun With Tweets *14*
2.2 Fun With Textual Data *15*
2.3 Fun With Visualizing Geographic Information *17*
2.4 Fun With Networks *19*

3 Programming Concepts for Data Analysis *23*
3.1 About Objects and Data Types *24*
3.1.1 Storing Single Values: Integers, Floating-Point Numbers, Booleans *25*
3.1.2 Storing Text *26*
3.1.3 Combining Multiple Values: Lists, Vectors, And Friends *28*
3.1.4 Dictionaries *32*
3.1.5 From One to More Dimensions: Matrices and n-Dimensional Arrays *33*
3.1.6 Making Life Easier: Data Frames *34*
3.2 Simple Control Structures: Loops and Conditions *35*
3.2.1 Loops *36*
3.2.2 Conditional Statements *37*
3.3 Functions and Methods *39*

4 How to Write Code *43*
4.1 Re-using Code: How Not to Re-Invent the Wheel *43*
4.2 Understanding Errors and Getting Help *46*
4.2.1 Error Messages *46*

4.2.2 Debugging Strategies *48*
4.3 Best Practice: Beautiful Code, GitHub, and Notebooks *49*

5 From File to Data Frame and Back *55*
5.1 Why and When Do We Use Data Frames? *56*
5.2 Reading and Saving Data *57*
5.2.1 The Role of Files *57*
5.2.2 Encodings and Dialects *59*
5.2.3 File Handling Beyond Data Frames *61*
5.3 Data from Online Sources *62*

6 Data Wrangling *65*
6.1 Filtering, Selecting, and Renaming *66*
6.2 Calculating Values *67*
6.3 Grouping and Aggregating *69*
6.3.1 Combining Multiple Operations *70*
6.3.2 Adding Summary Values *71*
6.4 Merging Data *72*
6.4.1 Equal Units of Analysis *72*
6.4.2 Inner and Outer Joins *75*
6.4.3 Nested Data *76*
6.5 Reshaping Data: Wide To Long And Long To Wide *78*
6.6 Restructuring Messy Data *79*

7 Exploratory Data Analysis *83*
7.1 Simple Exploratory Data Analysis *84*
7.2 Visualizing Data *87*
7.2.1 Plotting Frequencies and Distributions *88*
7.2.2 Plotting Relationships *92*
7.2.3 Plotting Geospatial Data *98*
7.2.4 Other Possibilities *99*
7.3 Clustering and Dimensionality Reduction *100*
7.3.1 *k*-means Clustering *101*
7.3.2 Hierarchical Clustering *102*
7.3.3 Principal Component Analysis and Singular Value Decomposition *106*

8 Statistical Modeling and Supervised Machine Learning *113*
8.1 Statistical Modeling and Prediction *115*
8.2 Concepts and Principles *117*
8.3 Classical Machine Learning: From Naïve Bayes to Neural Networks *122*
8.3.1 Naïve Bayes *122*
8.3.2 Logistic Regression *124*
8.3.3 Support Vector Machines *125*
8.3.4 Decision Trees and Random Forests *127*
8.3.5 Neural Networks *129*
8.4 Deep Learning *130*
8.4.1 Convolutional Neural Networks *131*

8.5 Validation and Best Practices *133*
8.5.1 Finding a Balance Between Precision and Recall *133*
8.5.2 Train, Validate, Test *137*
8.5.3 Cross-validation and Grid Search *138*

9 Processing Text *141*
9.1 Text as a String of Characters *142*
9.1.1 Methods for Dealing With Text *144*
9.2 Regular Expressions *145*
9.2.1 Regular Expression Syntax *146*
9.2.2 Example Patterns *147*
9.3 Using Regular Expressions in Python and R *150*
9.3.1 Splitting and Joining Strings, and Extracting Multiple Matches *151*

10 Text as Data *155*
10.1 The Bag of Words and the Term-Document Matrix *156*
10.1.1 Tokenization *157*
10.1.2 The DTM as a Sparse Matrix *159*
10.1.3 The DTM as a "Bag of Words" *162*
10.1.4 The (Unavoidable) Word Cloud *163*
10.2 Weighting and Selecting Documents and Terms *164*
10.2.1 Removing stop words *165*
10.2.2 Removing Punctuation and Noise *167*
10.2.3 Trimming a DTM *170*
10.2.4 Weighting a DTM *171*
10.3 Advanced Representation of Text *172*
10.3.1 *n*-grams *173*
10.2.3 Collocations *174*
10.3.3 Word Embeddings *176*
10.3.4 Linguistic Preprocessing *177*
10.4 Which Preprocessing to Use? *182*

11 Automatic Analysis of Text *184*
11.1 Deciding on the Right Method *185*
11.2 Obtaining a Review Dataset *187*
11.3 Dictionary Approaches to Text Analysis *189*
11.4 Supervised Text Analysis: Automatic Classification and Sentiment Analysis *191*
11.4.1 Putting Together a Workflow *191*
11.4.2 Finding the Best Classifier *194*
11.4.3 Using the Model *198*
11.4.4 Deep Learning *199*
11.5 Unsupervised Text Analysis: Topic Modeling *203*
11.5.1 Latent Dirichlet Allocation (LDA) *203*
11.5.2 Fitting an LDA Model *206*
11.5.3 Analyzing Topic Model Results *207*
11.5.4 Validating and Inspecting Topic Models *208*
11.5.5 Beyond LDA *209*

12 Scraping Online Data *212*
12.1 Using Web APIs: From Open Resources to Twitter *213*
12.2 Retrieving and Parsing Web Pages *219*
12.2.1 Retrieving and Parsing an HTML Page *219*
12.2.2 Crawling Websites *223*
12.2.3 Dynamic Web Pages *225*
12.3 Authentication, Cookies, and Sessions *228*
12.3.1 Authentication and APIs *228*
12.3.2 Authentication and Webpages *229*
12.4 Ethical, Legal, and Practical Considerations *230*

13 Network Data *233*
13.1 Representing and Visualizing Networks *234*
13.2 Social Network Analysis *241*
13.2.1 Paths and Reachability *242*
13.2.2 Centrality Measures *246*
13.2.3 Clustering and Community Detection *248*

14 Multimedia Data *258*
14.1 Beyond Text Analysis: Images, Audio and Video *259*
14.2 Using Existing Libraries and APIs *261*
14.3 Storing, Representing, and Converting Images *263*
14.4 Image Classification *270*
14.4.1 Basic Classification with Shallow Algorithms *272*
14.4.2 Deep Learning for Image Analysis *273*
14.4.3 Re-using an Open Source CNN *279*

15 Scaling Up and Distributing *283*
15.1 Storing Data in SQL and noSQL Databases *283*
15.1.1 When to Use a Database *283*
15.1.2 Choosing the Right Database *285*
15.1.3 A Brief Example Using SQLite *286*
15.2 Using Cloud Computing *286*
15.3 Publishing Your Source *290*
15.4 Distributing Your Software as Container *291*

16 Where to Go Next *293*
16.1 How Far Have We Come? *293*
16.2 Where To Go Next? *294*
16.3 Open, Transparent, and Ethical Computational Science *295*

Bibliography *297*
Index *303*

Preface

Why write another methods textbook? Aren't there enough textbooks already? And what about all the great online resources? We have been teaching computational analysis of communication for years for various universities and other organizations. These courses used different formats, ranged from semester-long courses to short workshops, used different techniques, and were taught at different levels – but we never found the book that really fit our audience. Regularly, students and colleagues ask us for book recommendations, and educators and administrators want to know which book to put on a reading list. And regularly, our answer was one along the lines of: Well, there is this great book on [R/Python/Neural Networks/...], but ....

The "but", in almost all cases, has to do with the audience: students of the social sciences who have at least some knowledge of and are interested in empirical research and quantitative methods, but have no experience in programming. They do want (or have to) learn programming to conduct the analyses they are interested in, but are not necessarily interested in programming for its own sake. They do not want to just push a button in some tool that limits their possibilities to what someone else has designed, but they also do not want to follow a whole Introduction to Computer Science with a comprehensive overview of programming concepts and paradigms that they might never need.

For years, we have therefore used our own materials to find a balance between teaching programming concepts where necessary but focussing on their application for answering questions that are of genuine interest to those studying various forms of communication. This book is our attempt to bring together and systematize this approach to teaching the analysis of communication.

A second driver for writing this book was to get over the "language war" that is sometimes visible in the field. In our own research and teaching, we find both R and Python to be great tools, each with their strengths and weaknesses. Too often, existing teaching materials focus on the language rather than the underlying concept. We believe that a good computational methods textbook should give practical instructions on the implementation of a concept in a given language, but put the concept rather than the language at the forefront. For that reason, we decided to use R and Python side by side, allowing students (and professors) to choose either – and to allow interested readers to view the differences and similarities between the languages.

Writing this book has also been an exercise in planning and coordination. With two of us being located in Amsterdam and one in Salamanca, we had many video calls to divide tasks and discuss each other's drafts. One can hardly call us tech-adverse, but nothing is as productive (and nice) as sitting together in a room, as we experienced during a writing weekend on the island of Texel. The COVID-19 pandemic, though, cancelled all plans for further in-person writing, and with suddenly many other unexpected priorities emerging, it took more time – and many more online meetings – for the final version of the book to see the light of the world.

This book would not have been possible without the continuous input we got over years – from students, colleagues, and others. They shaped our ideas on both how to analyze communication computationally, but also our ideas about how to teach this. It would also not have been possible without the patience of Nel, Rodrigo, and Sanne, when we again had to spend more hours than we thought on what at one point only became known as "the book".

Wouter van Atteveldt
Damian Trilling
Carlos Arcila Calderón
Amsterdam, Salamanca, Texel, & online

Acknowledgments

We would like to thank colleagues, friends, and students who provided feedback and input on earlier versions of parts of the manuscript: Dmitry Bogdanov, Andreu Casas, Modesto Escobar, Anne Kroon, Nicolas Mattis, Cecil Meeusen, Jesús Sánchez-Oro, Nel Ruigrok, Susan Vermeer, Mehdi Zamani, Rodrigo de la Barra, and Holli Semetko. We also want to thank the editors, copy-editors, and the other great people at Wiley as well as the initial reviewers for their help and confidence.

For an earlier version of the example for web scraping with Selenium, we would like to thank Marthe Möller.

And, of course, all others that we might have forgotten to mention here (sorry!).

1

Introduction

Abstract

This chapter explains how the methods outlined in this book are situated within the methodological and epistemological frameworks used by social scientists. It argues why the use of Python and R is fundamental for the computational analysis of communication. Finally, it shows how this book can be used by students and scholars.

Keywords computational social science, Python, R

- Understand the role of computational analysis in the social sciences
- Understand the choice between Python and/or R
- Know how to read this book

1.1 The Role of Computational Analysis in the Social Sciences

The use of computers is nothing new in the social sciences. In fact, one could argue that some disciplines within the social sciences have even been early adopters of computational approaches. Take the gathering and analyzing of large-scale survey data, dating back to the use of the Hollerith Machine in the 1890 US census. Long before every scholar had a personal computer on their desk, social scientists were using punch cards and mainframe computers to deal with such data. If we think of the analysis of *communication* more specifically, we already see attempts to automate content analysis in the 1960's (see, e.g. Scharkow, 2017).

However, something has profoundly changed in recent decades. The amount and type of data we can collect as well as the computational power we have access to have increased dramatically. In particular, digital traces that we leave when communicating online, from access logs to comments we place, have required new approaches (e.g., Trilling, 2017). At the same time, better computational facilities now allow us to ask questions we could not answer before.

González-Bailón (2017), for instance, argued that the computational analysis of communication now allows us to test theories that were formulated a century ago, such as Tarde's theory of social imitation. Salganik (2019) tells an impressive methodological story of continuity in showing how new digital research methods build on and relate

Computational Analysis of Communication: A Practical Introduction to the Analysis of Texts, Networks, and Images with Code Examples in Python and R, First Edition. Wouter van Atteveldt, Damian Trilling & Carlos Arcila Calderón.

to established methods such as surveys and experiments, while offering new possibilities by observing behavior in new ways.

A frequent misunderstanding, then, about computational approaches is that they would somehow be a-theoretical. This is probably fueled by clichés coined during the "Big Data"-hype in the 2010's, such as the infamous saying that in the age of Big Data, correlation is enough (Mayer-Schönberger and Cukier, 2013); but one could not be more wrong: as the work of Kitchin (2014a, b) shows, computational approaches can be well situated within existing epistemologies. For the field to advance, computational and theoretical work should be symbiotic, with each informing the other and with neither claiming superiority (Margolin, 2019). Thus, the computational scientists' toolbox includes both more data-driven and more theory-driven techniques; some are more bottom-up and inductive, others are more top-down and deductive. What matters here, and what is often overlooked, is in which stage of the research process they are employed. In other words, both inductive and deductive approaches as they are distinguished in more traditional social-science textbooks (e.g., Bryman, 2012) have their equivalent in the computational social sciences.

Therefore, we suggest that the data collection and data analysis process is thought of as a pipeline. To test, for instance, a theoretically grounded hypothesis about personalization in the news, we could imagine a pipeline that starts with scraping online news, proceeds with some natural-language processing techniques such as Named Entity Recognition, and finally tests whether the mentioning of persons has an influence on the placement of the stories. We can distinguish here between parts of the pipeline that are just necessary but not inherently interesting to us, and parts of the pipeline that answer a genuinely interesting question. In this example, the inner workings of the Named Entity Recognition step are not genuinely interesting for us – we just need to do it to answer our question. We do care about how well it works and especially which biases it may have that could affect our substantive outcomes, but we are not really evaluating any theory on Named Entity Recognition here. We are, however, answering a theoretically interesting question when we look at the pipeline as a whole, that is, when we apply the tools in order to tackle a social scientific problem. Of course, what is genuinely interesting depends on one's discipline: For a computational linguist, the inner workings of the named entity recognition may actually be the interesting part, and our research question just one possible "downstream task".

This distinction is also sometimes referred to as "building a better mousetrap" versus "understanding". For instance, Breiman (2001) remarked: "My attitude toward new and/or complicated methods is pragmatic. Prove that you've got a better mousetrap and I'll buy it. But the proof had better be concrete and convincing." (p. 230). In contrast, many social scientists are using statistical models to test theories and to understand social processes: they want to specifically understand how x relates to y, even if y may be better predicted by another (theoretically uninteresting) variable.

This book is to some extent about both building mousetraps and understanding. When you are building a supervised machine learning classifier to determine the topic of each text in a large collection of news articles or parliamentary speeches, you are building a (better) mousetrap. But as a social scientist, your work does not stop there. You need to use the mousetrap to answer some theoretically interesting question.

Actually, we expect that the contents of this book will provide a background that helps you to face the current research challenges in both academia and industry. On the one hand, the emerging field of Computational Social Science has become one of the most promising areas of knowledge and many universities and research institutes are looking for scholars with this profile. On the other hand, it is widely known that nowadays the computational skills will increase your job opportunities in private companies, public organizations or NGOs, given the growing interest in data-driven solutions.

When planning this book, we needed to make a couple of tough choices. We aimed to at least give an introduction to all techniques that students and scholars who want to computationally analyze communication will probably be confronted with. Of course, specific – technical – literature on techniques such as, for instance, machine learning can cover the subject in more depth, and the interested student may indeed want to dive into one or several of the techniques we cover more deeply. Our goal here is to offer enough working knowledge to apply these techniques and to know what to look for. While trying to cover the breadth of the field without sacrificing too much depth when covering each technique, we still needed to draw some boundaries. One technique that some readers may miss is agent-based modeling (ABM). Arguably, such simulation techniques are an important technique in the computational social sciences more broadly (Cioffi-Revilla, 2014), and they have recently been applied to the analysis of communication as well (Waldherr, 2014, Wettstein, 2020). Nevertheless, when reviewing the curricula of current courses teaching the computational analysis of communication, we found that simulation approaches do not seem to be at the core of such analyses (yet). Instead, when looking at the use of computational techniques in fields such as journalism studies (e.g., Boumans and Trilling, 2016), media studies (e.g., Rieder, 2017), or the text-as-data movement (Grimmer and Stewart, 2013), we see a core of techniques that are used over and over again, and that we have therefore included in our book. In particular, besides general data analysis and visualization techniques, these are techniques for gathering data such as web scraping or the use of API's; techniques for dealing with text such as natural language processing and different ways to turn text into numbers; supervised and unsupervised machine learning techniques; and network analysis.

1.2 Why Python and/or R?

By far most work in the computational social sciences is done using Python and/or R. Sure, for some specific tasks there are standalone programs that are occasionally used; and there are some useful applications written in other languages such as C or Java. But we believe it is fair to say that it is very hard to delve into the computational analysis of communication without learning at least either Python or R, and preferably both of them. There are very few tasks that you cannot do with at least one of them.

Some people have strong beliefs as to which language is "better" – we do not subscribe to that view. Most techniques that are relevant to us can be done in either language, and personal preference is a big factor. R started out as a statistical programming environment, and that heritage is still visible, for instance in the strong emphasis on vectors, factors, et cetera, or the possibility to estimate complex statistical models in just one line of code. Python started out as a general-purpose programming language, which means that some of the things we do feel a bit more "low-level" – Python abstracts away less of the underlying programming concepts than R does. This sometimes gives us more flexibility – at the cost of being more wordy. In recent years, however, Python and R have been growing closer to each other: with modules like *pandas* and *statsmodels*, Python now has R-like functionality handling data frames and estimating common statistical models on them; and with packages such as *quanteda*, handling of text – traditionally a strong domain of Python – has become more accessible in R.

This is the main reason why we decided to write this "bi-lingual" book. We wanted to teach techniques for the computational analysis of communication, without enforcing a specific implementation. We hope that the reader will learn from our book, say, how to transform a text into features and how to choose an appropriate machine learning model, but will find it of less importance in which language this happens.

However, sometimes, there are good reasons to choose one language above the other. For instance, many machine learning models in the popular *caret* package in R under the hood create a dense matrix, which severely limits the number of documents and features one can use; also, some complex web scraping tasks are maybe easier to realize in Python. On the other hand, R's data wrangling and visualization techniques in the *tidyverse* environment are known for their user-friendliness and quality. In the rare cases where we believe that R or Python is clearly superior for a given task, we indicate this; for the rest, we believe that it is up to the reader to choose.

1.3 How to Use This Book

This book differs from more technically oriented books on the one hand and more conceptual books on the other hand. We do cover the technical background that is necessary to understand what is going on, but we keep both computer science concepts and mathematical concepts to a minimum. For instance, if we had written a more technical book about programming in Python, we would have introduced rather earlier and in detail concepts such as classes, inheritence, and instances of classes. Instead, we decide to provide such information only as additional background where necessary and focus, rather pragmatically, on the application of techniques for the computational analysis of communication. Vice versa, if we had written a more conceptual book on new methods in our field, we would have given more emphasis to epistemological aspects, and had skipped the programming examples, which are now at the core of this book.

We do not expect much prior knowledge from the readers of this book. Sure, some affinity with computers helps, but there is no strict requirement on what you need to know. Also in terms of statistics, it helps if you have heard of concepts such as correlation or regression analysis, but even if your knoweldge here is rather limited, you should be able to follow along.

This also means that you may be able to skip chapters. For instance, if you already work with R and/or Python, you may not need our detailed instructions of the beginning. Still, the book follows a logical order in which chapters build on previous ones. For instance, when explaining supervised machine learning on textual data, we expect you to be familiar with previous chapters that deal with machine learning in general, or with the handling of textual data.

This book is designed in such a way that it can be used as a text book for introductory courses on the computational analysis of communications. Often, such courses will be on the graduate level, but it is equally possible to use this book in an undergraduate course; maybe skipping some parts that may go too deep. All code examples are not only printed in this book, but also available online. Students as well as social-scientists who want to brush up their skillset should therefore also be able to use this book for self-study, without a formal course around it. Lastly, this book can also be a reference for readers asking themselves: "How do I do this again?". In particular, if the main language you work in is R, you can look up how to do similar things in Python and vice versa.

Code examples

Regardless of the context in which you use this book, one thing is for sure: The only way to learn computational analysis methods is by practicing and playing around. For this reason, the code examples are probably the most important part of the book. Where possible, the examples use real world data that is freely available on the Internet. To make sure that the examples still work in five years' time, we generally provide a copy of this data on the book website, but we also provide a link to the original source.

One thing to note is that to avoid unnessecary repetition the examples are sometimes designed to continue on earlier snippets from that chapter. So, if you seem to be missing a data set, or if some package is not imported yet, make sure you run all the code examples from that chapter.

Note that although it is possible to copy-paste the code from the website accompaying this book[1], we would actually recommend typing the examples yourself. That way, you are more conscious about the commands you are using and you are adding them to your "muscle memory".

Finally, realize that the code examples in this book are just examples. There's often more ways to do something, and our way is not necessarily the only good (let alone the best) way. So, after you get an example to work, spend some time to play around with it: try different options, maybe try it on your own data, or try to achieve the same result in a different way. The most important thing to remember is: you can't break anything! So just go ahead, have fun, and if nothing works anymore you can always start over from the code example from the book.

1.4 Installing R and Python

R and Python are the most popular programming languages that data scientists and computational scholars have adopted to conduct their work. While many develop a preference for one or the other language, the chances are good that you will ultimately switch back and forth between them, depending on the specific task at hand and the project you are involved in.

---

[1] https://cssbook.net

Before you can start with analyzing data and communication in Python or R, you need to install interpreters for these languages (i.e., programs that can read code in these languages and execute it) on your computer. Interpreters for both Python and R are open source and completely free to download and use. Although there are various web-based services on which you can run code for both languages (such as Google Colab or RStudio Cloud), it is generally better to install an interpreter on your own computer.

After installing Python or R, you can execute code in these languages, but you also want a nice *Integrated Development Environment (IDE)* to develop your data analysis scripts. For R we recommend RStudio, which is free to install and is currently the most popular environment for working with R. For Python we recommend starting with JupyterLab or JupyterNotebook, which is a browser-based environment for writing and running Python code. All of these tools are available and well documented for Windows, MacOS, and Linux. After explaining how to install R and Python, there is a very important section on installing packages. If you plan to only use either R or Python (for now), feel free to skip the part about the other language.

If you are writing longer Python programs (as opposed to, for instance, short data analysis scripts) you probably want to install a full-blown IDE as well. We recommend PyCharm[2] for this, which has a free version that has everything you need, and the premium version is also free for students and academic or open source developers. See their website for download and installation instructions.

Anaconda

An alternative to installing R, Python, and optional libraries separately and as you need them (which we will explain later in this chapter) is to install the so called Anaconda Distribution, one of the most used and extensive platforms to perform data science. Anaconda is free and open-source, and is conceived to run Python and R code for data analysis and machine learning. Installing the complete Anaconda Distribution on your computer[3] provides you with everything that you need to follow the examples in this book and includes development environments such as Spyder, Jupyter, and RStudio. It also includes a large set of pre-installed packages often used in data science and its own package manager, *conda*, which will help you to install and update other libraries or dependencies. In short, Anaconda bundles almost all the important software to perform computational analysis of communication.

So, should you install Anaconda, or should you install all software separately as outlined in this chapter? It depends. On the pro side, by downloading Anaconda you have everything installed at once and do not have to worry about dependencies (e.g., Windows users usually do not have a C compiler installed, but some

[2] https://www.jetbrains.com/pycharm/

[3] https://www.anaconda.com/distribution/#download-section

packages may need it). On the con side, it is huge and also installs many things you do not need, you essentially get a non-standard installation, in which programs and packages are stored in different locations than those you (or your computer) may expect. Nowadays, as almost all computers actually already *have* some version of Python installed (even though you may not know it), you also end up in a possibly confusing situation where it may be unclear which version you are actually running, or for which version you installed a package. For this reason, our recommendation is to not use Anaconda unless it is already installed or you have a specific reason to do so (for example, if your professor requires you to use it).

1.4.1 Installing R and RStudio

Firstly, we will install R and its most popular IDE RStudio, and we will learn how to install additional packages and how to run a script. R is an object-based programming language orientated to statistical computing that can be used for most of the stages of computational analysis of communication. If you are completely new to R, but familiar with other popular statistical packages in social sciences (such as SPSS or STATA), you will find that you can perform in R many already-known statistical operations. If you are not familiar with other statistical packages, do not panic, we will guide you from the very beginning. Unlike much traditional software that requires just one complete and initial installation, when working with R, we will first install the raw programming language and then we will continue to install additional components during our journey. It might sound cumbersome, but in fact it will make your work more powerful and flexible, since you will be able to choose the best way to interact with R and especially you will select the packages that are suitable for your project.

Now, let us install R. The easiest way is to go to the RStudio CRAN page at https://cran.rstudio.com/.[4] Click on the link for installing R for your operating system, and install the latest version. If you use Linux, you may want to install R via your package manager. For Ubuntu linux, it is best to follow the instructions on https://cran.r-project.org/bin/linux/ubuntu/.

After installing R, let us immediately install RStudio Desktop (the free version). Go to https://rstudio.com/products/rstudio/download/#download and download and run the installer for your computer. If you open RStudio you should get a screen similar to Figure 1.1. If this is the first time you open RStudio you probably won't see the top left pane (the scripts), you can create that pane by creating a new *R Script* via the *file* menu or with the green plus icon in the top left corner.

Of the four panes in RStudio, you will probably spend most time in the top left pane, where you can view and edit your analysis *scripts*. A script is simply a list of commands that the computer should execute one after the other, for example: open your data, do some computations, and make a nice graph.

[4] *CRAN*, short for Comprehensive R Archive Network, is a network of websites on which R itself and various R packages are hosted.

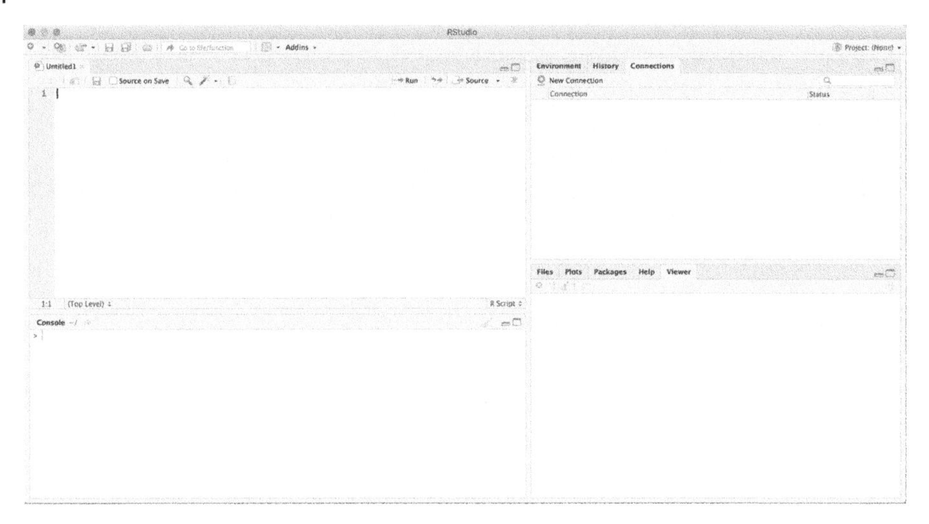

Figure 1.1 RStudio Desktop. *Source:* Rstudio, Inc.

To run a line of code, you can place your cursor anywhere on that line and click the *Run* icon or press control+Enter. To try that, type the following into your newly opened script:

```
print("Hello world")
```

Now, place your cursor on that line and press Run (or control+Enter). What happens is that the line is copied to the *Console* in the bottom left corner and executed. So, the results of your commands (and any error messages) will be shown in this console view.

In contrast to most traditional programming languages, the easiest way to run R code is line by line. You can simply place your cursor on the first line, and repeatedly press control+Enter, which executes a line and then places the cursor on the next line. You can also select multiple lines (or part of a line) to execute those commands together, but in general it is easier to check that everything is going as planned if you run the code line by line.

You can also write commands directly in the console and execute them (by pressing Enter). This can be useful for trying things out or to run things that only need to be run once, but in general we would strongly recommend typing all your commands in a script and then executing them. That way, the script serves as a log of the commands you used to analyze your data, so you (or a colleague) can read and understand how you did the analyses.

RStudio Projects

A very good idea to organize your data and code is to work with RStudio Projects. In fact, we would recommend you to now create a new empty project for the examples in this book. To do this, click on the *Project* button in the top right and select "New Project". Then, select New Directory and New Project and enter a name for this project and a parent folder for the project if you don't want it in your Documents. Using a project means that the scripts and data files for your project are all in the same location and you don't need to mess around with specifying the locations of files (which will probably be different for someone else or on a different computer). Moreover, RStudio remembers which files you were editing for each project, so if you are working on multiple projects it's very easy to switch between them. We recommend creating a project now for the book (and/or for any projects you are working on), and always switching to a project when you open RStudio.

On the right side of the RStudio workspace you will find two additional windows. In the top right pane there are two or more tabs: *environment* and *history*, and depending on additional packages you may have installed there may be some more. In *environment* you can manage your workspace (the set of elements you need to deploy for data analysis) and have a list of the objects you have uploaded to it. You may also import datasets with this tool. In the *history* tab you have an inventory of code executions, which you can save to a file, or move directly to console or to an R document.

Note that in the environment you can save and load your "workspace" (all data in the computer memory). However, relying on this functionality is often not a good idea: it will only save the state of your current session, whereas you will most likely want to save your R syntax file and/or your data instead. If you have your raw input data (e.g., as a csv file, see Chapter 5) and your analysis script, you can always reproduce what you have been doing. If you only have a snapshot of your workspace, you know the state in which you arrived, but cannot necessarily reproduce (or change) how you got there.

In the bottom right pane there are five additional useful tabs. In *files* you can explore your computer and manage all the files you may use for the project, including importing datasets. In *plots*, *help* and *viewer*, you can visualize the outputs, figures, documentation and general outcomes, respectively, that you have executed in your script. Finally, the tab for *packages* will be of great utility since it will let you install or update packages from CRAN or even from a file saved on your computer with a friendly interface.

1.4.2 Installing Python and Jupyter Notebook

Python is an object-orientated programming language and it is probably the favorite language of computational and data scientists in all disciplines around the world. There are different releases of Python, but the biggest difference used to be between Python 2 and Python 3. Fortunately, you will probably never need to install or use Python 2, and in fact, since January 2020 it is no longer supported. Thus, you can just use any recent Python 3 version for this book. When browsing through questions on online fora such as Stackoverflow or reading other people's code on Github (we will talk about that in Chapter 4), you still may come across legacy code in Python 2. Such code usually does not run directly in a Python 3 interpreter, but in most cases, only minor adaptions are necessary to make it work.

We will install and run Python and Jupyter Notebook using a *terminal* or command line interface. This is a tool that is installed on all computers that allows you to enter commands to the computer directly. First, create a project folder for this book using the File Explorer (Windows) or Finder (MacOS). Then, on Windows you can shift + Right click that folder and select "Open command Window here". On MacOS, after navigating to the folder you just created, you click on "Finder" in the menu at the top of the screen, then on "Services", then on "New Terminal at Folder." In both cases, this should open a new window (usually black or gray) that allows you to type commands.

Note that on most computers, Python is already installed by default. You can check this by typing the following command in your terminal:

```
python3 --version
```

On some versions of Windows, you may need to use `py` instead of `python3`:

```
py --version
```

In either case, the output of this command should be something like `Python 3.8.5`. If `python --version` also returns this version, you are free to use either command (but on older systems `python` can still refer to Python 2, so make sure that you are using Python 3 for this book!).

If Python is not installed on your system, go to https://www.python.org/downloads/windows/ or https://www.python.org/downloads/mac-osx/ and download and install the latest stable release (which at the time of writing is `3.9.0`).[5] After installing it, open a terminal again and run the command above to verify that it is installed correctly.

Included in any recent Python install is pip, the program that you will use for installing Python packages. You can check that pip is installed correctly by typing the following command on your terminal:

```
pip3 --version
```

Which should report something like `pip 20.0.2 from ... (python 3.8)`. Again, if `pip` reports the same version you can also use it instead of pip3. On some systems `pip3` will not work, so use `pip` in that case (but make sure to check that it points to Python 3).

Installing Jupyter Notebook. Next, we will install Jupyter Notebook, which you can use to run all the examples in this book and is a great environment for developing Python data analysis scripts. Jupyter Notebooks (which are also included in the IDE JupyterLab if you installed that), are run as a web application that allows you to create documents that contain code and inline text fragments. One of the nicest things about the Jupyter Notebook is that the code is inserted in fields (so-called "cells") that you can run one by one, getting its respective output, which, when added to the narrative text, will make your script more clean and reproducible. You can also add formatted text blocks (using a simple formatting language called *Markdown*) to explain to the reader what you are doing. In Section 4.3, we will address notebooks again as a good practice for a computational scientist.

You can install Jupyter notebook directly using pip using the following command (executed in a terminal):

```
pip3 install jupyter-notebook
```

---

[5] For linux, install python3 and pip using your package manager. For example, on ubuntu you can run `sudo apt install python3-pip`

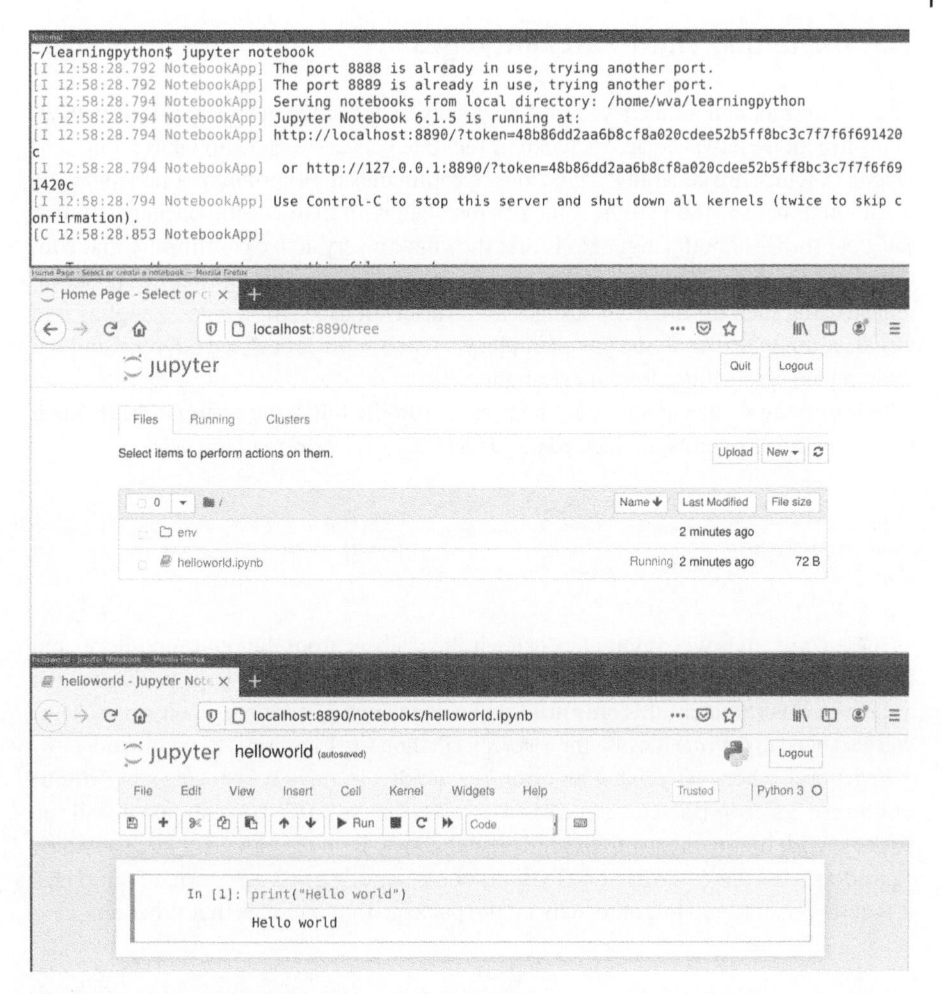

Figure 1.2 Jupyter Notebook. *Source:* Project Jupyter

Now, you can run Jupyter by executing the following command on the terminal:

```
jupyter notebook
```

This will print some useful information, including the URL at which you can access the notebook. However, it should also directly open this in a browser (e.g. Chrome) so you can directly start working. In your browser you should see the Jupyter main screen similar to the middle window in Figure 1.2. Create a new notebook by clicking on the *New* button in the top right and selecting Python 3. This should open a window similar to the bottom window in Figure 1.2.

In Jupyter, code is entered into cells. First, type `print("Hello World")` into the empty cell next to the `In[]:` prompt. Then, click the Run button or press control+Enter. This should execute your command and display the text `"Hello World"` in the output area right below the input cell. Note that you can create more cells using the plus icon or with the insert menu. You can also set the cell type via the Cell menu: select code for analysis scripts (which is the default), or Markdown for text fragments, which can be used to explain the code and/or interpret the results.

1.5 Installing Third-Party Packages

The `print` function used above is automatically included when you start R or Python. Many functions, however, are included in separate *packages* (also known as *libraries* or *modules*), which are generally collections of commands for a certain task or activity.

Although both R and Python come pre-installed with many useful packages, one of the great things of both languages is that they have a very active community that continuously develops, improves, and publishes new packages. Throughout this book, we will be using such third-party packages for a variety of tasks, from data wrangling and visualization to text analysis. For example, we will use the R package *tidyverse* and the Python packages *pandas* for data wrangling.

To install these packages on your computer, run the following commands: (Note: if you are using Anaconda, replace `pip3 install by conda install`)

Installing a package from Jupyter

```
!pip3 install pandas
# (On some systems, !pip install pandas)
```

Installing a package in R

```
install.packages("tidyverse")
```

These commands will automatically fetch the package from the right repository[6] and install them on your computer. This can take a while, especially for large packages such as tidyverse. Fortunately, this only needs to be done once. Every time you use a package, you also need to *activate* it using the `import` (Python) or `library` (R) command.

In general, whenever you get an error `No module named 'pandas'` (Python) or `there is no package called 'tidyverse'` (R), you can just install the package with that name using the code listed above. If you get an error such as `name 'pandas' is not defined` (Python) or `object 'ggplot' not found` (R), it is quite possible you forgot to activate the package that includes that function.

Packages used in each chapter

Some packages, like the *tidyverse* (R) and *pandas* (Python) packages for data handling are used in almost every chapter. Many chapters also introduce specific packages such as *igraph/networkx* for network analysis in Chapter 13. To make it easy to keep track of the packages needed for each chapter, every chapter that includes code in this book starts with a note like this that gives an overview of the main packages introduced in that chapter. It also includes the code needed to install these packages, which of course is only needed if you didn't install these packages before. Note again that if you are using Anaconda for Python, you should replace `!pip3 install` by `!conda install` in that code. On some systems, you may need to use `!pip install` instead of `!pip3 install`.

These notes also include a code block to import all the packages used for that chapter, which you need to run every time you use examples from that chapter.

---

[6] Similar to the App Store or Play Store, both R and Python have a centralized repository for third party packages. For R, this is the Comprehensive R Archive Network (CRAN) encountered earlier, while for Python this is the Python Package Index (PyPI) accessed by `pip`. Normally, all packages in these repositories are open source and safe to install.

2

Getting Started: Fun with Data and Visualizations

Abstract

This chapter is a lightning tour of some of the cool (and informative) things you can do with R and Python. Starting from a dataset of tweets about COVID-19, we show how you can analyze this data using text analysis, network analysis, and using geographic information. The goal of this chapter is not to teach you all these techniques in detail, rather, each of the examples showcases a possibility and guides you to the chapter where it will be explained in more detail. So don't worry too much about understanding every line of code, but relax and enjoy the ride!

Keywords basics of programming, data analysis

- Get an overview of the possibilities of R and Python for data analysis and visualization
- Understand how different aspects of data gathering, cleaning, and analysis work together
- Have fun with data and visualizations!

Packages used in this chapter

Since this chapter showcases a wide variety of possibilities, it relies on quite a number of third party packages. If needed, you can install these packages with the code below (see Section 1.4 for more details):

Python Code

```
!pip3 install pandas matplotlib geopandas
!pip3 install descartes shifterator
!pip3 install wordcloud gensim nltk networkx
```

R Code

```
install.packages(c("tidyverse", "igraph","maps",
    "quanteda", "quanteda.textplots",
    "quanteda.textstats", "topicmodels"))
```

After installing, you need to import (activate) the packages every session:

Computational Analysis of Communication: A Practical Introduction to the Analysis of Texts, Networks, and Images with Code Examples in Python and R, First Edition. Wouter van Atteveldt, Damian Trilling & Carlos Arcila Calderón.

Python Code

```python
import re
import pandas as pd
import matplotlib.pyplot as plt
from collections import Counter, defaultdict
from wordcloud import WordCloud
from gensim import corpora, models
import geopandas as gpd
import shifterator as sh
import nltk
from nltk.corpus import stopwords
import networkx as nx
```

R Code

```r
library(tidyverse)
library(lubridate)
library(quanteda)
library(quanteda.textplots)
library(quanteda.textstats)
library(topicmodels)
library(igraph)
library(maps)
```

2.1 Fun With Tweets

The goal of this chapter is to showcase how you can use R or Python to quickly and easily run some impressive analyses of real world data. For this purpose, we will be using a dataset of tweets about the COVID pandemic that is engulfing much of the world at the time this book is written. Of course, tweets are probably only representative for what is said on Twitter, but the data are (semi-)public and rich, containing text, location, and network characteristics. This makes them ideal for exploring the many ways in which we can analyze and visualize information with Python and R.

Example 2.1 shows how you can read this dataset into memory using a single command. Note that this does not retrieve the tweets from Twitter itself, but rather downloads our cached version of the tweets. In Chapter 12 we will show how you can download tweets and location data yourself, but to make sure we can get down to business immediately we will start from this cached version.

As you can see, the dataset contains almost 10000 tweets, listing their sender, their location and language, the text, the number of retweets, and whether it was a reply (retweet). You can read the start of the three most retweeted messages, which contain one (political) tweet from India and two seemingly political and factual tweets from the United States.

My first bar plot. Before diving into the textual, network, and geographic data in the dataset, let's first make a simple visualization of the date on which the tweets were posted. Example 2.2 does this in two steps: first, the number of tweets per hour is counted with an aggregation command. Next, a bar plot is made of this

Python Code

```python
tw=pd.read_csv("https://cssbook.net/d/covid.csv")
tw.head()
```

R Code

```r
tw = read_csv("https://cssbook.net/d/covid.csv")
head(tw)
```

R Output

```
# A tibble: 9,811 x 8
  status_id created_at          screen_name lang location text                retweet_count reply_to_screen.
  <dbl>     <dttm>              <chr>       <chr> <chr>    <chr>                      <dbl> <chr>
1  1.31e18 2020-09-25 16:50:33 ghulamabbas. en Lahore, P. "Secularism of #Gandhi and.   1203 NA
2  1.31e18 2020-09-25 22:49:07 GeoRebekah en Florida, . "On the day @GovRonDeSanti.     1146 NA
3  1.31e18 2020-09-25 19:39:16 AlexBerenson en New York "Updated @cgcgov figures: .      988 NA
# . with 9,808 more rows
```

Example 2.1 Retrieving cached tweets about COVID.

Python Code

```
tw.index=pd.DatetimeIndex(tw["created_at"])
tw["status_id"].groupby(pd.Grouper(freq="H")) \
    .count().plot(kind="bar")
# (note the use of \ to split a long line)
```

R Code

```
tweets_per_hour = tw %>%
  mutate(hour=round_date(created_at, "hour")) %>%
  group_by(hour) %>% summarize(n=n())
ggplot(tweets_per_hour, aes(x=hour, y=n)) +
  geom_col() + theme_classic() +
  xlab("Time") + ylab("# of tweets") +
  ggtitle("Number of COVID tweets over time")
```

R Output

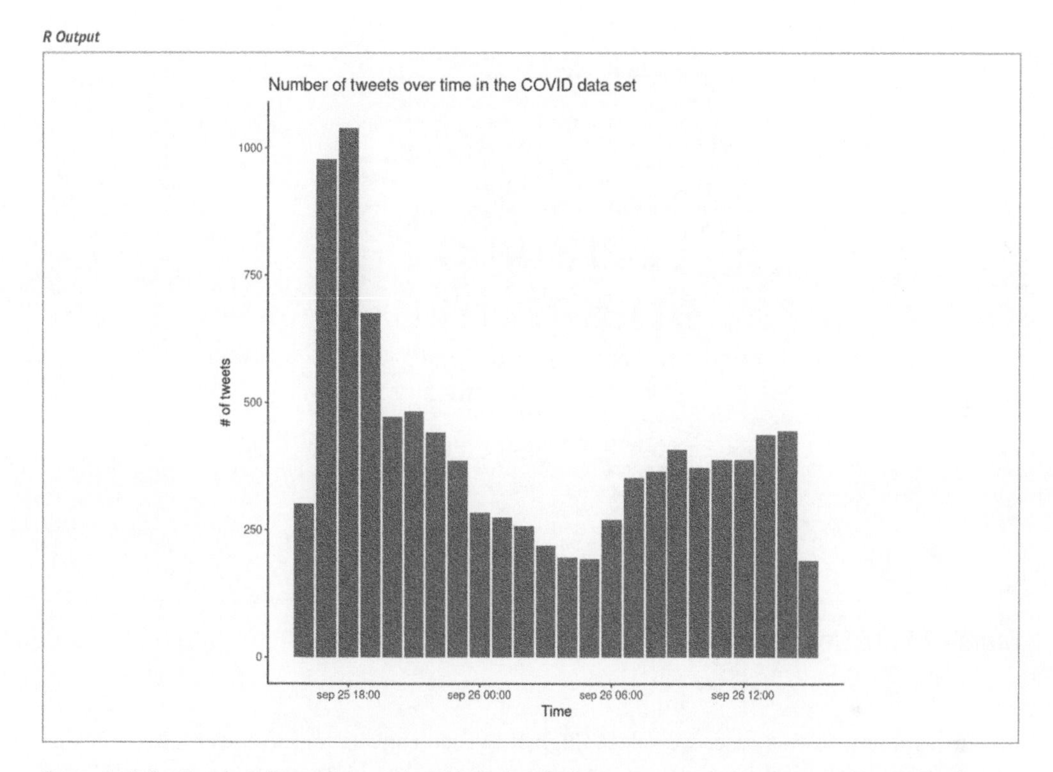

Example 2.2 Barplot of tweets over time. Note that Python output can vary due to different binning.

calculated value with some options to make it look relatively clean and professional. If you want to play around with this, you can for example try to plot the number of tweets per language, or create a line plot instead of a bar plot. For more information on visualization, please see Chapter 7. See Chapter 6 for an in-depth explanation of the aggregation command.

2.2 Fun With Textual Data

Corpus Analysis. Next, we can analyze which hashtags are most frequently used in this dataset. Example 2.3 does this by creating a *document-term matrix* using the package *quanteda* (in R) or by manually counting the words using a defaultdict (Python). The code shows a number of steps that are made to create the final results, each of which represent researcher choices about which data to keep and which to

Python Code

```python
freq = defaultdict(int)
for tweet in tw["text"]:
    for tag in re.findall("#\w+", tweet.lower()):
        if not re.search("#covid|#corona", tag):
            freq[tag]+=1
wc = WordCloud().generate_from_frequencies(freq)
plt.imshow(wc, interpolation="bilinear")
plt.axis("off")
```

R Code

```r
dtm_tags = filter(tw, lang=="en") %>%
  corpus() %>% tokens() %>%
  dfm(tolower = T) %>%
  dfm_select(pattern = "#*") %>%
  dfm_remove(c("#corona*", "#covid*"))
textplot_wordcloud(dtm_tags, max_words=100)
```

R Output

Example 2.3 My First Tag Cloud.

discard as noise. In this case, we select English tweets, convert text to lower case, remove stop words, and keep only words that start with #, while dropping words starting with `#corona` and `#covid`. To play around with this example, see if you can adjust the code to e.g. include all words or only at-mentions instead of the hashtags and make a different selection of tweets, for example Spanish language tweets or only popular (retweeted) tweets. Please see Chapter 10 if you want to learn more about corpus analysis, and see Chapter 6 for more information on how to select subsets of your data.

Topic Model. Where a word cloud (or tag cloud) shows which words occur most frequently, a *topic model* analysis shows which words co-occur in the same documents. Using the most common topic modeling algorithm, Latent Dirichlet Allocation or LDA, Example 2.4 explores the tweets by automatically clustering the tags selected earlier into 10 *topics*. Topic modeling is non-deterministic – if you run it again you can get slightly different topics, and topics are swapped around randomly as the topic numbers have no special meaning. By setting the computer's *random seed* you can ensure that if you run it again you get the same results. As you can see, some topics seem easily interpretable (such as topic 2 on social distancing and 8 on health care), it is always recommended that you inspect the clustered documents and edge cases in

Python Code

```
tags = [[tag.lower()
        for tag in re.findall("#\w+", tweet)]
       for tweet in tw["text"]]
voca = corpora.Dictionary(tags)
corpus = [voca.doc2bow(doc) for doc in tags]
m = models.LdaModel(corpus, num_topics=10,
    id2word=voca,
    distributed=False, random_state=1)
m.print_topics()
```

R Code

```
set.seed(1)
m = convert(dtm_tags, to="topicmodel") %>%
    LDA(10, method="Gibbs")
terms(m, 5)
```

R output

Topic 1	Topic 2	Topic 3	Topic 4	Topic 5	Topic 6	Topic 7	Topic 8	Topic 9	Topic 10
#health	#pandemic	#florida	#wfh	#pandemic	#sarscov2	#trump	#india	#trumpviru	#svaccine
#health-care	#masks	#staysafe	#sdusd	#canada	#business	#usa	#rss	#trump	#china
#maga	#virus	#socialdist	#anpcain-dgemic	#podcast	#economy	#biden	#islamopho	#bcilaim-atech	#anlogcek-down
#vote	#pence	#lockdown	#workfrom	#holemadeer-ship	#wearamas	#kuk	#gandhi	#american	#seducation
#wearamas	#kmask	#love	#remotewo	#rkmusic	#rip_india	#nmr2epdia	#nehru	#gopdeath	#cudltelhi

Example 2.4 Topic Model of the COVID tags. Note that Python output may look slightly different and contain different topics.

addition to the top words (or tags) as shown here. You can play around with this example, by using a different selection of words (modifying the code in Example 2.3) or changing the number of topics. You can also change (or remove) the random seed and see how running the same model multiple times will give different results. See Section 11.5 for more information about fitting, interpreting, and validating topic models.

2.3 Fun With Visualizing Geographic Information

For the final set of examples, we will use the location information contained in the Twitter data. This information is based on what Twitter users enter into their profile, and as such it is incomplete and noisy with many users giving a nonsensical location such as 'Ethereally here' or not filling in any location at all. However, if we assume that most users that do enter a proper location (such as Lahore or Florida in the top tweets displayed above), we can use it to map where most tweets are coming from.

The first step in this analysis is to resolve a name such as 'Lahore, Pakistan' to its geographical coordinates (in this case, about 31 degrees north and 74 degrees east). This is called geocoding, and both Google maps and Open Street Maps can be used to perform this automatically. As with the tweets themselves, we will use a cached version of the geocoding results here so we can proceed directly. Please see https://cssbook.net/datasets for the code that was used to create this file so you can play around with it as well.

Example 2.5 shows how this data can be used to create a map of Twitter activity. First, the cached user data is retrieved, showing the correct location for Lahore but also illustrating the noisiness of the data with the location "Un peu partout". Next, this data is *joined* to the Twitter data, so the coordinates are filled in where known. Finally, we plot this information on a map, showing tweets with more retweets as larger dots. See Chapter 7 for more information on visualization.

Python Code

```python
url = "https://cssbook.net/d/covid_users.csv"
users = pd.read_csv(url)
tw2 = tw.merge(users, on="screen_name", how="left")
world = gpd.read_file(
  gpd.datasets.get_path("naturalearth_lowres"))
gdf = gpd.GeoDataFrame(tw2,
  geometry=gpd.points_from_xy(tw2.long, tw2.lat))
ax = world.plot(color="white", edgecolor="black",
                figsize=(10,10))
gdf.plot(ax=ax, color="red", alpha=.2,
         markersize=tw["retweet_count"])
plt.show()
```

R Code

```r
url = "https://cssbook.net/d/covid_users.csv"
users = read_csv(url)
tw2 = left_join(tw, users)
ggplot(mapping=aes(x=long, y=lat)) +
  geom_polygon(aes(group=group),
    data=map_data("world"),
    fill="lightgray", colour = "white") +
  geom_point(aes(size=retweet_count,
                 alpha=retweet_count),
             data=tw2, color="red") +
  theme_void() + theme(aspect.ratio=1) +
  guides(alpha=FALSE, size=FALSE) +
  ggtitle("Location of COVID tweets",
          "Size indicates number of retweets")
```

R Output

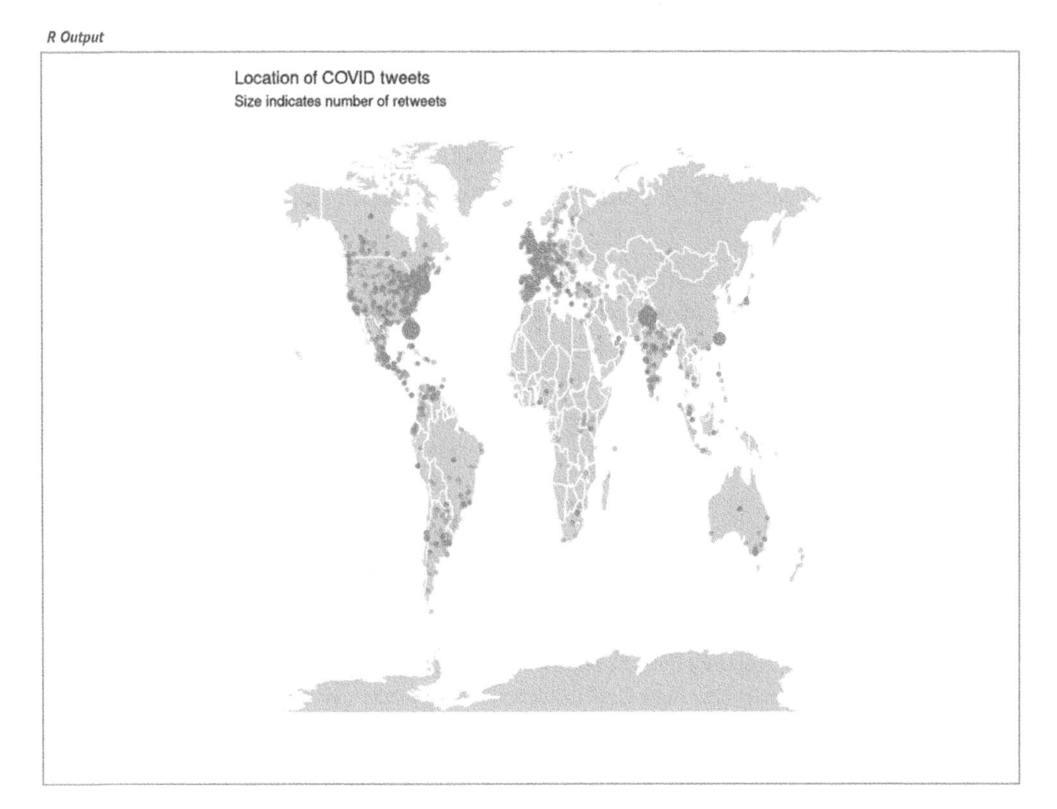

Example 2.5 Location of COVID tweets.

Combining textual and structured information. Since we know the location of a subset of our tweet's users, we can differentiate between e.g. American, European, and Asian tweets. Example 2.6 creates a very rough identification of North American tweets, and uses that to compute the relative frequency of words in those tweets compared to the rest. Not surprisingly, those tweets are much more about American politics, locations, and institutions. The other tweets talk about UK politics but also use a variety of names to refer to the pandemic. To play around with this, see if you can isolate e.g. Asian or South American tweets, or compare Spanish tweets from different locations.

Python Code

```python
nltk.download("stopwords")
cn = gdf.query("lang=='en'&(long<-60 & lat>25)")
cn = Counter(cn["text"].str.cat().lower().split())
cr = gdf.query("lang=='en' & (long>-60 | lat<25)")
cr = Counter(cr["text"].str.cat().lower().split())
for k in stopwords.words("english"):
    del cn[k]
    del cr[k]
key = sh.ProportionShift(type2freq_1=cn,
                         type2freq_2=cr)
key.get_shift_graph()
```

R Code

```r
dfm = tw2 %>% mutate(northamerica=ifelse(
    long < -60 & lat > 25,"N. America","Rest"))%>%
  filter(lang=="en") %>%
  corpus(docid_field="status_id") %>%
  tokens(remove_punct=T) %>%
  tokens_group(northamerica) %>%
  dfm(tolower=T) %>%
  dfm_remove(stopwords("en")) %>%
  dfm_select(min_nchar=4)
key = textstat_keyness(dfm, target="N. America")
textplot_keyness(key, margin=0.2) +
  ggtitle("Words preferred by North Americans",
          "(Only English-language tweets)") +
  theme_void()

```

R output

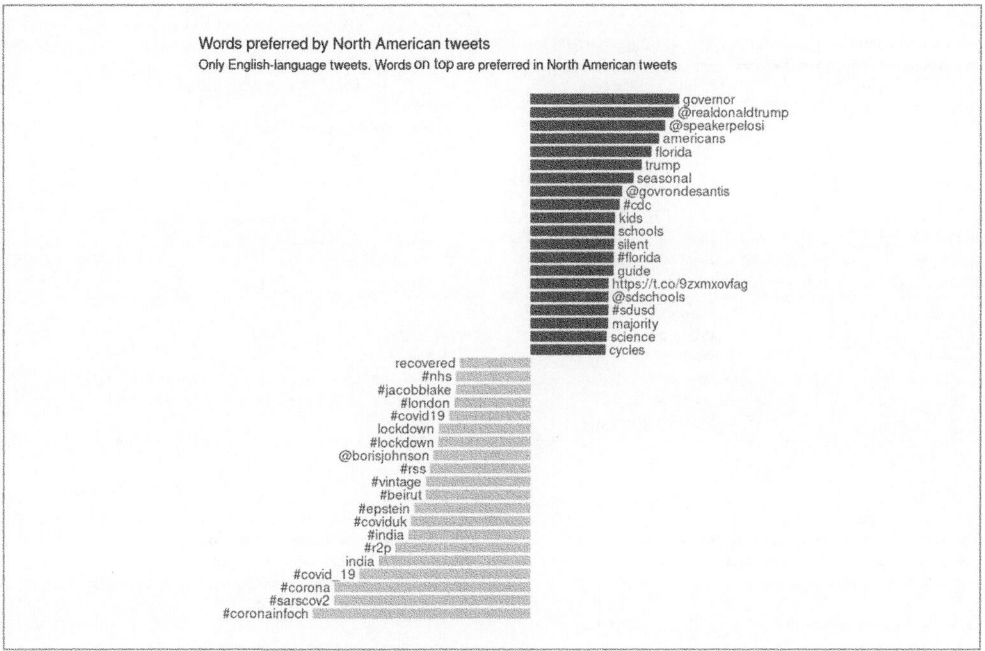

Words preferred by North American tweets
Only English-language tweets. Words on top are preferred in North American tweets

governor
@realdonaldtrump
@speakerpelosi
americans
florida
trump
seasonal
@govrondesantis
#cdc
kids
schools
silent
#florida
guide
https://t.co/9zxmxovfag
@sdschools
#sdusd
majority
science
cycles
recovered
#nhs
#jacobblake
#london
#covid19
lockdown
#lockdown
@borisjohnson
#rss
#vintage
#beirut
#epstein
#coviduk
#india
#r2p
india
#covid_19
#corona
#sarscov2
#coronainfoch

Example 2.6 Corpus comparison: North American tweets vs. the rest.

2.4 Fun With Networks

Twitter, of course, is a social network as well as a microblogging service: users are connected to other users because they follow each other and retweet and like each others' tweets. Using the `reply_to_screen_name` column, we can inspect the retweet network contained in the COVID tweet dataset. Example 2.7 first uses the data summarization commands from *tidyverse* (R) and *pandas* (Python) to create a data frame of connections or edges listing how often each user retweets each other user. The second code block shows how the *igraph* (R) and *networkx* (Python) packages are used to convert this edge list into a graph. From this graph, we select only the largest connected component and use a clustering algorithm to analyze which nodes (users) form cohesive subnetworks. Finally, a number of options are used to set the color and size of the edges, nodes, and labels, and the resulting network is plotted. As you can see, the central node is Donald Trump, who is retweeted by a large number of users, some of which are then retweeted by other users. You can play around with different settings for the plot options, or try to filter e.g. only tweets from a certain language. You could also easily compute social network metrics such as centrality on this network, and/or export the network for further analysis in specialized social network analysis software. See Chapter 13 for more information on network analysis, and Chapter 6 for the summarization commands used to create the edge list.

Python Code

```python
edges=tw2[["screen_name","reply_to_screen_name"]]
edges=edges.dropna().rename({"screen_name":"from",
  "reply_to_screen_name":"to"}, axis="columns")
edges.groupby(["from","to"]).size().head()
```

R Code

```r
edges = tw2 %>%
  select(from=screen_name,
         to=reply_to_screen_name) %>%
  filter(to != "") %>%
  group_by(to, from) %>%
  summarize(n=n())
head(edges)
```

R Output

to	from	n
<chr>	<chr>	<int>
_FutureIsUs	_FutureIsUs	1
JaylaS	AfronerdRadio	1
LoveMTB	ExpatriateNl	1
_nogueiraneto	ideobisium	1
NotFakeNews	panich52	1
_vikasupadhyay	SHADABMOHAMMAD7	4

Python Code

```python
g1 = nx.Graph()
g1.add_edges_from(edges.to_numpy())
largest = max(nx.connected_components(g1),key=len)
g2 = g1.subgraph(largest)

pos = nx.spring_layout(g2)
plt.figure(figsize=(20,20))
plt.axis("off")
sizes = [s*1e4 for s in
  nx.centrality.degree_centrality(g2).values()]
nx.draw_networkx_nodes(g2,pos, node_size=sizes)
nx.draw_networkx_labels(g2,pos)
nx.draw_networkx_edges(g2,pos)
plt.show()
```

R Code

```r
# create igraph and select largest component
g = graph_from_data_frame(edges)
components <- decompose.graph(g)
largest = which.max(sapply(components, gsize))
g2 = components[[largest]]
# Color nodes by cluster
clusters = cluster_spinglass(g2)
V(g2)$color = clusters$membership
V(g2)$frame.color = V(g2)$color
# Set node (user) and edge (arrow) size
V(g2)$size = degree(g2)^.5
V(g2)$label.cex = V(g2)$size/3
V(g2)$label = ifelse(degree(g2)<=1,"", V(g2)$name)
E(g2)$width = E(g2)$n
E(g2)$arrow.size= E(g2)$width/10
plot(g2)
```

R Output

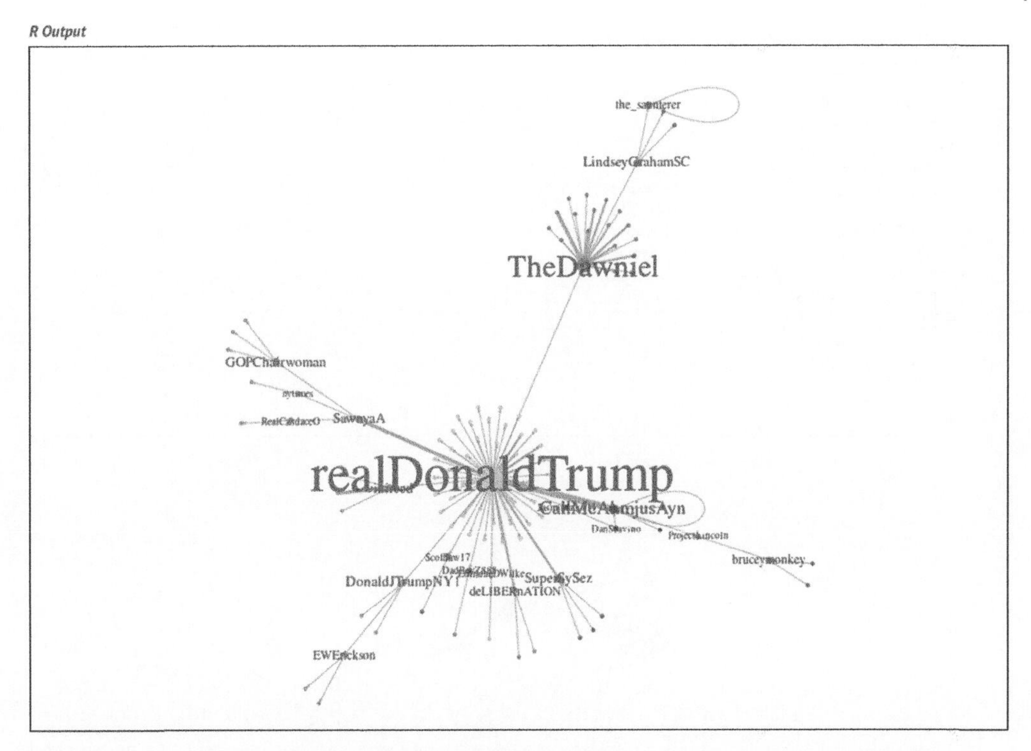

Example 2.7 Retweet network in COVID tweets. Note that Python output may look slightly different.

Geographic networks. In the final example of this chapter, we will combine the geographic and network information to show which regions of the world interact with each other. For this, in Example 2.8 we join the user information to the edges data frame created above twice: once for the sender, once for the replied-to user. Then, we adapt the earlier code for plotting the map by adding a line for each node in the network. As you can see, users in the main regions (US, EU, India) mostly interact with each other, with almost all regions also interacting with the US.

Python Code

```
u = users.drop(["location"], axis=1)
uf = u.rename({"screen_name":"from",
    "lat":"lat_from","long":"long_from"}, axis=1))
ut = u.rename({"screen_name":"to",
    "lat":"lat_to","long":"long_to"}, axis=1))
edges = edges.merge(uf).merge(ut).query(
    "long_to!=long_from & lat_to!=lat_from")

world = gpd.read_file(
    gpd.datasets.get_path("naturalearth_lowres"))
g_to = gpd.GeoDataFrame(edges.copy(),
    geometry=gpd.points_from_xy(edges.long_to,
                                edges.lat_to))
g_from = gpd.GeoDataFrame(edges.copy(),
    geometry=gpd.points_from_xy(edges.long_from,
                                edges.lat_from))

ax = world.plot(color="white", edgecolor="black",
                figsize=(10,10))
```

R Code

```
edges2 = edges %>%
  inner_join(users, by=c("from"="screen_name"))%>%
  inner_join(users, by=c("to"="screen_name"),
             suffix=c("", ".to")) %>%
  filter(lat != lat.to | long != long.to )
ggplot(mapping=aes(x = long, y = lat)) +
  geom_polygon(aes(group=group), map_data("world"),
    fill="lightgray", colour = "white") +
  geom_point(aes(size=retweet_count,
  alpha=retweet_count), data=tw2, color="red")+
  geom_curve(aes(xend=long.to, yend=lat.to, size=n),
             edges2, curvature=.1, alpha=.5) +
  theme_void() + guides(alpha=FALSE, size=FALSE) +
  ggtitle("Retweet network of COVID tweets",
  "Bubble size indicates total no. of retweets")
```

```
g_from.plot(ax=ax, color="red", alpha=.2)                 20
g_to.plot(ax=ax, color="blue", alpha=.2)                  21
                                                          22
e=g_from.join(g_to,lsuffix="_from",rsuffix="_to")         23
e = e[["geometry_from", "geometry_to"]]                   24
px = lambda point: point.x                                25
py = lambda point: point.y                                26
x_values = list(zip(e["geometry_from"].map(px),           27
                    e["geometry_to"].map(px)))            28
y_values = list(zip(e["geometry_from"].map(py),           29
                    e["geometry_to"]. map(py)))           30
plt.plot(x_values, y_values, linewidth = 1,               31
    linestyle = "-", color = "green", alpha=.3)           32
plt.show()                                                33
```

R Output

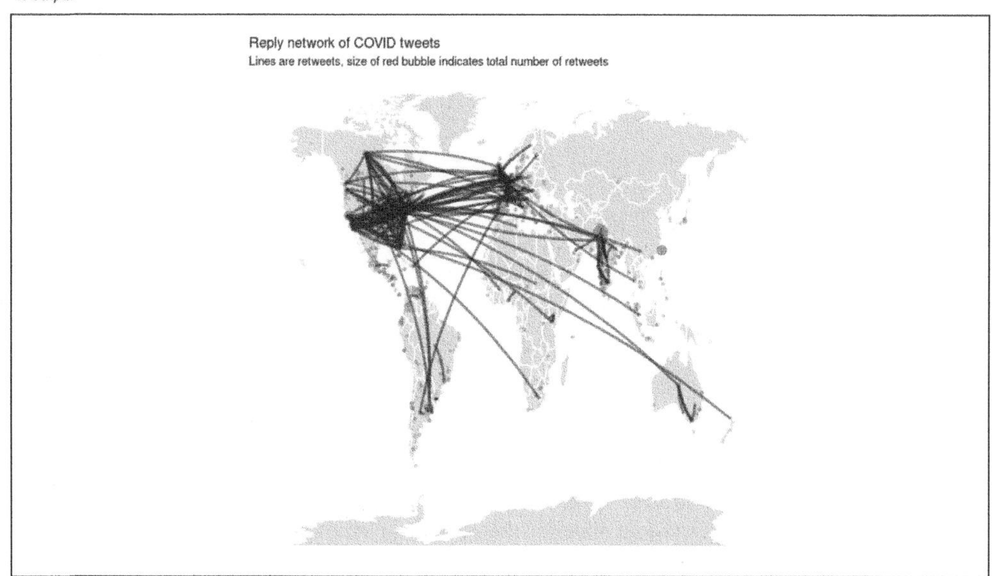

Example 2.8 Reply Network of Tweets.

3

Programming Concepts for Data Analysis

Abstract

This chapter introduces readers to the basics of programming in Python and R. It explains how to deal with objects, statements, expressions, variables and different types of data, and shows how to create and understand simple control structures such as loops and conditions.

Keywords basics of programming, data types, control strucutres, functions

- Understand objects and data types
- Write control structures
- Use functions and methods

Packages used in this chapter

This chapter focuses on the built-in capabilities of Python and R, so it does not rely on many packages. For R, only *glue* is used (which allows nice text formatting). For Python, we only use the packages *numpy* and *pandas* for data frame support. If needed, you can install these packages with the code below (see Section 1.4 for more details).

Python Code

```
!pip3 install numpy pandas
```

R Code

```
install.packages("glue")
```

After installing, you need to import (activate) the packages every session:

Python Code

```
import numpy as np
import pandas as pd
```

R Code

```
library(glue)
```

Computational Analysis of Communication: A Practical Introduction to the Analysis of Texts, Networks, and Images with Code Examples in Python and R, First Edition. Wouter van Atteveldt, Damian Trilling & Carlos Arcila Calderón.

3.1 About Objects and Data Types

Now that you have seen what R and Python can do in Chapter 2, it is time to take a small step back and learn more about how it all actually works under the hood. In both languages, you write a *script* or *program* containing the commands for the computer. But before we get to some real programming and exciting data analyses, we need to understand how data can be represented and stored.

No matter whether you use R or Python, both store your data in memory as *objects*. Each of these objects has a name, and you create them by assigning a value to a name. For example, the command `x = 10` creates a new object[1], named `x`, and stores the value 10 in it. This object is now stored in memory and can be used in later commands. Objects can be simple values such as the number 10, but they can also be pieces of text, whole data frames (tables), or analysis results. We call this distinction the *type* or *class* of an object.

Objects, pointers, and variables

In programming, a distinction is often made between an object (such as the number 10) and the variable in which it is stored (such as `x`). The latter is also called a "pointer". However, this distinction is not very relevant for most of our purposes. Moreover, in statistics, the word variable often refers to a column of data, rather than to the name of, for instance, the object containing the whole data frame (or table). For that reason, we will use the word *object* to refer to both the actual object or value and its name. (If you want some extra food for thought and want to challenge your brain a bit, you can see the relationship between the idea of a pointer and the discussion about mutable and immutable objects below.)

Let us create an object that we call a (an arbitrary name, you can use whatever you want), assign the value 100 to it, and use the class function (R) or type function (Python) to check what kind of object we created (Example 3.1). As you can see, R reports the type of the number as "numeric", while Python reports it as "int", short for integer or whole number. Although they use different names, both languages offer very similar data types. Table 3.1 provides an overview of some common basic data types.

Python Code

```
a = 100
print(type(a))
```

R Code

```
a = 100
print(class(a))
```

Python Output

```
<class 'int' >
```

R Output

```
[1] "numeric"
```

Example 3.1 Determining the type of an object.

[1] In both R and Python, the equals sign (=) can be used to assign values. In R, however, the traditional way of doing this is using an arrow (<-). In this book we will use the equals sign for assignment in both languages, but remember that for R, `x = 10` and `x <-10` are essentially the same.

Table 3.1 Most used basic data types in Python and R.

Python		R		Description
Name	Example	Name	Example	
int	`1`	integer	`1L`	whole numbers
float	`1.3`	numeric	`1.3`	numbers with decimals
str	`"Spam", 'ham'`	character	`"Spam", 'ham'`	textual data
bool	`True, False`	logical	`TRUE, FALSE`	the truth values

Let us have a closer look at the code in Example 3.1 above. The first line is a command to create the object *a* and store its value 100; and the second is illustrative and will give you the class of the created object, in this case "numeric". Notice that we are using two native functions of R, `print` and `class`, and including `a` as an argument of `class`, and the very same `class(a)` as an argument of `print`. The only difference between R and Python, here, is that the relevant Python function is called `type` instead of `class`.

Once created, you can now perform multiple operations with `a` and other values or new variables as shown in Example 3.2. For example, you could transform `a` by multiplying `a` by 2, create a new variable `b` of value 50 and then create another new object `c` with the result of `a + b`.

3.1.1 Storing Single Values: Integers, Floating-Point Numbers, Booleans

When working with numbers, we distinguish between integers (whole numbers) and floating point numbers (numbers with a decimal point, called "numeric" in R). Both Python and R automatically determine the data type when creating an object, but differ in their default behavior when storing a number that can be represented as an int: R will store it as a float anyway and you need to force it to do otherwise, for Python it is the other way round (Example 3.3). We can also convert between types later on, even though converting a float to an int might not be too good an idea, as you truncate your data.

So why not just always use a float? First, floating point operations usually take more time than integer operations. Second, because floating point numbers are stored as a combination of a coefficient and an exponent (to the base of 2), many decimal fractions can only approximately be stored as a floating point number. Except for specific domains (such as finance), these inaccuracies are often not of much practical importance. But it explains why calculating `6*6/10` in Python returns 3.6, while `6*0.6` or `6*(6/10)` returns 3.599 999 999 999 999 6. Therefore, if a value can logically only be a whole number (anything that is countable, in fact), it makes sense to restrict it to an integer.

We also have a data type that is even more restricted and can take only two values: true or false. It is called "logical" (R) or "bool" (Python). Just notice that boolean values are case sensitive: while in R you must capitalize the whole value (`TRUE`, `FALSE`), in Python we only capitalize the first letter: `True`, `False`. As you can see in Example 3.3, such an object behaves exactly as an integer that is only allowed to be 0 or 1, and it can easily be converted to an integer.

Python Code

```python
a = 100
a = a*2    # equivalent to (shorter) a*=2
b = 50
c = a + b
print(a, b, c)
```

R Code

```r
a = 100
a = a*2
b = 50
c = a + b
print(a)
print(b)
print(c)
```

Python Output

```
200 50 250
```

R Output

```
[1] 200
[1] 50
[1] 250
```

Example 3.2 Some simple operations.

Python Code

```python
d = 20
print(type(d))
# forcing python to treat 20 as a float
d2 = 20.0
print(type(d2))

e = int(20.7)
print(type(e))
print(e)

f = True
print(type(f))
print(int(f))
print(int(False))
```

R Code

```r
d = 20
print(class(d))
# forcing R to treat 20 as an int
d2 = 20L
print(class(d2))

e = as.integer(20.7)
print(class(e))
print(e)

f = TRUE
print(class(f))
print(as.integer(f))
print(as.integer(FALSE))
```

Python Output

```
<class 'int'>
<class 'float'>
<class 'int'>
20
<class 'bool'>
1
0
```

R Output

```
[1] "numeric"
[1] "integer"
[1] "integer"
[1] 20
[1] "logical"
[1] 1
[1] 0
```

Example 3.3 Floating point numbers, integers, and boolean values.

3.1.2 Storing Text

As a computational analyst of communication you will usually work with text objects or strings of characters. Commonly simply known as "strings", such text objects are also referred to as "character vector objects" in R. Every time you want to analyze a social-media message, or any other text, you will be dealing with such strings.

As you see in Example 3.4, you can create a string by enclosing text in quotation marks. You can use either double or single quotation marks, but you need to use the same mark to begin and end the string. This can be useful if you want to use quotation marks within a string, then you can use the other type to denote the beginning and end of the string. If you need to use a single quotation mark within a single-quoted string, you can *escape* the quotation mark by prepending it with a backslash (`\'`), and

Python Code

```python
text1 = "This is a text"
print(f"Type of text1: {type(text1)}")
text2 = "Using 'single' and \"double\" quotes"
text3 = 'Using \"single\" and "double" quotes'
print(f"Are text2 and text3 equal?{text2==text3}")
```

R Code

```r
text1 = "This is a text"
glue("Class of text1: {class(text1)}")
text2 = "Using 'single' and \"double\" quotes"
text3 = 'Using \'single\' and "double" quotes'
glue("Are text2 and text3 equal? {text2==text3}")
```

Python Output

```
Type of text1: <class 'str'>
Are text2 and text3 equal? True
```

R Output

```
Class of text1: character
Are text2 and text3 equal? TRUE
```

Python Code

```python
somebytes= text1.encode("utf-8")
print(type(somebytes))
print(somebytes)
```

R Code

```r
somebytes= charToRaw(text1)
print(class(somebytes))
print(somebytes)
```

Python Output

```
<class 'bytes'>
b'This is a text'
```

R Output

```
[1] "raw"
[1] 54 68 69 73 20 69 73 20 61 20 74 65 78 74
```

Example 3.4 Strings and bytes.

similarly for double-quoted strings. To include an actual backslash in a text, you also escape it with a backslash, so you end up with a double backslash (\\).

The Python example also shows a concept introduced in Python 3.6: the f-string. These are strings that are prefixed with the letter `f` and are *formatted* strings. This means that these strings will automatically insert a value where curly brackets indicate that you wish to do so. This means that you can write: `print(f"The value of i is {i}")` in order to print "The value of i is 5" (given that `i` equals 5). In R, the *glue* package allows you to use an f-string-like syntax as well: `glue("The value of i is {i}")`.

Although this will be explained in more detail in Section 5.2.2, it is good to introduce how computers store text in memory or files. It is not too difficult to imagine how a computer internally handles *integers*: after all, even though the number may be displayed as a decimal number to us, it can be trivially converted and stored as a binary number (effectively, a series of zeros and ones) — we do not have to care about that. But when we think about text, it is not immediately obvious how a string should be stored as a sequence of zeros and ones, especially given the huge variety of writing systems used for different languages.

Indeed, there are several ways of how textual characters can be stored as bytes, which are called *encodings*. The process of moving from bytes (numbers) to characters is called decoding, and the reverse process is called encoding. Ideally, this is not something you should need to think of, and indeed strings (or character vectors) already represent decoded text. This means that often when you read from or write data to a file, you need to specify the encoding (usually UTF-8). However, both Python and R also allow you to work with the raw data (e.g. before decoding) in the form of *bytes* (Python) or *raw* (R) data, which is sometimes necessary if there are encoding problems. This is shown briefly in the bottom part of Example 3.4. Note that while R shows the underlying hexadecimal byte values of the raw data (so 54 is `T`, 68 is `h` and so on) and Python displays the bytes as text characters, in both cases the underlying data type is the same: raw (non-decoded) bytes.

3.1.3 Combining Multiple Values: Lists, Vectors, And Friends

Until now, we have focused on the basic, initial data types or "vector objects", as they are called in R. Often, however, we want to group a number of these objects. For example, we do not want to manually create thousands of objects called tweet0001, tweet0002, ..., tweet9999 – we'd rather have one list called tweets that contains all of them. You will encounter several names for such combined data structures: lists, vectors, arrays, series, and more. The core idea is always the same: we take multiple objects (be it numbers, strings, or anything else) and then create one object that combines all of them (Example 3.5).

As you see, we now have one name (such as `scores`) to refer to all of the scores. The Python object in Example 3.5 is called a *list*, the R object a *vector*. There are more such combined data types, which have slightly different properties that can be important to know about: first, whether you can mix different types (say, integers and strings); second, what happens if you change the array. We will discuss both points below and show how this relates to different specific types of arrays in Python and R which you can choose from. But first, we will show how to work with them.

Operations on vectors and lists. One of the most basic operations you can perform on all types of one-dimensional arrays is *indexing*. It lets you locate any given element or group of elements within a vector using its or their positions. The first item of a vector in R is called 1, the second 2, and so on; in Python, we begin counting with 0. You can retrieve a specific element from a vector or list by simply putting the index between square brackets `[]` (Example 3.6).

In the first case, we asked for the score of the 5th student ("9"); in the second we asked for the 1st and 10th position ("8" "5"); and finally for all the elements between the 1st and 4th position ("8" "8" "7" "6"). We can directly indicate a range by using a `:`. After the colon, we provide the index of the last element (in R), while Python stops just *before* the index.[2] If we want to pass multiple single index values instead of a range in R, we need to create a vector of these indices by using `c()` (Example 3.6). Take a moment to compare the different ways of indexing between Python and R in Example 3.6!

Python Code

```python
scores = [8, 8, 7, 6, 9, 4, 9, 2, 8, 5]
print(type(scores))
countries = ["Netherlands", "Germany", "Spain"]
print(type(countries))
```

R Code

```r
scores = c(8, 8, 7, 6, 9, 4, 9, 2, 8, 5)
print(class(scores))
countries = c("Netherlands", "Germany", "Spain")
print(class(countries))
```

Python Output

```
<class 'list'>
<class 'list'>
```

R Output

```
[1] "numeric"
[1] "character"
```

Example 3.5 Collections arrays (such as vectors in R or lists in Python) can contain multiple values.

[2] This is related to the reason why Python starts counting with zero. If you are interested in this, have a look at https://www.cs.utexas.edu/users/EWD/transcriptions/EWD08xx/EWD831.html

Python Code

```
scores = ["8","8","7","6","9","4","9","2","8","5"]

print(scores[4])
print([scores[0], scores[9]])
print(scores[0:4])

# Convert the first 4 scores into numbers
# Note the use of a list comprehension [.. for ..]
# This will be explained in the section on loops
scores_new = [int(e) for e in scores[1:4]]
print(type(scores_new))
print(scores_new)
```

R Code

```
scores=c("8","8","7","6","9","4","9","2","8","5")

scores[5]
scores[c(1, 10)]
scores[1:4]

# Convert the first 4 scores into numbers
scores_new = as.numeric(scores[1:4])
class(scores_new)
scores_new
```

Python Output

```
9
['8', '5']
['8', '8', '7', '6']
<class 'list'>
[8, 7, 6]
```

R Output

```
[1] "9"
[1] "8" "5"
[1] "8" "8" "7" "6"
[1] "numeric"
[1] 8 8 7 6
```

Example 3.6 Slicing vectors and converting data types.

Indexing is very useful to access elements and also to create new objects from a part of another one. The last line of our example shows how to create a new array with just the first four entries of `scores` and store them all as numbers. To do so, we use *slicing* to get the first four scores and then either change its class using the function as.numeric (in R) or convert the elements to integers one-by-one (Python) (Example 3.6).

We can do many other things like adding or removing values, or creating a vector from scratch by using a function (Example 3.7). For instance, rather than just typing a large number of values by hand, we often might wish to create a vector from an operator or a function, without typing each value. Using the operator `:` (R) or the functions `seq` (R) or `range` (Python), we can create numeric vectors with a range of numbers.

Can we mix different types?. There is a reason that the basic data types (numeric, character, etc.) we described above are called "vector objects" in R: The vector is a very important structure in R and consists of these objects. A vector can be easily created with the `c` function and can only combine elements of the same type (numeric, integer, complex, character, logical, raw). Because the data types within a vector correspond to only one class, when we create a vector with for example numeric data, the `class` function will display "numeric" and not "vector".

If we try to create a vector with two different data types, R will force some elements to be transformed, so that all elements belong to the same class. For example, if you re-build the vector of scores with a new student who has been graded with the letter *b* instead of a number (Example 3.8), your vector will become a character vector. If you print it, you will see that the values are now displayed surrounded by `"`.

In contrast to a vector, a *list* is much less restricted: a list does not care whether you mix numbers and text. In Python, such lists are the most common type for creating a

Python Code

```
# Appending a new value to a list:
scores.append(7)

# Create a new list instead of overwriting:
scores4 = scores + [7]

# Removing an entry:
del scores[-10]

# Creating a list containing various ranges
list(range(1,21))
list(range(-5,6))

# A range of fractions: 0, 0.2, 0.4, ... 1.0
# Because range only handles integers, we first
# make a range of 0, 2, etc, and divide by 10
my_sequence = [e/10 for e in range(0,11,2)]
```

R Code

```
1  # appending a new value to a vector
2  scores = c(scores, 7)
3
4  # Create a new list instead of overwriting:
5  scores4 = c(scores, 7)
6
7  # removing an entry from a vector
8  scores = scores[-10]
9
10
11 # Creating a vector containing various ranges
12 range1 = 1:20
13 range2 = -5:5
14
15 # A range of fractions: 0, 0.2, 0.4, ... 1.0
16 my_sequence = seq(0,1, by=0.2)
17
```

Example 3.7 Some more operations on one-dimensional arrays.

R Code

```
1 scores2 = c(8, 8, 7, 6, 9, 4, 9, 2, 8, 5, "b")
2 print(class(scores2))
3 print(scores2)
```

Output

```
[1] "character"
[1] "8" "8" "7" "6" "9" "4" "9" "2" "8" "5" "b"
[1] 8 8 7 6 9 4 9 2 8 5
```

Example 3.8 R enforces that all elements of a vector have the same data type.

one-dimensional array. Because they can contain very different objects, running the `type` function on them does not return anything about the objects inside the list, but simply states that we are dealing with a list (Example 3.5). In fact, lists can even contain other lists, or any other object for that matter.

In R you can also use lists, even though they are much less popular in R than they are in Python because vectors are better if all objects are of the same type. R lists are created in a similar way as vectors, except that we have to add the word `list` before declaring the values. Let us build a list with four different kinds of elements, a numeric object, a character object, a square root function (`sqrt`), and a numeric vector (Example 3.9). In fact, you can use any of the elements in the list through indexing – even the function `sqrt` that you stored in there to get the square root of 16!

Python users often like the fact that lists give a lot of flexibility, as they happily accept entries of very different types. But also Python users sometimes may want a stricter structure like R's vector. This may be especially interesting for high-performance calculations, and therefore, such a structure is available from the *numpy* (which stands for Numbers in Python) package: the *numpy* array. This will be discussed in more detail when we deal with data frames in Chapter 5.

Object references and mutable objects

A subtle difference between Python and R is how they deal with copying objects. Suppose we define x containing the numbers 1, 2, 3 (`x=[1,2,3]` in Python or `x=c(1,2,3)` in R) and then define an object *y* to equal *x* (`y=x`). In R, both objects are kept separate, so changing *x* does not affect *y*, which is probably what you expect. In Python, however, we now have two variables (names) that both point to or *reference* the same object, and if we change *x* we also change *y* and vice versa, which can be quite unexpected. Note that if you really want to copy an object in Python, you can run `x.copy()`. See Example 3.10 for an example. Note that this is only important for *mutable* objects, that is, objects that can be changed. For example, lists in Python and R and vectors in R are mutable because you can replace or append members. Strings and numbers, on the other hand, are immutable: you cannot change a number or string, a statement such as `x=x*2` creates a new object containing the value of `x*2` and stores it under the name `x`.

Sets and Tuples. The *vector* (R) and *list* (Python) are the most frequently used collections for storing multiple objects. In Python there are two more collection types you are likely to encounter. First, *tuples* are very similar to lists, but they cannot be changed after creating them (they are *immutable*). You can create a tuple by replacing the square brackets by regular parentheses: `x=(1,2,3)`.

Second, in Python there is an object type called a *set*. A set is a mutable collection of *unique* elements (you cannot repeat a value) with no order. As it is not properly ordered, you cannot run any indexing or slicing operation on it. Although R does not have an explicit set type, it does have functions for the various set operations, the most

Python Code

```
my_list = [33, "Twitter", np.sqrt, [1,2,3,4]]
print(type(my_list))

# this resolves to sqrt(16):
print(my_list[2](16))
```

R Code

```
my_list = list(33, "Twitter", sqrt, c(1,2,3,4))
class(my_list)

# this resolves to sqrt(16):
my_list[[3]](16)
```

Python Output

```
<class 'list'>
4.0
```

R Output

```
[1] "list"
[1] 4
```

Example 3.9 Lists can store very different objects of multiple data types and even functions.

Python Code

```
x = [1,2,3]
y = x
y[0] = 99
print(x)
```

R Code

```
x = c(1,2,3)
y = x
y[1] = 99
print(x)
```

Python Output

```
[99, 2, 3]
```

R Output

```
[1] 1 2 3
```

Example 3.10 The (unexpected) behavior of mutable objects..

useful of which is probably the function `unique` which removes all duplicate values in a vector. Example 3.11 shows a number of set operations in Python and R, which can be very useful, e.g. finding all elements that occur in two lists.

3.1.4 Dictionaries

Python *dictionaries* are a very powerful and versatile data type. Dictionaries contain unordered[3] and mutable collections of objects that contain certain information in another object. Python generates this data type in the form of {`key : value`} pairs in order to map any object by its key and not by its relative position in the collection. Unlike in a list, in which you index with an integer denoting the position in a list, you can index a dictionary using the key. This is the case shown in Example 3.12, in which we want to get the values of the object "positive" in the dictionary *sentiments* and of the object "A" in the dictionary *grades*. You will find dictionaries very useful in your journey as a computational scientist or practitioner, since they are flexible ways to store and retrieve structured information. We can create them using the curly brackets {} and including each key-value pair as an element of the collection (Example 3.12).

In R, the closest you can get to a Python dictionary is to use lists with named elements. This allows you to assign and retrieve values by key, however the key is

Python Code

```python
a = {3, 4, 5}
my_list = [3, 2, 3, 2, 1]
b = set(my_list)
print(f"Set a: {a}; b: {b}")
print(f"intersect:  a & b = {a & b}")
print(f"union:      a | b = {a | b}")
print(f"difference: a - b = {a - b}")

```

R Code

```r
a = c(3, 4, 5)
my_vector = c(3, 2, 3, 2, 1)
b = unique(my_vector)
print(b)
print(intersect(a,b))
print(union(a,b))
print(setdiff(a,b))

```

Python Output

```
Set a: {3, 4, 5}; b: {1, 2, 3}
intersect: a & b = {3}
union: a | b = {1, 2, 3, 4, 5}
difference: a - b = {4, 5}
```

R Output

```
[1] 3 2 1
[1] 3
[1] 3 4 5 2 1
[1] 4 5
```

Example 3.11 Sets.

Python Code

```python
sentiments = {"positive":1, "neutral" : 0,
              "negative" : -1}
print(type(sentiments))
print("Sentiment for positive:",
      sentiments["positive"])

grades = {}
grades["A"] = 4
grades["B"] = 3
grades["C"] = 2
grades["D"] = 1

print(f"Grade for A: {grades['A']}")
print(grades)
```

R Code

```r
sentiments = list(positive=1, neutral=0,
                  negative=-1)
print(class(sentiments))
print(glue("Sentiment for positive: ",
           sentiments$positive))

grades = ()
grades$A = 4
grades$B = 3
grades$C = 2
grades$D = 1
# Note: grades[["A"]] is equivalent to grades$A
print(glue("Grade for A: {grades[['A']]}"))
print(glue("Grade for A: {grades$A}"))
print(grades)
```

[3] Newer versions of Python actually do remember the order in which items are inserted into a dictionary. However, for the purpose of this introduction, you can assume that you hardly ever care about the order of elements in a dictionary

Python Output

```
<class 'dict'>
Sentiment for positive: 1
Grade for A: 4
{'A': 4, 'B': 3, 'C': 2, 'D': 1}
```

R Output

```
[1] "list"
Sentiment for positive: 1
Grade for A: 4
$A
[1] 4

$B
[1] 3

$C
[1] 2

$D
[1] 1
```

Example 3.12 Key-value pairs in Python dictionaries and R named lists.

restricted to names, while in Python most objects can be used as keys. You create a named list with `d = list(name=value)` and access individual elements with either `d$name` or `d[["name"]]`.

A good analogy for a dictionary is a telephone book (imagine a paper one, but it actually often holds true for digital phone books as well): the names are the keys, and the associated phone numbers the values. If you know someone's name (the key), it is *very easy* to look up the corresponding values: even in a phone book of thousands of pages, it takes you maybe 10 or 20 seconds to look up the name (key). But if you know someone's phone number (the value) instead and want to look up the name, that's very inefficient: you need to read the whole phone book until you find the number.

Just as the elements of a list can be of *any* type, and you can have lists of lists, you can also nest dictionaries to get dicts of dicts. Think of our phone book example: rather than storing just a phone number as value, we could store another dict with the keys "office phone", "mobile phone", etc. This is very often done, and you will come across many examples dealing with such data structures. You have one restriction, though: the keys in a dictionary (as opposed to the values) are not allowed to be mutable. After all, imagine that you could use a list as a key in a dictionary, and if at the same time, some other pointer to that very same list could just change it, this would lead to a quite confusing situation.

3.1.5 From One to More Dimensions: Matrices and *n*-Dimensional Arrays

Matrices are two-dimensional rectangular datasets that include values in rows and columns. This is the kind of data you will have to deal with in many analyses shown in this book, such as those related to machine learning. Often, we can generalize to higher dimensions.

In Python, the easiest representation is to simply construct a list of lists. This is, in fact, often done, but has the disadvantage that there are no easy ways to get, for instance, the dimensions (the shape) of the table, or to print it in a neat(er) format. To get all that, one can transform the list of lists into an `array`, a datastructure provided by the package *numpy* (see Chapter 5 for more details).

To create a matrix in R, you have to use the function `matrix` and create a vector of values with the indication of how many rows and columns will be on it. We also have to tell R if the order of the values is determined by the row or not. In Example 3.13, we create two matrices in which we vary the `byrow` argument to be TRUE and FALSE, respectively, to illustrate how it changes the values of the matrix, even when the shape (2 x 3) remains identical. As you may imagine, we can operate with matrices, such as adding up two of them.

3.1.6 Making Life Easier: Data Frames

So far, we have discussed the general built-in collections that you find in most programming languages such as the list and array. However, in data science and statistics you are very likely to encounter a specific collection type that we haven't discussed yet: the *data frame*. Data frames are discussed in detail in Chapter 5, but for completeness we will also introduce them briefly here.

Data frames are user-friendly data structures that look very much like what you find in SPSS, Stata, or Excel. They will help you in a wide range of statistical analysis. A data frame is a tabular data object that includes rows (usually the instances or cases) and columns (the variables). In a three-column data frame, the first variable can be *numeric*, the second *character* and the third *logical*, but the important thing is that each variable is a vector and that all these vectors must be of the same length. We create data frames from scratch using the data.frame() function. Let's generate a simple data frame of three instances (each case is an author of this book) and three variables of the types numeric (*age*), character (*country* where they obtained their master degree) and logic (*living abroad*, whether they currently live outside the the country in which they were born) (Example 3.14). Notice that you have the label of the variables at the top of each column and that it creates an automatic numbering for indexing the rows.

Python Code

```
matrix = [[1, 2, 3], [4, 5, 6], [7,8,9]]
print(matrix)

array2d = np.array(matrix)
print(array2d)
```

R Code

```
my_matrix = matrix(c(0, 0, 1, 1, 0, 1),
    nrow = 2, ncol = 3, byrow = TRUE)
print(dim(my_matrix))
print(my_matrix)

my_matrix2 = matrix(c(0, 0, 1, 1, 0, 1),
    nrow = 2, ncol = 3, byrow = FALSE)
print(my_matrix2)
```

Python Output

```
[[1, 2, 3], [4, 5, 6], [7, 8, 9]]
[[1 2 3]
 [4 5 6]
 [7 8 9]]
```

R Output

```
[1] 2 3
     [,1] [,2] [,3]
[1,]    0    0    1
[2,]    1    0    1
     [,1] [,2] [,3]
[1,]    0    1    0
[2,]    0    1    1
```

Example 3.13 Working with two- or *n*-dimensional arrays.

Python Code

```
authors = pd.DataFrame({"age": [38, 36, 39],
  "countries": ["Netherlands","Germany","Spain"],
  "living_abroad": [False, True, True]})
print(authors)
```

R Code

```
authors = data.frame(age = c(38, 36, 39),
  countries = c("Netherlands","Germany","Spain"),
  living_abroad= c(FALSE, TRUE, TRUE))
print(authors)
```

R Output

```
  age countries   living_abroad
0 38  Netherlands False
1 36  Germany     True
2 39  Spain       True
```

Example 3.14 Creating a simple data frame.

3.2 Simple Control Structures: Loops and Conditions

Control structures in Python and R

This section and the next explain the working of control structures such as loops, conditions, and functions. These exist (and are very useful) in both Python and R. In R, however, you do not need them as much because most functions can work on whole columns in one go, while in Python you often run things on each row of a column and sometimes do not use data frames at all. Thus, if you are primarily interested in using R you could consider skipping the remainder of this chapter for now and returning later when you are ready to learn more. If you are learning Python, we strongly recommend continuing this chapter as control structures are used in many of the examples in the book.

Having a clear understanding of objects and data types is a first step towards comprehending how object-orientated languages such as R and Python work, but now we need to get some literacy in writing code and *interacting* with the computer and the objects we created. Learning a programming language is just like learning any new language. Imagine you want to speak Italian or you want to learn how to play the piano. The first thing will be to learn some words or musical notes, and to get familiarized with some examples or basic structures – just as we did in Chapter 2. In the case of Italian or the piano, you would then have to learn some grammar: how to form sentences, how play some chords; or, more generally, how to reproduce patterns. And this is exactly how we now move on to acquiring computational literacy: by learning some rules to make the computer do exactly what you want.

Remember that you can interact with R and Python directly on their consoles just by typing any given command. However, when you begin to use several of these commands and combine them you will need to put all these instructions into a script that you can then run partially or entirely. Recall Section 1.4, where we showed how IDEs such as RStudio (and Pycharm) offer both a console for directly typing single commands and a larger window for writing longer scripts.

Both R and Python are *interpreted* languages (as opposed to *compiled* languages), which means that interacting with them is very straightforward: You provide your computer with some *statements* (directly or from a script), and your computer reacts. We call a sequence of these statements a *computer program*. When we created objects by writing, for instance, `a = 100`, we already dealt with a very basic statement, the *assignment statement*. But of course the statements can be more complex.

In particular, we may want to say more about how and when statements need to be executed. Maybe we want to repeat the calculation of a value for each item on a list, or maybe we want to do this only if some condition is fulfilled.

Both R and Python have such *loops* and *conditional statements*, which will make your coding journey much easier and with more sophisticated results because you can control the way your statements are executed. By controlling the flow of instructions you can deal with a lot of challenges in computer programming such as iterating over unlimited cases or executing part of your code as a function of new inputs.

In your script, you usually indicate such loops and conditions visually by using *indentation*. Logical empty spaces – two in R and four in Python – depict blocks and sub-blocks on your code structure. As you will see in the next section, in R, using indentation is optional, and curly brackets will indicate the beginning ({) and end (}) of a code block; whereas in Python, indention is mandatory and tells your interpreter where the block starts and ends.

3.2.1 Loops

Loops can be used to repeat a block of statements. They are executed once, indefinitely, or until a certain condition is reached. This means that you can operate over a set of objects as many times as you want just by giving one instruction. The most common types of loops are *for*, *while*, and *repeat* (do-while), but we will be mostly concerned with so-called for-loops. Imagine you have a list of headlines as an object and you want a simple script to print the length of each message. Of course you can go headline by headline using indexing, but you will get bored or will not have enough time if you have thousands of cases. Thus, the idea is to operate a loop in the list so you can get all the results, from the first until the last element, with just one instruction. The syntax of the for-loop is:

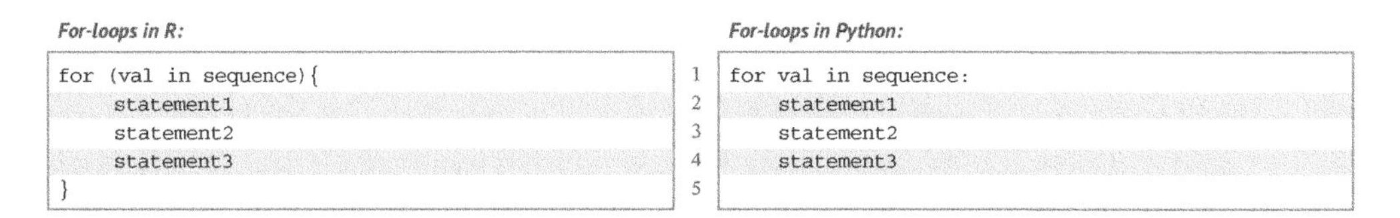

For-loops in R:

```
for (val in sequence){
      statement1
      statement2
      statement3
}
```

For-loops in Python:

```
for val in sequence:
     statement1
     statement2
     statement3
```

As Example 3.15 illustrates, every time you find yourself *repeating* something, for instance printing each element from a list, you can get the same results easier by *iterating* or *looping* over the elements of the list, in this case. Notice that you get the same results, but with the loop you can automate your operation writing few lines of code. As we will stress in this book, a good practice in coding is to be efficient and harmonious in the amount of code we write, which is another justification for using loops.

Don't repeat yourself!

You may be used to copy-pasting syntax and slightly changing it when working with some statistics program: you run an analysis and then you want to repeat the same analysis with different datasets or different specifications. But this is error-prone and hard to maintain, as it involves a lot of extra work if you want to change something. In many cases where you find yourself pasting multiple versions of your code, you would probably be better using a for-loop instead.

Another way to iterate in Python is using list comprehensions (not available natively in R), which are a stylish way to create list of elements automatically even with conditional clauses. This is the syntax:

```
newlist = [expression for item in list if conditional]
```

In Example 3.16 we provide a simple example (without any conditional clause) that creates a list with the number of characters of each headline. As this example

Python Code

```python
headlines = ["US condemns terrorist attacks",
    "New elections forces UK to go back to the UE",
    "Venezuelan president is dismissed"]
# Manually counting each element
print("manual results:")
print(len(headlines[0]))
print(len(headlines[1]))
print(len(headlines[2]))
#and the second is using a for-loop
print("for-loop results:")
for x in headlines:
    print(len(x))
```

R Code

```r
headlines = list ("US condemns terrorist attacks",
    "New elections forces UK to go back to the UE",
    "Venezuelan president is dismissed")
# Manually counting each element
print("manual results: ")
print(nchar(headlines[1]))
print(nchar(headlines[2]))
print(nchar(headlines[3]))
# Using a for-loop
print("for-loop results:")
for (x in headlines){
    print(nchar(x))
}
```

Python Output

```
manual results:
40
44
33
for-loop results:
40
44
33
```

R Output

```
[1] "manual results: "
[1] 40
[1] 44
[1] 33
[1] "for-loop results:"
[1] 40
[1] 44
[1] 33
```

Example 3.15 For-loops let you repeat operations.

Python Code

```python
len_headlines= [len(x) for x in headlines]
print(len_headlines)

# Note: the "list comprehension" above is
#   equivalent to the more verbose code below:
len_headlines = []
for x in headlines:
    len_headlines.append(len(x))
print(len_headlines)
```

Output

```
[40, 44, 33]
[40, 44, 33]
```

Example 3.16 List comprehensions are very popular in Python.

illustrates, list comprehensions allow you to essentially write a whole for-loop in one line. Therefore, list comprehensions are very popular in Python.

3.2.2 Conditional Statements

Conditional statements will allow you to control the flow and order of the commands you give the computer. This means you can tell the computer to do this or that, depending on a given circumstance. These statements use logic operators to test *if* your condition is met (True) or not (False) and execute an instruction accordingly. Both in R and Python, we use the clauses *if*, *else if* (*elif* in Python), and *else* to write the syntax of the conditional statements. Let's begin showing you the basic structure of the conditional statement:

If-statement in R:

```
if (condition) {
    statement_1
} else if (other_condition) {
    statement_2
} else {
    statement_3
}
```

If-statement in Python:

```
if (condition) {
    Statement1
elif other_condition:
    Statement2
else:
Statement3
```

Suppose you want to print the headlines of Example 3.15 only if the text is less than 40 characters long. To do this, we can include the conditional statement in the loop, executing the body only if the condition is met (Example 3.17)

We could also make it a bit more complicated: first check whether the length is smaller than 40, then check whether it is exactly 44 (`elif` / `else if`), and finally specify what to do if none of the conditions was met (`else`).

In Example 3.18, we will print the headline if it is shorter than 40 characters, print the string What a coincidence! if it is exactly 44 characters, and print "Too low" in all other cases. Notice that we have included the clause *elif* in the structure (in R it is noted *else if*). *elif* is a combination of *else* and *if*: if the previous condition is not satisfied, this condition is checked and the corresponding code block (or *else* block) is executed. This avoids having to nest the second *if* within the *else*, but otherwise the reasoning behind the control flow statements remains the same.

Python Code

```
for x in headlines:
    if len(x)<40:
        print(x)
```

R Code

```
for (x in headlines){
    if (nchar(x)<40) {
        print(x)}
    }
```

Output

```
Venezuelan president is dismissed
```

Example 3.17 A simple conditional control structure.

Python Code

```
for x in headlines:
    if len(x)<30:
        print(x)
    elif len(x) == 44:
        print("What a coincidence!")
    else :
        print("Too low")
```

R Code

```
for (x in headlines) {
    if (nchar(x)<30) {
        print(x)
    } else if (nchar(x)==44){
        print("What a coincidence!")
    } else {
        print("Too low")
}
}
```

Output

```
US condemns terrorist attacks
What a coincidence!
Too low
```

Example 3.18 A more complex conditional control structure.

3.3 Functions and Methods

Functions and *methods* are fundamental concepts in writing code in object-orientated programming. Both are objects that we use to store a set of statements and operations that we can use later without having to write the whole syntax again. This makes our code simpler and more powerful.

We have already used some built-in functions, such as `length` and `class` (R) and `len` and `type` (Python) to get the length of an object and the class to which it belongs. But, as you will learn in this chapter, you can also write your own functions. In essence, a function takes some input (the *arguments* supplied between brackets) and returns some output. Methods and functions are very similar concepts. The difference between them is that the functions are defined independently from the object, while methods are created based on a class, meaning that they are associated with an object. For example, in Python, each string has an associated method `lower`, so that writing `'HELLO'.lower()` will return 'hello'. In R, in contrast, one uses a function, `tolower('HELLO')`. For now, it is not really important to know why some things are implemented as a method and some are implemented as a function; it is partly an arbitrary choice that the developers made, and to fully understand it, you need to dive into the concept of `classes`, which is beyond the scope of this book.

Tab completion

Because methods are associated with an object, you have a very useful trick at your disposal to find out which methods (and other properties of an object) there are: TAB completion. In Jupyter, just type the name of an object followed by a dot (e.g., `a.<TAB>` in case you have an object called a) and hit the TAB key. This will open a drop-down menu to choose from.

We will illustrate how to create simple functions in R and Python, so you will have a better understanding of how they work. Imagine you want to create two functions: one that computes the 60% of any given number and another that estimates this percentage only if the given argument is above the threshold of 5. The general structure of a function in R and Python is:

Defining and calling a function in Python:

```
def f(par1, par2=0):
    statements
    return return_value

result = f(arg1, arg2)
result = f(par1=arg1, par2=arg2)
result = f(arg1, par2=arg2)
result = f(arg1)
```

Defining and calling a function in R:

```
1  f = function (par1, par2=0) {
2    statements
3    return_value
4  }
5  result = f(arg1, arg2)
6  result = f(par1=arg1, par2=arg2)
7  result = f(arg1, par2=arg2)
8  result = f(arg1)
```

In both cases, this defines a function called `f`, with two *arguments*, `arg_1` and `arg_2`. When you call the function, you specify the values for these parameters (the arguments) between brackets after the function name. You can then store the result of the function as an object as normal.

As you can see in the syntax above, you have some choices when specifying the arguments. First, you can specify them *by name* or *by position*. If you include the name (`f(param1=arg1)`) you explicitly bind that argument to that parameter. If you don't include the name (`f(arg1, arg2)`) the first argument matches the first parameter and so on. Note that you can mix and match these choices, specifying some parameters by name and others by position.

Second, some functions have *optional parameters*, for which they provide a default value. In this case, `par2` is optional, with default value `0`. This means that if you don't specify the parameter it will use the default value instead. Usually, the mandatory parameters are the main objects used by the function to do its work, while the optional parameters are additional options or settings. It is recommended to generally specify these options by name when you call a function, as that increases the readability of the code. Whether to specify the mandatory arguments by name depends on the function: if it's obvious what the argument does, you can specify it by position, but if in doubt it's often better to specify them by name.

Finally, note that in Python you explicitly indicate the result value of the function with `return value`. In R, the value of the last expression is automatically returned, although you can also explicitly call `return(value)`.

Example 3.19 shows how to write our function and how to use it.

The power of functions, though, lies in scenarios where they are used repeatedly. Imagine that you have a list of 5 (or 5 million!) scores and you wish to apply the function `perc_60_cond` to all the scores at once using a loop. This costs you only two extra lines of code (Example 3.20).

So far you have taken your first steps as a programmer, but there are many more advanced things to learn that are beyond the scope of this book. You can find a lot of literature, online documentation and even wonderful Youtube tutorials to keep learning. We can recommend the books by Crawley (2012) and VanderPlas (2016) to have more insights into R and Python, respectively. In the next chapter, we will go deeper

Python Code

```python
#The first function just computes 60% of the value
def perc_60(x):
    return x*0.6
print(perc_60(10))
print(perc_60(4))

# The second function only computes 60% it the
# value is bigger than 5
def perc_60_cond(x):
    if x>5:
        return x*0.6
    else:
        return x
print(perc_60_cond(10))
print(perc_60_cond(4))
```

R Code

```r
#The first function just computes 60% of the value
perc_60 = function(x) x*0.6

print(perc_60(10))
print(perc_60(4))

# The second function only computes 60% it the
# value is bigger than 5
perc_60_cond = function(x) {
  if (x>5) {
    return(x*0.6)
  } else {
    return(x)
  }
}
print(perc_60_cond(10))
print(perc_60_cond(4))
```

Output

```
6.0
2.4
6.0
4
```

Example 3.19 Writing functions.

Python Output

```
# Apply the function in a for-loop
scores = [3,4,5,7]
for x in scores:
    print(perc_60_cond(x))
```

R Output

```
1 # Apply the function in a for-loop
2 scores = list(3,4,5,6,7)
3 for (x in scores) {
4   print(perc_60_cond(x))
5 }
```

Output

```
3
4
5
4.2
```

Example 3.20 Functions are particular useful when used repeatedly.

into the world of code in order to learn how and why you should re-use existing code, what to do if you become stuck during your programming journey and what are the best practices when coding.

> A specific type of Python function that you may come across at some point (for instance, in Section 12.2.2) is the `generator`. Think of a function that returns a list of multiple values. Often, you do not need all values at once: you may only need the *next* value at a time. This is especially interesting when calculating the whole list would take a lot of time or a lot of memory. Rather than waiting for all values to be calculated, you can immediately begin processing the first value before the next arrives; or you can work with data so large that it doesn't all fit into your memory at the same time. You recognize a generator by the `yield` keyword instead of a `return` keyword (Example 3.21)

Python Code

```
1 mylist = [35,2,464,4]
2
3 def square1(somelist):
4     listofsquares = []
5     for i in somelist:
6         listofsquares.append(i**2)
7     return(listofsquares)
8
9 def square2(somelist):
10    for i in somelist:
11        yield i**2
12
13 print("As a list:")
14 mysquares = square1(mylist)
15 for mysquare in mysquares:
16     print(mysquare)
17 print(type(mysquares))
18 print(f"The list has {len(mysquares)} entries")
19
20
21 print("\nAs a generator:")
22
23 mysquares = square2(mylist)
24 for mysquare in mysquares:
25     print(mysquare)
26 print(type(mysquares))
27 # This throws an error (generators have no length)
28 print(f"mysquares has {len(mysquares)} entries")
```

Output

```
As a list:
1225
4
215296
16
<class 'list'>
The list has 4 entries
As a generator:
1225
4
215296
16
<class 'generator'>
```

Example 3.21 Generators behave like lists in that you can iterate (loop) over them, but each element is only calculated when it is needed. Hence, they do not have a length.

4

How to Write Code

Abstract

Programming is no longer a solitary activity, and almost all questions, problems, and error messages have been encountered and solved before. This chapter explains the most common forms of collaboration and sources of outside help, as well as outlining best practice on how to write and share code yourself.

Keywords package, library, errors, computational hygiene, notebooks

---

- Understand the importance of re-using code when programming
- Help beginning coders to avoid getting stuck
- Explain "computational hygiene" and show best practices in R and Python to write and share code

---

In Chapter 3, you have learned how to write your first lines of code. You created objects of different types, used and wrote some functions, and explored the major control structures. You are probably eager to write your first longer piece of code and produce some interesting data processing or analysis script. In this chapter, we prepare you to do this with as little frustration as possible. You will be introduced to some best practices so that you can implement them right from the start and to some tools that will make your life easier.

First, we will answer the question "how do you avoid reinventing the wheel": when is it appropriate to simply use someone else's existing code, and when do you need to write your own code from scratch? And is there a middle ground? Second, we will discuss how to turn error messages – which you will inevitably see a lot – from a frustrating annoyance into a helpful tool. Finally, we will discuss some best practices when writing code.

4.1 Re-using Code: How Not to Re-Invent the Wheel

Just as in any human language, programming languages also consist of a vocabulary, syntax rules, and expressions. Using the proper words and grammar, you can build from scratch any idea your imagination allows. That's a wonderful thing! But, let's be honest: the language itself, the expressions, ideas, and all the abstract constructs

Computational Analysis of Communication: A Practical Introduction to the Analysis of Texts, Networks, and Images with Code Examples in Python and R, First Edition. Wouter van Atteveldt, Damian Trilling & Carlos Arcila Calderón.

seldom come originally from you. And in fact, that's great as well: otherwise, you'd have to deeply think of every element before talking and expressing any thought. Instead, you use pre-existing rules, ideas, perceptions, and many different narratives to create your own messages to interact with the world. It's the same with coding: you never start from scratch.

Of course you *can* code anything you want from the very beginning, even just using 0's and 1's! When reading through the previous chapters, maybe you even started to think that complex operations will be exhausting and will take a really long time. After all, from the basic operations we did to a useful statistical model seems like a long way to go.

Luckily, this is not the case. There is almost no project in which computational scientists, data analysts, or developers do not re-use earlier code in order to achieve their goals more quickly and efficiently. The more common a task is, the greater the chance that you do not have to re-invent the wheel. Of course, you have to give credit where credit is due, but it is not uncommon to paste code snippets from others into your own code and adapt them. This is especially true for standard operations, for which there are only so many ways to achieve the desired result.

There are different ways to re-use earlier code. One is to copy and adapt raw lines of code written by someone else or by yourself in the past. In fact, there are many online repositories such as GitHub or BitBucket that contain many programs and well-documented code examples (see Section 4.3). When conducting computational analyses, you will spend a significant part of your time in such repositories trying to understand what others have done and figuring out how you can use it in your own work. Of course, make sure that the license of the code allows you to use it in the way you want. Also, give credit where credit is due: at the very least, place a comment with a link in your code to indicate what inspired you.

Another way is to build or import functions that summarize many lines of code into a simpler command, as we explained in Section 3.3. Functions are indeed powerful strategies to re-use the code since you do not have to write the same code over and over again if you need it in multiple places. Packages are probably the most elegant approach to recycle the work done by other colleagues. In Section 1.4 you already learned how to install a package and you probably noticed how easy it is to bring many pre-built functionalities onto your workspace. You can also write and publish your own package in the future to help your colleagues to write less code and to be more efficient in their daily job (see also Section 15.3)!

Many questions can arise here: what to re-use? When to use a function written by someone else instead of writing the code yourself? Which scripts and sources are trustworthy? Which is the best package to choose? How many packages should we use within the same project? Should we care about package versions? And must we know every package that is released in our field? There are of course multiple answers to these questions and it will be probably a matter of practice how to obtain the most appropriate ones. In general, we can say that one premise is to re-use and share code as much as you can. This idea is limited by constraints of quality, availability, parsimony, updates, and expertise. In other words, when recycling code we should think of the reputation of the source, the difficulty of accessing it, the risk of having an excessive and messy number of inputs, the need to share the last developments with your colleagues, and the fact that you will never be able to know everything.

Let's take an example. Imagine you want to compute the Levenshtein distance between two strings. That's a pretty straightforward metric that answers the question: "How many edits (removing/changing/inserting characters) do I need to apply to transform string1 into string2?" It can be used for plagiarism detection, but may be interesting for us to determine, for instance, whether a newspaper copied some content from somewhere else, even if small changes have been applied. You could now try to write some code to calculate that (and we are sure you could do that if you invested some time in it!), but it is such a common problem that it has been solved multiple times before. You could, for instance, look up some functions that are known to solve the problem and copy-paste them into your code. You can find a large number of different implementations for both Python and R here: https://en.wikibooks.org/wiki/Algorithm_Implementation/Strings/Levenshtein_distance. You can then choose and copy-paste the one which is most appropriate for you. One that is very fast, because you want to compare a huge set of strings? One that is easy to understand? One that uses only a few lines of code to not distract the reader? Alternatively, if you look for available packages for Python and R, you see that there are multiple packages that you can install with `install.packages` (R) or `pip` (Python) and then import. If you go for that route, you don't need to care about the internal workings and can "abstract away" and outsource the problem – on the other hand, the users of your code now have one more dependency to install before they can use your code.

In the case of package selection, we understand it can be quite overwhelming, with so many different packages from different contributors. In fact, sometimes the same task, such as topic modeling, can be done using multiple different packages. So, how to find and choose the best package? Besides resources like this book, the most important guide is probably the community around you: using packages that a lot of other people also use means that the package is probably well maintained and documented, and that there is a community to ask for help if needed. Since all packages on Pypi and CRAN are free to download and install, however, you can also shop around and see what the various packages do. When comparing different packages, it is always good to check their documentation and their GitHub page: packages that are well documented and that are updated frequently are often a good choice.

For example, the authors of this book had several intensive discussions of which packages to mention and use in the proposed exercises, an issue that became complex given the variety of topics addressed in this book. In the case of text analysis, a library such as `NLTK` for Python was incredibly popular among computational analysts until a few years ago becoming a package of reference in the field, but it has – at least for some applications – been overpassed by friendly and sophisticated new packages for natural language processing like `SpaCy`. So, which should we have included in this book? The one which is well-known (with excellent documentation by the way) and still used by thousands of practitioners and students around the world, or the one which is penetrating the market because of its easiness and advantages? Moreover, when choosing the second option, are we sure a more trendy package is going to be stable in time or is it going to be superseded by a different one in just few months?

There isn't the one golden way of how to re-use code and packages, but this dynamic scenario also depicts an exciting and provocative field that forces us to keep ourselves up to date.

4.2 Understanding Errors and Getting Help

Even though re-using code makes writing programs easier and less error-prone, every programmer makes mistakes. Programming can be a frustrating endeavor, and you will encounter error messages, bugs, and problems that you don't know how to fix. This section shows how error messages are useful rather than scary and lists the main error messages encountered in the beginning. It explains how to search for help within the R/Python documentation, how to use online resources, and how to formulate questions to your instructor or community so you get a useful answer.

If you tried out some of the concepts in Chapter 3, you have probably already come across some typical or basic errors in programming. Maybe you tried to call a function from a library that you forgot to load before, or maybe you tried to multiply a string with a float. There are thousands of errors that you will encounter, and there is no exhaustive list of them, so you won't find a complete structured catalogue to solve your problems when coding. This might seem a rough road for any scientist but in fact you will get used to finding the answers by different means.

4.2.1 Error Messages

There are two common strategies to *avoid getting stuck* and move on with your task: one is to understand the *type* of error you are getting, and the other is to know *where* to go to obtain valuable help. We would add a third one: be patient and do not despair!

Both R and Python produce warning or error messages when something is wrong in your code. Beginning computational researchers may sometimes feel afraid, confused, or even frustrated when they get such a *painful* message (we have all felt this) and some then would become so anxious that they don't pay enough attention to the text of the error message thinking it will not be helpful to solve the problem and blaming themselves for not being a perfect programmer. But the more you code the more you realize that getting these error messages is just part of the routine and that it is very useful to carefully read the warning instead of skipping it.

In most cases, the error message in your console will tell you exactly where the problem is: a specific line or operation within your code. With this information in many cases you will quickly identify what the problem is about and you will know how to solve it. One of the most common causes for errors is just very silly typos!

Next to the location (the line number), the error message will also tell you more about the problem. For example, when trying to multiply the float object `a` by the string object `b` you will get "Error in a * b : non-numeric argument to binary operator" in R or "TypeError: can't multiply sequence by non-int of type 'float'" in Python. As intimidating as this language may sound in the first place, if you re-read it, you will realize that it, in fact, explains exactly what went wrong. This helps you to understand what you did wrong and enable you to fix it.

If you get a warning or error that you don't understand or get an incorrect result in your code you have three options to get more information: use the `help` commands to know more about any object or function (`help(object)` in both R and Python); read the documentation of base R, base Python or of any individual package (there are plenty of them online!); and look at the wonderful community of worldwide coders, read what they have discussed so far or even pose a question to *challenge* their minds.

Let's consider this third option. Imagine you read the text of an error message and you feel you don't understand it. It may be because the wording is too complex or because it just gives an "error code" (i.e. "error 401 - Unauthorized" when trying to connect to the Twitter API). If your first thought is to try searching for it in Google, then this is completely correct: it might take you to code documentation, or better to an online discussion in sites such as Stack Overflow, which is a useful question and answer website for coders (see Figure 4.1). It is very likely that some colleagues have already posed a question about the meaning of that error and others have already provided an answer to what it means and especially help with *how to* fix it.

Depending on the complexity and novelty of your problem you might find a helpful answer in a few minutes or it might take you hours. Never get desperate if you visit many discussions without understanding everything directly: you may have to come back to some of them after reading all. Moreover, some answers will include the exact code you need (ready for copy-and-paste), code to be adapted (i.e. changing the name of your variables) or in pseudocode (informal description of the code). In all of the cases you will be the responsible for making sense of the huge (and sometimes messy) amount of sources you will come across.

It is of course possible that you don't get what you need in previous discussions. In that case you will be able to create your own question and wait for someone to reply.

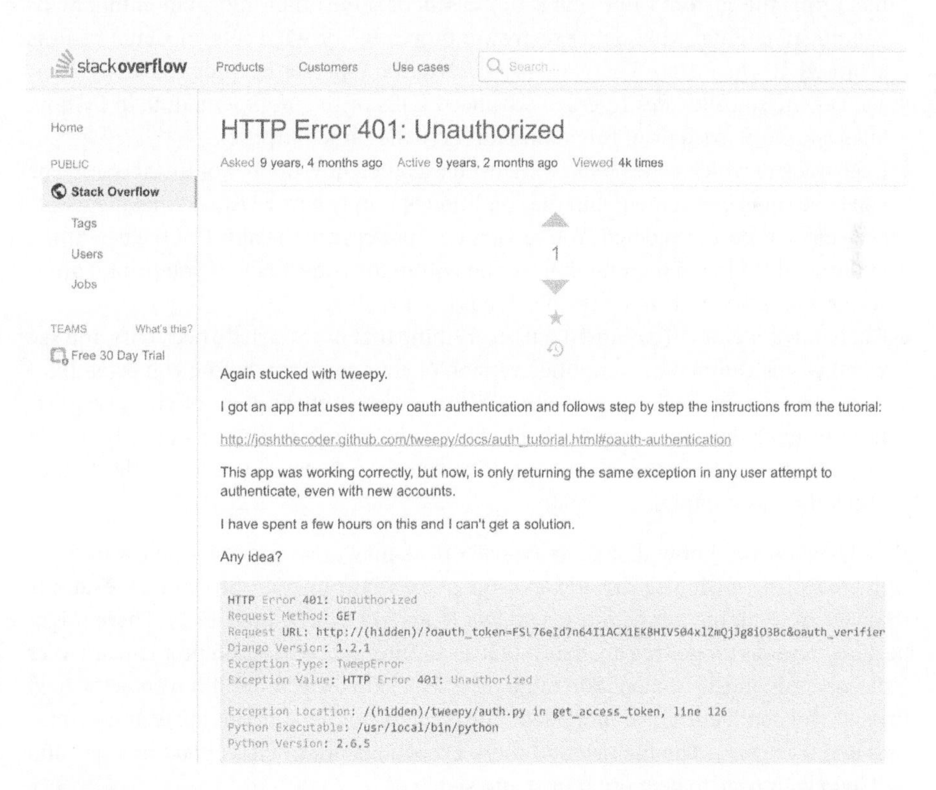

Figure 4.1 A online discussion in Stackover Flow about a warning message. *Source*: Stack Exchange Inc.

If you decide to do this, take some advice into account. First, be sure that the answer is not elsewhere within the same website (a first answer could be just a link to a previous post!). Second, don't worry that your question is silly or too basic: you will find in the community all kinds of coders, from those who are taking their first steps to those who are very advanced. Third, be clear, specific, and focus on what you need to solve. This is probably the most important advice since it is necessary that other coders understand what you need in a few words (not philosophical discussions or previous elaborated rationales) so they can decide to spend some minutes of their time and help you. It is a very common practice that you copy in the questions the warning message or the code you are having trouble with because peers can even fix it themselves and give the solution right away. Do not worry if your post receives a lot of replies after getting what you needed. This thread might also help others in the future!

4.2.2 Debugging Strategies

It's not always straightforward to understand what is going wrong. Maybe your script does not even produce an error message, but just produces some unexpected result.

Of course, every program is different and there is not one way to solve every issue, but there are some simple strategies that help you debugging your code. The underlying core principle is to better understand what exactly is happening.

- Print more. For example, if you have a for-loop, just add a print statement to the loop that prints the current value that is processed, or some other information that helps you understanding what data exactly are processed, or what intermediate result is achieved at which step. There are more advanced tools for keeping track of values, such as the so-called *debuggers* in advanced IDEs or the *logging* module in Python, but a couple of extra print functions can serve the same purpose.
- Keep track of which code blocks have been executed how often. For instance, maybe you have some *if* statement, but the condition is simply never True, so that the whole code block is never executed. You can create an integer with value 0 at the beginning of the code, and then increment it by one within the code block. If you print it afterwards, you know how often the block has been visited.
- Cut it down. Remove (comment out) everything that is not strictly necessary and see whether you can make a simplified version of your code run, before you extend it.
- Add consistency checks. For instance, if from a theoretical point of view, two lists need to have the same length, check it with the length function; similarly, if you know that an object must have a specific value (e.g., because you know the result), check this assumption.

Finally, when you know that some typical errors may arise and you don't want your script to stop or crash, you can add an `exception` in your code. Suppose for example that you are building a function to connect to an API (see Section 12.1). There might be many reasons for getting an error, such as an Internet connection problem, a server issue, or a missing document. You might decide to skip the error and continue the next lines or you could even give more detail instructions of what to do (i.e. wait five minutes and try again). The inclusion of these exceptions are in fact a good practice and will help your code to be more robust and stable.

Python Code

```python
headlines = ("US condemns terrorist  attacks",
     None, "Venezuelan president is dismissed")

for x in headlines:
     try:
          # Getting len of None will raise an error
          if len(x)<40:
               print(x)
     except:
          print(f"{x} is not a valid headline")
```

R Code

```r
headlines = list ("US condemns terrorist attacks",
  NA, "Venezuelan president is dismissed")

for (x in headlines){
     tryCatch(
          # Getting nchar of NA will raise an error
          if (nchar(x)<40) print(x),
          error=function(error_condition) {
            print(paste(x, "is not a valid headline"))
          }
     )
}
```

Python Output

```
None is not a valid headline
Venezuelan president is dismissed
```

R Output

```
[1] "NA is not a valid headline"
[1] "Venezuelan president is dismissed"
```

Example 4.1 Error handling.

Let's make Example 3.17 from the previous chapter more robust so that it does not fail if an invalid headline is passed. For instance, in Python, the object `None` has no defined length; and in R, it is illegal to calculate the number of characters of NA. It is a good idea to think about how you want to deal with this: either you want your script to just fail (and clean up the data), or you may want to deal with the error in some way. Especially if you have little control over the input data and/or if the process you are dealing with takes a long time, you may want to handle these errors rather than having your script fail. In Example 4.1, we show how to use such a try/except construction: you indicate which code block you want to try (e.g., run as normal); and in the next block, you indicate what should happen if that results in an error.

Note that using *try ... except* statements like this is fairly common in Python code, in R it is not needed as frequently. In many cases where a Python function like `int` raises an exception if the input cannot be converted to an integer, the R function `as.numeric` just returns a missing value. Thus, in R you normally only encounter these statements when using external resources, for example when using an API or scraping a web page. See Chapter 12 for more details on these topics.

4.3 Best Practice: Beautiful Code, GitHub, and Notebooks

This section gives a brief explanation of "computational hygiene": how to structure your code so you can understand it later, the importance of naming and documentation, the use of versioning and online repositories such as GitHub, and the use of literate programming (such as through the use of RMarkdown or Jupyter notebooks) to explain, share, and publish code.

Coding is more than learning the basic rules and creating a message. If you want to use code to communicate ideas and to work with peers you have to take care of many content and shape details in order to guarantee the comprehension and reproducibility of the scripts. It even applies to the code you write for "private use" because it is highly likely that you will forget your original thoughts from one day to another, or that you later realize that you need to share it with someone else to ask for help. Thus instead of writing personal, hidden and illegible code without adopting any social conventions, you should dedicate some extra effort to make your scripts easy and ready to share.

The first step of the computational hygiene is within the code itself. Every time you create an object, a variable or a function, you have to take many apparently unimportant decisions such as giving a name, separating words, lines or blocks, and including comments. These decisions are personal but should mostly depend on social conventions in order to be useful. As you may imagine, there are many of these conventions for general programming and specially for specific languages. To mention just a few, you can find an "official" style guide for Python[1] or Google's R style guide[2]. Some of these guides are extensive (they cover every detail) and some are more general or abstract. You do not have to see them as a "bible" that needs to be strictly adhered to in each and every situation, but they offer very good guidance for best practice. In fact, even when you find them useful it is true that you will probably learn more of these practices from reading good examples and especially from interacting with a specific community and its rules.

We will mention some general guidelines that apply for both R and Python. If it is the first time you are venturing into the world of code you will find this advice useful, but if you are a more advanced learner you will probably get more specific knowledge in the more detailed sources for each language and community.

In the case of naming we encourage you to use meaningful names or standard abbreviations to objects, using lower-case or mixed-case (remember both Python and R are case-sensitive!), avoiding special characters and operators (such as &, @ or %), and not exceeding 32 characters. You normally begin with a letter[3] (an upper-case when defining a class), followed by other letters or numbers, and using underscores to separate the words if necessary (i.e. `data_2020` or `Filter_Text`). Some suggest that variable names should be nouns and function names should be verbs, which seems logical if you think of the nature of these objects.

When writing code, please also take into consideration white space and indentations, because you should use them to give proper structure to the code by creating the block statements. In the case of R, also pay attention to the use of curly braces: the convention is that the opening curly brace begins after some code and is always followed by a new line; and the closing curly brace is in its own line except if there are more instructions in the block. Do not write very long lines (more than 80 characters) to help your code fit the screen and avoid lateral scrolling. Good separation of words, lines and blocks will make your script more readable!

Now, if you want to make your code highly understandable and shareable, you have to include *documentation*. This is probably a very basic dimension of coding but unfortunately some authors forget to take the few minutes it takes to describe what their script does (and why), making your journey more difficult. An essential good practice

---

[1] https://www.python.org/dev/peps/pep-0008/

[2] https://google.github.io/styleguide/Rguide.html

[3] An exception are so-called private identifiers – identifiers that are not supposed to be directly addressed. They conventionally begin with an underscore.

in coding is to include enough information to clarify your code when it is not clear by itself. You can do this in different ways (even by writing a separate codebook), but the most natural and straightforward manner is to include some *comments* in the code. These comments should be included both at the beginning of the script to give an overview or introduction to the code, and within the script (in independent lines or at the the end of a line) to give specific orientations. In many cases, you will need to read your code later (for example when you need to revise an article or analysis), and a short time spent documenting your code will save you a lot of time later.

R and Python use the hash sign # to create these comments. Notice that the comment will always begin after the hash, so this part will not be executed. If the first character in your line is a # all the text included will be considered as a comment; but if you have already written some code in a line and include a # after the code, the initial code will be executed and you will always see the comment by its side. You will normally combine these two ways of documenting your script. As a rule of thumb, insert a comment if the code itself is not obvious, and explain the choices and intentions of the code. So, if a line says `df = df - 1`, a comment like *Decrease df by one* is not very useful (as that is obvious from the code), but a comment like *Remove one degree of freedom since we estimated the mean* does help, as it makes it clear why we are subtracting one from the `df` object.

Additionally, Python and R encourage the use of so-called `docstrings`: In Python, place a string surrounded by triple quotation marks at the start of a function; in R, place a comment `#'` right above the function.[4] In this documentation, you can explain what the function does and what parameters it requires. The nice thing is that if properly used, docstrings are automatically displayed in help functions and automatically generated documentation.

Another way to make your code more beautiful and, crucially, easier to re-use by others and yourself is to make your code as generic as possible. For instance, imagine you need to calculate the sum of the length of two texts, "Good morning!" and "Goodbye!". You could just write `r = 13 + 8`. But what if the strings change in the future? And how to remember what `13 + 8` was supposed to mean? Instead of using such *hardcoded* values, you can therefore write it better as `r = len("Good morning!") + len("Goodbye")` (for R, replace `len` by `nchar`). But the strings themselves are still hardcoded, so you can create these strings and assign them the names `s1` and `s2` first, and then just calculate `r = len(s1) + len(s2)`. In practice, these types of generalization often involve the uses of functions (Section 3.3) and loops (Section 3.2.1). So, don't use hard-coded values or "magic numbers": `circumference=6.28*r` is much less clear than `PI=3.14; circumference= 2*PI*r`.

Moreover, you must be aware that your code is *dynamic* and it will normally evolve over time. For example, you may have different files (.py or .R) containing different versions of your script, though this is normally inefficient and chaotic. In order to have a more powerful control of versions and to track changes, coders usually use online repositories to host their scripts for private use and especially to share them. And there

---

[4] For more information, see https://www.python.org/dev/peps/pep-0257/#what-is-a-docstring and https://cran.r-project.org/web/packages/roxygen2/vignettes/roxygen2.html,respectively

are many of these sites, but we believe that GitHub[5] is the most well known and is preferred by data scientists (Figure 4.2 shows the repository we used to write this book).

Once you upload (or *commit* and *push*) your code to GitHub you can access it from anywhere and will be able to track the historical changes, which in practice will allow you to have multiple versions in the very same place. You will decide if you make the code public or keep it private, and who to invite to edit the repository. When working collaboratively you it will feel like editing a *wiki* of code, while having a *webpage* for your project and a *network of friends* (followers), will be similar to social media. You can work locally or even from a web interface, and then synchronize the changes. When you allow colleagues to download (or *clone*) your repository you are then making a good contribution to the community of developers and you can also monitor your impact. In addition to code, you can also upload other kinds of files, including notebooks, and organize them in folders, just as you have on your own computer.

One extended good practice when sharing code is the use of *literate programming*, which is an elegant, practical, and pedagogic way to document and execute the base code. We have already mentioned in this section the importance of including documentation within your code (i.e. using the # sign and docstrings), but you also have the opportunity to extend this documentation (with formatted texts, images and even equations!) and put everything together to present in a logical structure everything necessary to understand the code and to run the executable lines step by step.

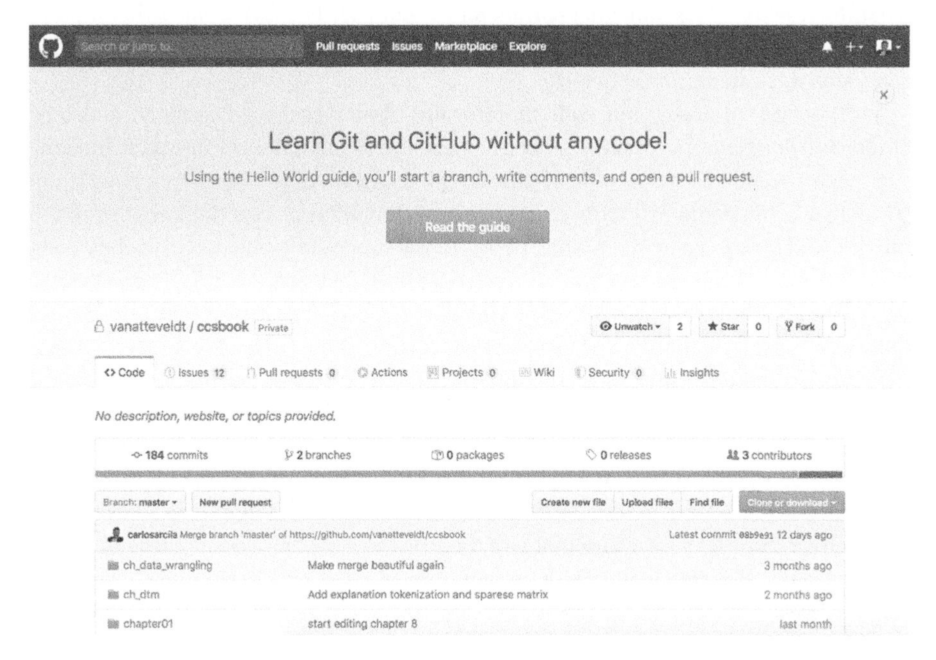

Figure 4.2 The online repository GitHub. *Source:* GitHub, Inc.

[5] https://github.com/

There are different approaches to implement this literate programming in web and local environments, but the standard in R and Python is the use of *notebooks*. In a notebook you can alternate a text processor with an executable cell to place formatted text between blocks of code. By doing this you can include complete documentation of your scripts, and even more important you can execute each cell one step at a time (loading the results in memory while the notebook is open). This last point allows you avoid the risk of executing the whole script at once, and also gives you more control of the intermediate outputs produced in your code. Once you get used to notebooks, you will probably never write code for data analysis in a basic editor again!

The usual tool in R is the R Markdown Notebook and in Python the Jupyter Notebook (see Figure 4.3), but in practice you can also deploy Python in Markdown and R in Jupyter. Both tools can help you with similar tasks to organize your script, though their internal technical procedures are quite different. We have chosen Jupyter to develop the examples in this book because it is a web-based interactive tool. Moreover, there are several services such as Google Colab[6] (Figure 4.4), that allow you to remotely run these notebooks online without installing anything on your computer, making the code highly reproducible.

So far you have seen many of the possibilities that the world of code offers you from a technical and collaboration perspective. We will come back to ethical and normative considerations throughout the book, in particular in Section 12.4 and Section 16.3.

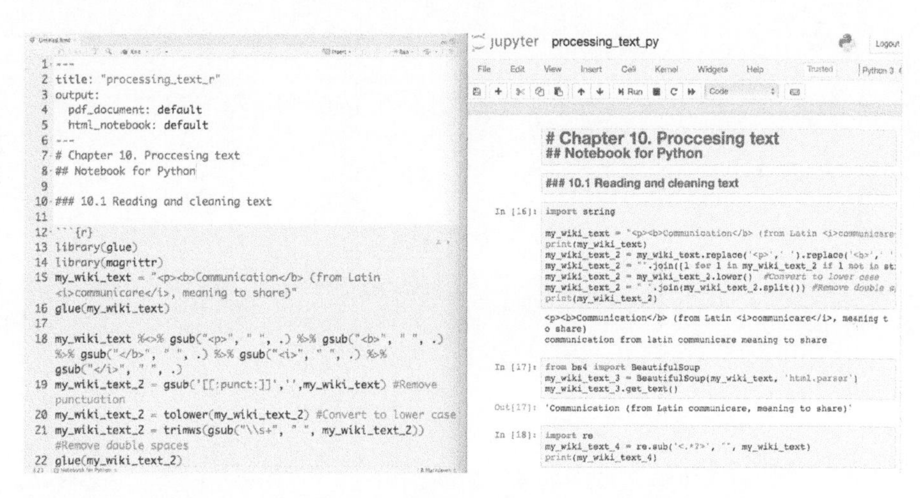

Figure 4.3 Markdown (left) and Jupyter (right) Notebooks. *Source*: Project Jupyter

---

[6] https://colab.research.google.com

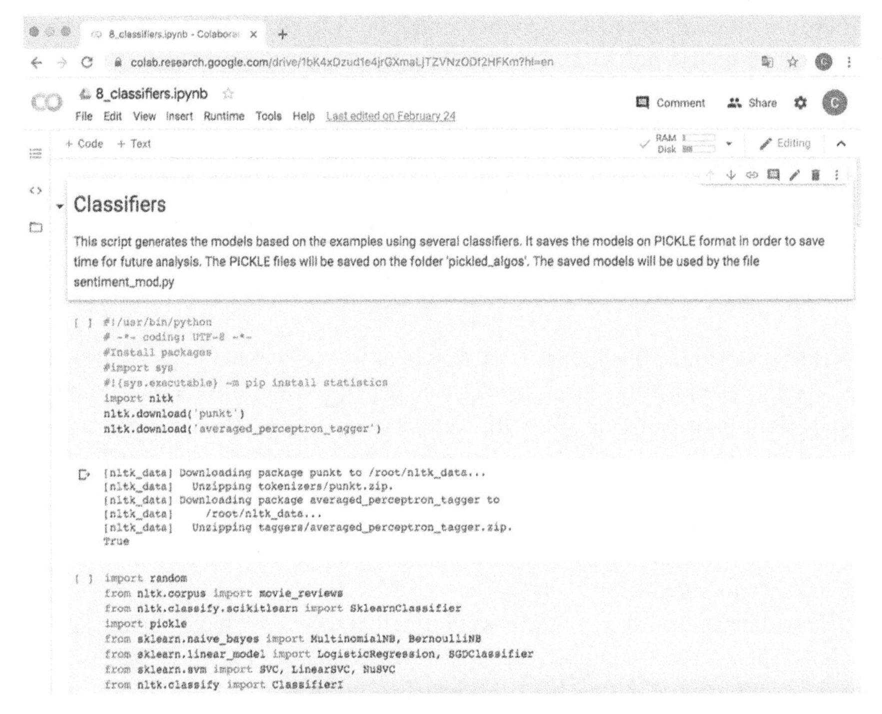

Figure 4.4 Jupyter notebook in Google Colab. *Source*: Project Jupyter

5

From File to Data Frame and Back

Abstract

This chapter teaches you the basics of file handling, such as different file formats and encodings. It introduces csv files, json files, plain text files, and binary file formats. We discuss different approaches to organizing data in files, and how to write data frames to and read them from these files. Finally, we provide guidance for retrieving example datasets.

Keywords file formats, encodings, reading and writing files, data frames, datasets

- Know how to handle different encodings and dialects
- Make an informed choice for a file format
- Know how to access existing datasets

Packages used in this chapter

This chapter relies mostly on the *pandas* (Python) and *tidyverse* (R) functionality to read and write files. Additionally, *haven* is used to read data from other tools such as SPSS. Finally, we show how to use existing data from packages such as *sotu* (R) and *nltk* and *scikit-learn* (Python). If needed, you can install these packages with the code below (see Section 1.4 for more details):

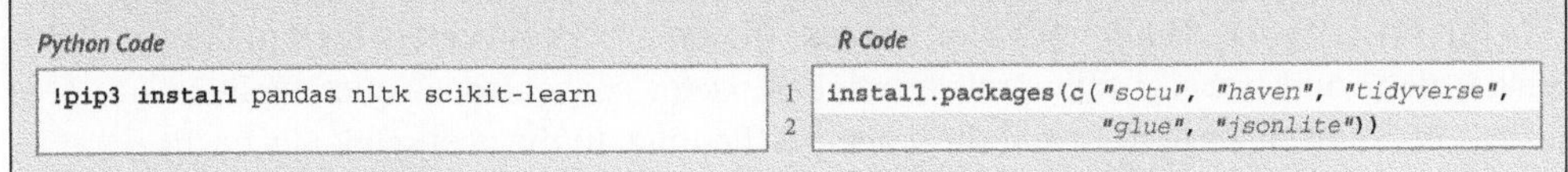

Python Code

```
!pip3 install pandas nltk scikit-learn
```

R Code

```
install.packages(c("sotu", "haven", "tidyverse",
                   "glue", "jsonlite"))
```

After installing, you need to import (activate) the packages every session:

Python Code

```
import json
import urllib
import pandas as pd
import nltk
from nltk.corpus import state_union
from sklearn.datasets import fetch_20newsgroups
```

R Code

```
library(tidyverse)
library(haven)
library(sotu)
library(glue)
library(jsonlite)
```

Computational Analysis of Communication: A Practical Introduction to the Analysis of Texts, Networks, and Images with Code Examples in Python and R, First Edition. Wouter van Atteveldt, Damian Trilling & Carlos Arcila Calderón.

5.1 Why and When Do We Use Data Frames?

In Section 3.1, we introduced basic data types: strings (which contain text), integers (which contain whole numbers, or numbers without anything "behind the dot"), floats (floating point numbers; numbers with decimals), and bools (boolean values, True or False). We also learned that a series of multiple values (e.g., multiple integers, multiple strings) can be stored in what we call a vector (R) or a list (Python).

In most social-scientific applications, however, we do not deal with isolated series of values. We rather want to link multiple values to each other. One way to achieve this is by the use of dictionaries (see Section 3.1). Such data structures are really useful for nested data: For example, if we do not want to store people's age only but also want to store their addresses, we could store a dict within a dict. In fact, as we will see later in this chapter, many data that computational social scientists use to work with come in such a format. For instance, data about an online product can contain many reviews which in turn have various pieces of information on the review author.

But ultimately, for many social-scientific analyses, a tabular data format is preferred. We are used to thinking of observations (cases) as rows with columns containing information or measurements about these observations (e.g., age, gender, days per week of newspaper reading, ...). It also simplifies how we can run many statistical analyses later on.

We could simply construct a list of lists to achieve such a tabular data format. In fact, this list-of-lists technique is often used to store tabular data or matrices, and you will probably encounter it in some examples in this book or elsewhere. The list-of-lists approach is very low-level, though: if we wanted, for instance, to insert a column or a row at a specific place, writing the code to do so could be cumbersome. There are also no things like column headers, and no consistency checks: nothing would warn us if one row actually contained more "columns" than another, which should not be the case in a rectangular table.

To make our lives easier, we can therefore use a data structure called a data frame. Data frames can be generated from list-of-list structures, from dictionaries, and many others. One way of doing this is shown in Example 5.1, but very often, you'd rather read data from a file or an online resource directly into a data frame (see Section 5.2).

In this book, we use data frames a lot, because they are very convenient for handling tabular data, and because they provide a lot of useful convenience functionalities, instead of requiring us to re-invent the wheel all the time. In the next section, we will discuss some of them.

Of course, there are some situations when data frames are *not* a good choice to organize your data:

- Your data are one-dimensional. Think, for example, of resources like a list of stop-words, or a list of texts without any meta-information.
- Your data do not have a tabular structure. Think, for example, of deeply nested data, network data or of very messy data.
- Your data are so large that you cannot (or do not want to) load it into memory. For instance, if you want to process the text of all articles on Wikipedia, you probably want to process them one-by-one instead of loading all articles at the same time.

Python Code

```python
# Create two lists that will be columns
list1 = ["Anna", "Peter", "Sarah", "Kees"]
list2 = [40, 33, 40, 77]

# or we could have a list of lists instead
mytable = [["Anna", 40],
           ["Peter", 33],
           ["Sarah", 40],
           ["Kees", 77]]


# Convert an array to a dataframe
df=pd.DataFrame(mytable)

# or create the data frame directly from vectors
df2=pd.DataFrame.from_records(zip(list1,list2))

# No. of rows, no. of columns, and shape
print(f"{len(df)} rows x {len(df.columns)} cols")
print(f"Its shape is {df.shape}")

print("Element-wise equality of df and df2:")
print(df == df2)
```

R Code

```r
# Create two vectors that will be columns
vector1 <- c("Anna","Peter","Sarah","Kees")
vector2 <- c(40,33,40,77)

# Create an array of four rows and two columns
myarray <- array(c(vector1,vector2), dim=c(4,2))





# Convert an array to a dataframe
df1=data.frame(myarray)

# or create the data frame directly from vectors
df2=data.frame(vector1, vector2)

# No. of rows, no. of columns, and dimension
print(glue("{ncol(df1)} rows x {nrow(df1)} cols"))
print(dim(df1))

print("Element-wise equality of df1 and df2:")
print(df1 == df2)
```

Python Output

```
4 rows x 2 cols
Its shape is (4,2)
Elemnent-wise equality of df and df2:
      0     1
0  True  True
1  True  True
2  True  True
3  True  True
```

R Output

```
2 rows x 4 cols
[1] 4 2
[1] "Element-wise equality of df1 and df2:"
       X1   X2
[1,] TRUE TRUE
[2,] TRUE TRUE
[3,] TRUE TRUE
[4,] TRUE TRUE
```

Example 5.1 Creating a data frame from other data structures.

Therefore, you will come across (and we will introduce you to) examples in which we do *not* use data frames to organize our data. But in most cases we will, because they make our life easier: once we constructed our data frame, we have a range of handy functions at our disposal, that allow us to select rows or columns, add new rows or columns, apply functions to them, and so on. We will discuss these in Chapter 6.

But how do we – toy examples like those in Example 5.1 aside – get data into and out of data frames?

5.2 Reading and Saving Data

5.2.1 The Role of Files

In statistical software like SPSS or Stata, or in all typical office applications for that matter, you *open* a file, do some work on it, and then *save* the changes to the same file once you are done. You basically "work on that file".

That's not how your typical workflow in R or Python looks like. Here, you work on one or multiple data frames (or some other data structures). That means that you might start by *reading* the contents of some file into a data frame, but once that is done, there is no link between the data frame and that file any more. Once your work is done, you can save your data frame to a file, of course, but it is a good practice not to overwrite your input file, so that you can always go back to where you started. A typical workflow would look like this:

1. Read raw data from file `myrawdata.csv` into data frame `df`
2. Do some operations and analyses on `df`
3. Save `df` to file `myfinaldata.csv`

Note that the last step is not even necessary, but may be handy if running the script takes very long, or if you want to re-distribute the resulting file.

The format in which we read files into a data frame and the format to which we save our final data frame by no means need to be identical. We can, for example, read data created by someone else in Stata's proprietary `.dta` format into a data frame and later save it to a `.csv` table.

While we sometimes do not have the choice in which format we get our input data, we have a range of options regarding our output data. We usually prefer formats that are *open* and *interoperable* for this, which ensures that they can be used by as many people as possible, and that they are not tied to any specific (proprietary) software tool which might not be available to everyone and can be discontinued in the future.

The most common file formats that are relevant to us are listed in Table 5.1. `txt` files are particularly useful for long texts (think of one file containing one newspaper article or even a whole book), but they are bad for storing associated meta data. `csv` files are the default choice for tabular data, and `json` files allow us to store nested data in a dictionary-like format.

For the sake of completeness, we also listed the native Python and R formats pickle, RDS, and RDA. Because of their lack of interoperability, they are not very suitable for long-term storage or for sharing data, but they can have a place in a workflow as an intermediate step to solve the issue that none of the other formats are able of store all properties of a data frame (e.g., the csv file cannot store whether a given column in an R data frame is to be understood as containing strings such as "man", "woman", "non-binary" or a factor with the three levels man, woman, non-binary). If it is important to store an object (such as a data frame) exactly as-it-is, we can use these formats. One of the rare instances where we use these formats is in Example 11.8, where we store machine learning models for later reuse.

Table 5.1 Basics of data frame handling.

	Used for?	**open**	**interoperable?**
txt	plain text	yes	yes
csv	tabular data	yes	yes
json	nested data, key-value pairs	yes	yes
pickle	Python objects	yes	no
RDS/RDA	R objects	yes	no

5.2.2 Encodings and Dialects

Plain `txt` files, `csv` files, and `json` files are all files that are based on text. Unlike binary file formats, you can read them in any text editor. Try it yourself to understand what is going on under the hood.

Download a csv file (such as https://cssbook.net/d/gun-polls.csv) and open it in a text editor of your choice. Some people swear that their preferred editor is the best (google to learn about the vi versus emacs war for some entertainment), but if you have no strong feeling, then Notepad++, Atom, or Sublime may be good choices that you may want to look into.

As you will see (Figure 5.1), a csv file internally just looks like a bunch of text in which each line represents a row and in which the columns are separated by a comma (hence the name comma separated values (csv)). Looking at the data in a text editor is a very good way to find out what happens if reading your files into a data frame does not work as expected – which can happen more frequently than you would expect.

Mostly due to historical reasons, not every text based file (which, as we have seen, includes csv files) is internally stored in the same way. For a long time, it was common to *encode* in such a way that one character mapped to one byte. That was easy from a programming perspective (after all, the *n*th character of a text can be directly read from and written to the *n*th byte of a file) and was also storage-efficient. But given that a byte consists of 8 bits, that means that there are only 256 possible characters. All letters in the alphabet in uppercase, again in lowercase, numbers, punctuation, some control characters – and you are out of characters. Due to this limitation, there were different encodings or codepages for different languages that told a program which value should be interpreted as which character.

We all know the phenomenon of garbled special characters, like German umlauts or Scandinavian characters like ø, å, or œ being displayed as something completely different. This happens when files are read with a different encoding than the encoding that was used for creating them.

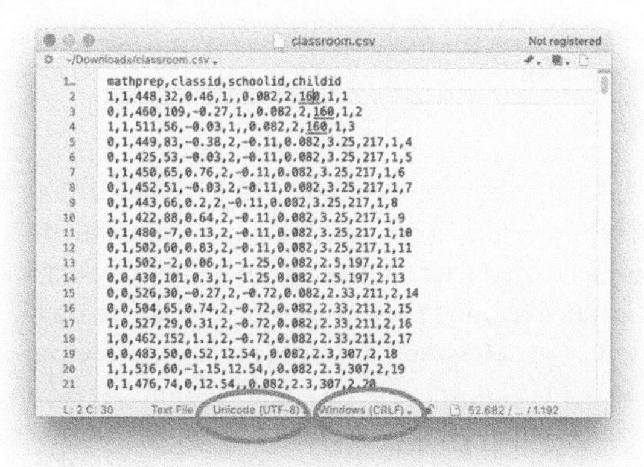

Figure 5.1 A csv file opened in a text editor, illustrating that the columns are separated by commas, and showing the encoding and the line endings. *Source:* Apple Inc.

In principle, this issue has been solved due to the advent of Unicode. Unicode allows all characters from all scripts to be handled, including emoticons, Korean and Chinese characters, and so on. The most popular encoding for Unicode characters is called UTF-8, and it has been around for decades.

To avoid any data loss, it is advisable to make sure that your whole workflow uses UTF-8 files. Most modern applications support UTF-8, even though some still by default use a different encoding (e.g., "Windows-1252") to store data. As Figure 5.1 illustrates, you can use a text editor to find out what encoding your data has, and many editors also offer an option to change the encoding. However, you cannot recover what has been lost (e.g., if at one point you saved your data with an encoding that only allows 256 different characters, it follows logically that you cannot recover that information).

As we will show in the practical code examples below, you can also force Python and R to use a specific encoding, which can come in handy if your data arrives in a legacy encoding.

Related to the different encodings a file can have, but less problematic, are different conventions of how a *line ending* is denoted. Windows-based programs have been using a Carriage Return followed by a Line Feed (denoted as `\r\n`), very old versions of MacOS used a Carriage Return only (`\r`), and newer versions of MacOS as well as Linux use a Line Feed only (`\n`). In our field, the Linux (or Unix) style line endings have become most dominant, and Python 3 even automatically converts Windows style line endings to Unix style line endings when reading a file – even on Windows itself.

A third difference is the use of so-called *byte-order markers* (BOM). In essence, a BOM is an additional byte added to the beginning of a text file to indicate that it is a UTF-encoded file and to indicate in which order the bytes are to be read (the so-called endianness). While informative, this can cause trouble if your program does not expect that byte to be there. In that case, you might either want to remove it or explicitly specify the encoding as such. For instance, in Example 5.3, you can add an argument such as encoding="utf-8" or encoding="utf8-bom" to the open (Python) or scan (R) command.

In short, the most standard form in which you probably want to encode your data is in UTF-8 with Linux-style line endings without the use of a byte-order marker.

In the case of reading and writing csv files, we thus need to know the encoding, and potentially also the line ending conventions and the presence of a byte-order marker. However, there are also some additional variations that we need to consider. There is no single definition of what a csv file needs to look like, and there are multiple dialects that are widely used. They mainly differ in two aspects: the delimiter that is chosen, and the quoting and/or escaping of values.

First, even though csv stands for comma separated values, one could use other characters instead of a comma to separate the columns. In fact, because many countries use a comma instead of a dot as a decimal separator ($10.30 versus 10,30€), in these countries a semicolon (;) is used instead of a comma as column delimiter. To avoid any possible confusion, others use a tab character (`\t`) to separate columns. Sometimes, these files are then called a tab-separated file, and instead of .csv, they may have a file extension such as `.tsv`, `.tab`, or even `.txt`. However, this does not change the way how you can read them – but what you need to know is whether your columns are separated by `,`, `;`, or `\t`.

Second, there may be different ways to deal with strings as values in a csv file. For instance, it may be that a specific value contains the same character that is also used as a delimiter. These cases are usually resolved by either putting all strings into quotes, putting only strings that contain such ambiguities in quotes, or by prepending the ambiguous character with a specific escape character. Most likely, all of this is just handled automatically under the hood, but in case of problems, you might want to look into this and check out the documentation of the packages you are using on how to specify which strategy is to be used.

Let's get practical and try out reading and writing files into a data frame (Example 5.2).

Of course, we can read more than just csv files. In the Python example, you can use tabcompletion to get an overview of all file formats Python supports: type `pd.read` and then press the TAB key to get a list of all supported files. For instance, you could `pd.read_excel('test.xlsx')`, `df3 = pd.read_stata('test.dta')`, or `df4 = pd.read_json('test.json')` Similarily, for R, you can hit TAB after typing `haven::` to get an overview over functions such as `read_spss`.

Python Code

```python
url = "https://cssbook.net/d/media.csv"
# Directly read a csv file from internet
df = pd.read_csv(url)

# We can also explicitly specify delimiter etc.
df = pd.read_csv(url, delimiter = ",")
# Note: use help(pd.read_csv) to see all options

# Save dataframe to a csv:
df.to_csv("mynewcsvfile.csv")
```

R Code

```r
url = "https://cssbook.net/d/media.csv"
# Directly read a csv file from internet
df = read_csv(url)

# We can also explicitly specify delimiter etc.
df = read_delim(url, delim = ",")
# Note: use ?read_csv to see all options

# Save dataframe to a csv:
write_csv(df, "mynewcsvfile.csv")
```

Example 5.2 Reading files into a data frame.

5.2.3 File Handling Beyond Data Frames

Data frames are a very useful data structure for organizing and analyzing data, and will occur in many examples in this book. However, not all things that we might want to read from a file needs to go into a data frame. Imagine if we have a list of words that we later want to remove from some texts (so-called stopwords, see Chapter 9). We could make a list (or vector) of such words directly in our code. But if we have more than a couple of such words, it is easier and more readable to keep them in an external file. We could create a file `stopwords.txt` in a text editor with one of such words per line:

```
and
or
a
an
```

If you do not wish to create this list yourself, you could also download one from https://cssbook.net/d/stopwords.txt and save it in the same directory as your Python or R script.

Then, you can read this file into a vector or list (see Example 5.3).

Python Code

```python
# Define stopword list in the code itself
stopwords = ["and","or","a","an","the"]


# Better idea: Download stopwords file and read it
url = "https://cssbook.net/d/stopwords.txt"
urllib.request.urlretrieve(url, "stopwords.txt")
with open("stopwords.txt") as f:
        stopwords = [w.strip() for w in f]
stopwords
```

R Code

```r
# Define stopword list in the code itself
stopwords = c("and", "or", "a", "an", "the")

# Better idea: Download stopwords file and read it
url = "https://cssbook.net/d/stopwords.txt"
download.file(url, "stopwords.txt")
stopwords = scan("stopwords.txt", what="string")
stopwords
```

Example 5.3 Reading files without data frames.

Example 5.4 provides you with some more elaborate code examples that allows us to dig a bit deeper into the general way of handling files.

In the Python example, we can open a file and assign a handle to it that allows us to refer to it (the name of the handle is arbitrary, let's just call it `f` here). Then, we can use a for loop to iterate over all lines in the file and add it to a list.

Python Code

```python
# Modify the stopword list and save it:
stopwords += ["somenewstopword", "andanotherone"]
with open("newstopwords.txt", mode = "w") as f:
    f.writelines(stopwords)



# Use json to read/write dictionaries
somedict = {"label":"Report",
            "entries":[1,2,3,4]}

with open("test.json", mode = "w") as f:
    json.dump(somedict, f)

with open("test.json", mode = "r") as f:
    d = json.load(f)
print(d)
```

R Code

```r
# Modify the stopword list and save it:
stopwords = c(stopwords,
              "somenewstopword", "andanotherone")
fileConn<-file("newstopwords.txt")
writeLines(stopwords, fileConn)
close(fileConn)

# Use json to read/write named lists
somedict = list(label="Report",
                entries=c(1,2,3,4))

write_json(somedict, "/tmp/x.json", auto_unbox=T)


d=read_json("/tmp/x.json", simplifyVector = T)
print(d)
```

Example 5.4 More examples of reading from and writing to files.

The `mode = 'r'` specifies that we want to read from the file. `mode = 'w'` would open the file for writing, create it if necessary, and immediately deletes all content that may have been in there if the file already existed (!). Note that the `.strip()` is necessary to remove the line ending itself, and also any possible whitespace at the beginning or end of a line. If we want to save our stopwords, we can do this in a similar way: we first open the file (this time, for writing), and then use the file handle's methods to write to it. We are not limited to plain text files, here. For instance, we can use the same approach to read json files into a Python dict or to store a Python dict into a json file.

We could also combine this with a for loop that goes over all files in a dictionary. Imagine we have a folder full of positive movie reviews, and another one full of negative movie reviews that we want to use to train a machine learning classifier (see Section 11.4). Let's further assume that all these reviews are saved as `.txt` files. We can iterate over all of them, as shown in Example 11.1. If you want to read text files into a data frame in R, the *readtext* package may be interesting for you.

5.3 Data from Online Sources

Many data that are interesting to those analyzing communication are nowadays gathered online. In Chapter 12, you will learn how to use APIs to retrieve data from web services, and how to write your own web scraper to automatically download large numbers of web pages and extract relevant information. For instance, you might want to retrieve customer reviews from a website or articles from news sites.

In this section, however, we will focus on how to re-use existing datasets that others have made available online. For instance, the open science movement has led to more and more datasets being shared openly using repositories such as Dataverse, Figshare, or others. Re-using existing data can be very good for several reasons: first, to confirm (or not) the conclusions drawn by others; second, to avoid wasting resources by re-collecting very similar or even identical data all over again; and third, because gathering a large, high-quality dataset might just not be feasible with your means. This is especially true when you need annotated (i.e., hand-coded) data for supervised machine learning purposes (Chapter 8).

We can distinguish between two types of existing online datasets: datasets that are inherently interesting, and so-called toy datasets.

Toy datasets may include made-up data, but often, contain real data. However, they are not analyzed to gain scientific insights (any more), as they may be too small, outdated, or already analyzed all-over again. These provide a great way, though, to learn and explore new techniques: after all, the results and the characteristics of the data are already known. Hence, such toy datasets are often even included in R and Python packages. Some of them are really well-known in teaching (e.g., the iris dataset containing measurements of some flowers; or the titanic dataset containing statistics on survival rates of passengers on the Titanic; MINIST for image classification; or the MPG dataset on car fuel consumption). Many of these are included in packages like *scikit-learn*, *seaborn*, or *ggplot2*– and you can have a look at their documentation.

For instance, the 20 Newsgroups dataset contains 18 846 posts from newsgroups plus the groups where they were posted (Example 5.5). This can be an interesting resource for practising with natural language processing, unsupervised, and supervised machine learning. Other interesting resources are collections of political speeches, such as the state-of-the-union speeches from the US, which are available in multiple packages (Example 5.6). Other interesting datasets with large collections of textual data may be the Financial News dataset compiled by Chen (2017) or the political news dataset compiled by Horne et al. (2018).

Python Code

```
# Note: use fetch_20newsgroups? for more options
d=fetch_20newsgroups(
    remove=("headers", "footers", "quotes"))
df=pd.DataFrame(zip(d["data"], d["target_names"]))
df.head()
```

R Code

```
url = "https://cssbook.net/d/20_newsgroups.csv"
d = read_csv(url)
head(d)
```

Python Output

```
    0                                                  1
I was wondering if anyone out there could enli...   alt.atheism
A fair number of brave souls who upgraded thei...    comp.graphics
well folks, my mac plus finally gave up the gh...   comp.os.ms-windows.misc
\nDo you have Weitek's address/phone number? ...   comp.sys.ibm.pc.hardware
From article <C5owCB.n3p@world.std.com>, by to...   comp.sys.mac.hardware
```

Example 5.5 In Python, *scikit-learn* has a convenience function to automatically download the 20 newsgroup dataset and automatically clean it up. In R, you can download the raw version (there are multiple copies floating around on the internet) and perform the cleaning yourself.

Python Code

```python
# Note: download is only needed once...
nltk.download("state_union")
sentences = state_union.sents()
print(f"There are {len(sentences)} sentences.")
```

R Code

```r
speeches = sotu_meta
# show only first 50 characters
speeches %>%
    mutate(text = substr(sotu_text,0,50)) %>%
    head()
```

R Output

president	year	years_active	party	sotu_type	text
<chr>	<int>	<chr>	<chr>	<chr>	<chr>
George Washington	1790	1789-1793	Nonpartisan	speech	entatives: I embrace with great satisfaction the
George Washington	1790	1789-1793	Nonpartisan	speech	resentatives: In meeting you again I feel much sa
George Washington	1791	1789-1793	Nonpartisan	speech	resentatives: "In vain may we expect peace with
George Washington	1792	1789-1793	Nonpartisan	speech	entatives: It is some abatement of the satisfacti
George Washington	1793	1793-1797	Nonpartisan	speech	resentatives: Since the commencement of the term
George Washington	1794	1793-1797	Nonpartisan	speech	resentatives: When we call to mind the gracious i

Example 5.6 A collection of US state-of-the-union speeches is available in multiple packages in various forms.

There are also some more generic resources that you may want to consider for finding more datasets to play around with. On https://datasetsearch.research.google.com, you can search for datasets of all kinds, both really interesting ones and toy datasets. Another great research is https://kaggle.com, a site that hosts data science competitions.

6

Data Wrangling

Abstract

This chapter shows you how to do "data wrangling" in R and Python. Data wrangling is the process of transforming raw data into a shape that is suitable for analysis. The sections of this chapter first take you through the normal data wrangling pipeline of filtering, changing, grouping, and joining data. Finally, the last section shows how you can reshape data.

Keywords data wrangling, data cleaning, filtering, merging, reshaping

- Filter rows and columns in data frames
- Compute new columns and summary statistics for data frames
- Reshape and merge data frames

Packages used in this chapter

This chapter uses the *readxl* package for reading Excel files and various parts of the *tidyverse* including *ggplot2*, *dplyr*, and *tidyr* (which are installed automatically when you install tidyverse). In Python we rely mostly on the *pandas* package, but we also use *scipy* package for statistics and the *xlrd* for reading Excel files. You can install these packages with the code below if needed (see Section 1.4 for more details):

Python Code

```
!pip3 install pandas scipy seaborn xlrd
```

R Code

```
install.packages("tidyverse")
```

After installing, you need to import (activate) the packages every session:

Python Code

```
import pandas as pd
import seaborn as sns
import scipy.stats
import re
```

R Code

```
library(tidyverse)
library(readxl)
```

Computational Analysis of Communication: A Practical Introduction to the Analysis of Texts, Networks, and Images with Code Examples in Python and R, First Edition. Wouter van Atteveldt, Damian Trilling & Carlos Arcila Calderón.

6.1 Filtering, Selecting, and Renaming

Selecting and renaming columns. A first clean up step we often want to do is removing unnecessary columns and renaming columns with unclear or overly long names. In particular, it is often convenient to rename columns that contain spaces or non-standard characters, so it is easier to refer to them later.

Selecting rows. As a next step, we can decide to filter certain rows. For example, we might want to use only a subset of the data, or we might want to remove certain rows because they are incomplete or incorrect.

As an example, FiveThirtyEight published a quiz about American public opinion about guns, and were nice enough to also publish the underlying data[1]. Example 6.1 gives an example of loading and cleaning this dataset, starting with the function `read_csv` (included in both *tidyverse* and *pandas*) to load the data directly from the Internet. This dataset contains one poll result per row, with a *Question* column indicating which question was asked, and the columns listing how many Americans (adults or registered voters) were in favor of that measure, in total and for Republicans and Democrats. Next, the columns *Republican* and *Democratic Support* are renamed to shorten the names and remove the space. Then, the URL column is dropped using the *tidyverse* function `select` in R or the *pandas* function `drop` in Python. Notice that the result of these operations is assigned to the same object `d`. This means that the original `d` is overwritten.

Note: In R, the *tidyverse* function `select` is quite versatile. You can specify multiple columns using `select(d, column1, column2)` or by specifying a range of columns: `select(d, column1:column3)`. Both commands keep only the specified columns. As in the example, you can also specify a negative selection with the minus sign: `select(d, -column1)` drops `column1`, keeping all other columns. Finally, you can rename columns in the select command as well: `select(d, column1=col1, column2)` renames `col` to `column1`, keeps that column and `column2`, and drops all other columns.

We then filter the dataset to list only the polls on whether teachers should be armed (you can understand this is close to our heart). This is done by comparing the value of the *Question* column to the value `'arm-teachers'`. This comparison is done with a double equals sign (==). In both Python and R, a single equals sign is used for assignment, and a double equals sign is used for comparison. A final thing to notice is that while in R we used the *dplyr* function (`filter`) to filter out rows, in Python we *index* the data frame using square brackets on the *pandas* DataFrame attribute `loc`(ation): `d.loc[]`.

[1] https://projects.fivethirtyeight.com/guns-parkland-polling-quiz/; see https://github.com/fivethirtyeight/data/tree/master/poll-quiz-guns for the underlying data.

Note that we chose to assign the result of this filtering to d2, so after this operation we have the original full dataset d as well as the subset d2 at our disposal. In general, it is your choice whether you overwrite the data by assigning to the same object, or create a copy by assigning to a new name[2]. If you will later need to work with a different subset, it is smart to keep the original so you can subset it again later. On the other hand, if all your analyses will be on the subset, you might as well overwrite the original. We can always re-download it from the internet (or reload it from our harddisk) if it turns out we needed the original anyway.

Python Code

```
url="https://cssbook.net/d/guns-polls.csv"
d=pd.read_csv(url)
d = d.rename(columns={"Republican Support": "rep",
    "Democratic Support": "dem"})
d = d.drop(columns="URL")
# alternatively, we can write:
# d.drop(columns="URL", inplace=True)
d2 = d.loc[d.Question == "arm-teachers"]
d2
```

R Code

```
url="https://cssbook.net/d/guns-polls.csv"
d = read_csv(url)
d = rename(d, rep= 'Republican Support',
           dem= 'Democratic Support')
d = select(d, -URL)


d2 = filter(d, Question == "arm-teachers")
d2
```

R Output

Question <chr>	Start <chr>	End <chr>	Pollster <chr>	Population <chr>	Support <dbl>	rep <dbl>	dem <dbl>
arm-teachers	2/23/18	2/25/18	YouGov/Huffpost	Registered Voters	41	69	20
arm-teachers	2/20/18	2/23/18	CBS News	Adults	44	68	20
arm-teachers	2/27/18	2/28/18	Rasmussen	Adults	43	71	24
arm-teachers	2/27/18	2/28/18	NPR/Ipsos	Adults	41	68	18
arm-teachers	3/3/18	3/5/18	Quinnipiac	Registered Voters	40	77	10
arm-teachers	2/26/18	2/28/18	Survey Monkey	Registered Voters	43	80	11

Example 6.1 Filtering.

6.2 Calculating Values

Very often, we need to calculate values for new columns or change the content of existing columns. For example, we might wish to calculate the difference between two columns, or we may need to clean a column by correcting clerical errors or converting between data types.

In these steps, the general pattern is that a column is assigned a new value based on a calculation that generally involves other columns. In both R and Python, there are two general ways to accomplish this. First, you can simply assign the new values to an existing or new column, using the column selection notation discussed in Section 3.1: `df["column"] = ...` in Python, or `df$column = ...` in R.

---

[2] Keep in mind that in Python, `df2=df1` does *not* create a copy of a data frame, but a pointer to the same memory location (see the discussion on mutable objects in Section 3.1). This may often not be of practical importance, but if you really need to be sure that a copy is created, use `df2=df1.copy()`.

Both Python and R also offer a function that allows multiple columns to be changed, returning a new copy of the data frame rather than changing the original data frame. In R, this is done using the *tidyverse* function `mutate`, which is the recommended way to compute values. The Python equivalent, *pandas* function `assign`, is used more rarely as it does not offer many advantages over direct assignment.

In either case, you can use arithmetic: e.g. `rep - dem` to compute the difference between these columns. This works directly in R `mutate`, but in Python or in R direct assignment you also need to specify the name of the data frame. In Python, this would be `d["rep"] - d["dem"]`[3], while in R this is `d$rep - d$dem`.

In many cases, however, you want to use various functions to perform tasks like cleaning and data conversion (see Section 3.3 for a detailed explanation of built-in and custom functions). For example, to convert a column to numeric you would use the base R function `as.numeric` in R or the *pandas* function `to_numeric` in Python. Both functions take a column as argument and convert it to a numeric column.

Almost all R functions work on whole columns like that. In Python, however, many *functions* work on individual values rather than columns. To apply a function on each element of a column `col`, you can use `df.col.apply(my_function)` (where `df` and `col` are the names of your data frame and column). In contrast, *Pandas* columns have multiple useful *methods* that – because they are methods of that column – apply to the whole column[4]. For example, the method `df.col.fillna` replaces missing values in the column `col`, and `df.col.str.replace` conducts a find and replace. Unlike functions that expect individual values rather than columns as an input, there is no need to explicitly `apply` such a method. As always, you can use tab completion (pressing the TAB key after writing `df.col.`) to get a menu that includes all available methods.

To illustrate some of the many possibilities, Example 6.2 has code for cleaning a version of the guns polls in which we intentionally introduced two problems: we added some typos to the *rep* column and introduced a missing value in the *Support* column. To clean this, we perform three steps: First, we remove all non-numeric characters using a regular expression (see Section 9.2 for more information on text handling and regular expressions). Next, we need to explicitly convert the resulting column into a numeric column so we can later use it in calculations. Finally, we replace the missing value by the column mean (of course, it is doubtful that that is the best strategy for imputing missing values here, we do it mainly to show how one can deal with missing values technically. You will find some more discussion about missing values in Section 7.1).

The cleaning process is actually performed twice: lines 5-10 use direct assignment, while lines 12-19 use the `mutate/assign` function. Finally, lines 21-27 show how you can define and apply a custom function to combine the first two cleaning steps. This can be quite useful if you use the same cleaning steps in multiple places, since it reduces the repetition of code and hence the possibility of introducing bugs or inconsistencies.

Note that all these versions work fine and produce the same result. In the end, it is up to the researcher to determine which feels most natural given the circumstances.

---

[3] You can also write `d.rep - d.dem`, which is shorter, but does not work if your column names contain, for instance, spaces.

[4] See Section 3.3 for a refresher on methods and functions.

Python Code

```python
# version of the guns polls with some errors
url="https://cssbook.net/d/guns-polls-dirty.csv"
d2=pd.read_csv(url)

# Option 1: clean with direct assignment
# Note that when creating a new column,
# you have to use df["col"] rather than df.col
d2["rep2"] = d2.rep.str.replace("[^0-9\\.]", "")
d2["rep2"] = pd.to_numeric(d2.rep2)
d2["Support2"]=d2.Support.fillna(d.Support.mean())

# Alternatively, clean with .assign
# Note the need to use an anonymous function
# (lambda) to chain calculations
cleaned = d2.assign(
    rep2 = d2.rep.str.replace("[^0-9\\.]", ""),
    rep3 = lambda d2: pd.to_numeric(d2.rep2),
    Support2=d2.Support.fillna(d2.Support.mean()))


# Finally, you can create your own function
def clean_num(x):
    x = re.sub("[^0-9\\.]", "", x)
    return int(x)

cleaned["rep3"] = cleaned.rep.apply(clean_num)
cleaned.head()
```

R Code

```r
# version of the guns polls with some errors
url="https://cssbook.net/d/guns-polls-dirty.csv"
d2 = read_csv(url)

# Option 1: clean with direct assignment.
# Note the need to specify d2$ everywhere
d2$rep2=str_replace_all(d2$rep, "[^0-9\\.]", "")
d2$rep2 = as.numeric(d2$rep2)
d2$Support2 = replace_na(d2$Support,
                         mean(d2$Support, na.rm=T))

# Alternative, clean with mutate
# No need to specify d2$,
# and we can assign to a new or existing object
cleaned = mutate(d2
    rep2 = str_replace_all(rep, "[^0-9\\.]", ""),
    rep2 = as.numeric(rep2),
    Support2 = replace_na(Support,
        mean(Support, na.rm=TRUE)))

# Finally, you can create your own function
clean_num = function(x) {
    x = str_replace_all(x, "[^0-9\\.]", "")
    as.numeric(x)
}
cleaned = mutate(cleaned, rep3 = clean_num(rep))
head(cleaned)
```

R Output

Question <chr>	Start <chr>	End <chr>	Pollster <chr>	Population <chr>	Support <dbl>	rep <chr>	dem <dbl>	rep2 <dbl>	Support2 <dbl>	rep3 <dbl>
arm-teachers	2/23/18	2/25/18	YouGov/Huffpost	Registered Voters	41	69	20	69	41.0	69
arm-teachers	2/20/18	2/23/18	CBS News	Adults	NA	68	20	68	41.6	68
arm-teachers	2/27/18	2/28/18	Rasmussen	Adults	43	71d	24	71	43.0	71
arm-teachers	2/27/18	2/28/18	NPR/Ipsos	Adults	41	68	18	68	41.0	68
arm-teachers	3/3/18	3/5/18	Quinnipiac	Registered Voters	40	77	10	77	40.0	77
arm-teachers	2/26/18	2/28/18	SurveyMonkey	Registered Voters	43	80	11	80	43.0	80

Example 6.2 Mutate.

As noted above, in R we would generally prefer `mutate` over direct assignment, mostly because it fits nicely into the *tidyverse* workflow and you do not need to repeat the data frame name. In Python, we would generally prefer the direct assignment, unless you want to make a copy of the data with these changes, in which case `assign` can be more useful.

6.3 Grouping and Aggregating

The functions we used to change the data above operated on individual rows. Sometimes, however, we wish to compute summary statistics of groups of rows. This essentially shifts the unit of analysis to a higher level of abstraction. For example, we could compute per-school statistics from a data file containing information per student; or we could compute the average number of mentions of a politician per day from data file containing information per articles (each date might have multiple articles and each article multiple mentions to politicians!).

In data analysis, this is called *aggregation*. In both Python and R, it consists of two steps: First, you define which rows are *grouped* together to form a new unit by specifying which column identifies these groups. In the previous examples, this would be the school name or the date of each article. It is also possible to group by multiple columns, for example to compute the average per day per news source.

The next step is to specify one or more summary (or *aggregation*) functions to be computed over the desired value columns. These functions compute a summary value, like the mean, sum, or standard deviation, over all the values belonging to each group. In the example, to compute average test scores per school we would apply the average (or mean) function to the test score value column. In general, you can use multiple functions (e.g. mean and variance) and multiple columns (e.g. mean test score and mean parental income).

The resulting dataset is reduced both in rows and in columns. Each row now represents a group of previuos cases (e.g. school or date), and the columns are now only the grouping columns and the computed summary scores.

Example 6.3 shows the code in R and Python to define groups and compute summary values. First, we group by poll *question*; and for each question, we compute the average and standard deviation. The syntax is a little different for R and Python, but the idea is the same: first we create a new variable `groups` that stores the grouping information, and then we create the aggregate statistics. In this example, we do not store the result of the computation, but print it on the screen. To store the results, simply assign it to a new object as normal.

In R, you use the *dplyr* function `group_by` to define the groups, and then call the function `summarize` to compute summary values by specifying `name=function(value)`.

In Python, the grouping step is quite similar. In the summarization step, however, you specify which summaries to compute in a dictionary[5]. The keys of the dictionary list the value columns to compute summaries of, and the values contain the summary functions to apply, so `{'value': function}` or `{'value': [list of functions]}`.

6.3.1 Combining Multiple Operations

In the examples above, each line of code (often called a *statement*) contained a single operation, generally a call to a function or method (see Section 3.3). The general shape of each line in R was `data = function(data, arguments)`, that is, the data is provided as the first argument to the function. In Python, we often used methods that "belong to" objects such as data frames or columns. Here, we therefore specify the object itself followed by a period and its method that is to be called, i.e. `object = object.method(arguments)`.

Although there is nothing wrong with limiting each line to a single operation, both languages allow multiple operations to be chained together. Especially for grouping

[5] See Section 3.1 for more information on working with dictionaries.

Python Code

```python
groups = d.groupby("Question")
groups.agg({"Support": ["mean", "std"]})
```

R Code

```r
groups = group_by(d, Question)
summarize(groups, m=mean(Support), sd=sd(Support))
```

R Output

Question <chr>	m <dbl>	sd <dbl>
age-21	75.85714	6.011893
arm-teachers	42.00000	1.549193
background-checks	87.42857	7.322503
ban-assault-weapons	61.75000	6.440285
ban-high-capacity-magazines	67.28571	3.860669
mental-health-own-gun	85.83333	5.455884
repeal-2nd-amendment	10.00000	NA
stricter-gun-laws	66.45455	5.145165

Example 6.3 Aggregation. Note that in the Python example, we can specify often-used functions such as `"mean"` simply as a string, but instead, we could also pass functions directly, such as numpy's `np.mean`.

and summarizing, it can make sense to link these operations together as they can be thought of as a single "data wrangling" step.

In Python, this can be achieved by adding the second `.method()` directly to the end of the first statement. Essentially, this calls the second method on the result of the first method: `data = data.method1(arguments).method2(arguments)`. In R, the data needs, of course, to be included in the function arguments. But we can also chain these function calls. This is done using the *pipe operator* (`%>%`) from the (cutely named) *magrittr* package. The pipe operator inserts the result of the first function as the first argument of the second function. More technically, `f1(d) %>% f2()` is equivalent to `f2(f1(d))`. This can be used to chain multiple commands together, e.g. `data = data %>% function1(arguments) %>% function2(arguments)`.

Python Code

```python
d.groupby("Question").agg(
    {"Support": ["mean", "std"]})
```

R Code

```r
d %>% group_by(Question) %>%
    summarize(m=mean(Support), sd=sd(Support))
```

Example 6.4 Combining multiple functions or methods. The result is identical to Example 6.3.

Example 6.4 shows the same operation as in Example 6.3, but chained into a single statement.

6.3.2 Adding Summary Values

Rather than reducing a data frame to contain only the group-level information, it is sometimes desirable to add the summary values to the original data. For example, if we add the average score per school to the student-level data, we can then determine whether individual students outperform the school average.

Of course, the summary scores are the same for all rows in the same group: all students in the same school have the same school average. So, these values will be repeated for these rows, essentially mixing individual and group level variables in the same data frame.

Python Code

```
# Note the use of ( ) to split a long line
d["mean"] = (d.groupby("Question")["Support"]
             .transform("mean"))
d["deviation"] = d["Support"] - d["mean"]
d.head()
```

R Code

```
d = d %>% group_by(Question) %>%
  mutate(mean = mean(Support),
         deviation=Support - mean)
head(d)
```

R Output

Question <chr>	Start <chr>	End <chr>	Pollster <chr>	Population <chr>	Support <dbl>	rep <dbl>	dem <dbl>	mean <dbl>	deviation <dbl>
age-21	2/20/18	2/23/18	CNN/SSRS	Registered Voters	72	61	86	75.85714	-3.857143
age-21	2/27/18	2/28/18	NPR/Ipsos	Adults	82	72	92	75.85714	6.142857
age-21	3/1/18	3/4/18	Rasmussen	Adults	67	59	76	75.85714	-8.857143
age-21	2/22/18	2/26/18	Harris Interactive	Registered Voters	84	77	92	75.85714	8.142857
age-21	3/3/18	3/5/18	Quinnipiac	Registered Voters	78	63	93	75.85714	2.142857
age-21	3/4/18	3/6/18	YouGov	Registered Voters	72	65	80	75.85714	-3.857143

Example 6.5 Adding summary values to individual cases.

Example 6.5 shows how this can be achieved in Python and R, computing the mean support per question and then calculating how each poll deviates from this mean.

In R, the code is very similar to Example 6.4 above, simply replacing the *dplyr* function `summarize` by the function `mutate` discussed above. In this function you can mix summary functions and regular functions, as shown in the example: first the mean per group is calculated, followed by the deviation of this mean.

The Python code also uses the same syntax used for computing new columns. The first line selects the *Support* column on the grouped dataset, and then calls the *pandas* method transform of that column to compute the mean per group, adding it as a new column by assigning it to the column name. The second line uses the regular assignment syntax to create the deviation based on the support and calculated mean.

6.4 Merging Data

In many cases, we need to combine data from different sources or data files. For example, we might have election poll results in one file and socio-economic data per area in another. To test whether we can explain variance in poll results from factors such as education level, we would need to combine the poll results with the economic data. This process is often called merging or joining data.

6.4.1 Equal Units of Analysis

The easiest joins are when both datasets have the same unit of analysis, i.e. the rows represent the same units. For example, consider the data on public and private capital ownership published by Piketty (2014) alongside his landmark book *Capital in the 21st Century*. As shown in Example 6.6, he released separate files for public and private capital ownership. If we wished to analyze the relationship between these (for example to recreate Figure 3.6 on page 128 of that book), we first need to combine them into a single data frame.

To combine these data frames, we use the *pandas* data frame method `merge` in Python or the *dplyr* method `full_join` in R. Both methods join the data frames on one or more *key* columns. The key column(s) identify the units in both data frames, so in this case the *Year* column. Often, the key column is some sort of identifier, like a respondent or location ID. The resulting data frame will contain the shared key column(s), and all other columns from both joined data frames.

In both Python and R, all columns that occur in both data frames are by default assumed to be the key columns. In many cases, this is the desired behavior as both data frames may contain e.g. a *Year* or *RepondentID* column. Sometimes, however, this is not the case. Possibly, the key column is called differently in both data frames, e.g. *respID* in one and *Respondent* in the other. It is also possible that the two frames contain columns with the same name, but which contain actual data that should not be used as a key. For example, in the Piketty data shown above the key column is called *Year* in both frames, but they also share the columns for the countries which are data columns.

Python Code

```python
url = "https://cssbook.net/d/private_capital.csv"
private = pd.read_csv(url)
private.tail()
```

R Code

```r
url = "https://cssbook.net/d/private_capital.csv"
private = read_csv(url)
tail(private)
```

R Output

Year <dbl>	U.S. <dbl>	Japan <dbl>	Germany <dbl>	France <dbl>	U.K. <dbl>	Italy <dbl>	Canada <dbl>	Australia <dbl>	Spain <dbl>
2005	4.70	5.74	3.84	5.00	4.99	6.24	3.73	5.22	7.24
2006	4.88	5.83	3.78	5.34	5.19	6.37	3.88	5.32	7.69
2007	4.94	5.79	3.79	5.53	5.23	6.42	4.02	5.55	7.92
2008	4.36	5.87	3.90	5.53	4.91	6.61	3.83	5.44	7.86
2009	4.06	6.19	4.15	5.63	5.04	6.91	4.13	5.04	7.89
2010	4.10	6.01	4.12	5.75	5.22	6.76	4.16	5.18	7.55

Python Code

```python
url = "https://cssbook.net/d/private_capital.csv"
public = pd.read_csv(url)
public.tail()
```

R Code

```r
url = "https://cssbook.net/d/private_capital.csv"
public = read_csv(url)
tail(public)
```

R Output

Year <dbl>	U.S. <dbl>	Japan <dbl>	Germany <dbl>	France <dbl>	U.K. <dbl>	Italy <dbl>	Canada <dbl>	Australia <dbl>	Spain <dbl>
2005	0.48	0.34	0.04	0.28	0.32	-0.56	-0.16	0.67	0.13
2006	0.51	0.36	0.02	0.37	0.32	-0.54	-0.10	0.69	0.20
2007	0.54	0.38	0.06	0.46	0.32	-0.52	-0.03	0.69	0.26
2008	0.49	0.34	0.08	0.43	0.28	-0.52	0.00	0.71	0.25
2009	0.36	0.24	0.07	0.35	0.19	-0.65	-0.02	0.71	0.14
2010	0.21	0.14	0.04	0.31	0.06	-0.68	-0.04	0.67	0.05

Example 6.6 Private and Public Capital data (source: Piketty 2014).

In these cases, it is possible to explicitly specify which columns to join on (using the `on=` (Python) / `by=` (R) argument). However, we would generally recommend preprocessing the data first and select and/or rename columns such that the only shared columns are the key columns. The reason for that is that if columns in different data frames mean the same thing (i.e. *respID* and *Respondent*), they should generally have the same name to avoid confusion. In the case of "accidentally" shared column names, such as the country names in the current example, it is also better to rename them so it is obvious which is which in the resulting dataset: if shared columns are not used in the join, by default they get ".x" and ".y" (R) or "_x" and "_y" (Python) appended to their name, which is not very meaningful. Even if the key column is the only shared column, however, it can still be good to explicitly select that column to make it clear to the reader (or for yourself in the future) what is happening.

Python Code

```python
private_fr = private[["Year", "France"]].rename(
    columns={"France":"fr_private"})
public_fr = public[["Year", "France"]].rename(
columns={"France":"fr_public"})
    capital_fr = pd.merge(private_fr, public_fr)
# Data for Figure 3.6 (Piketty, 2014, p 128)
capital_fr.head()
```

R Code

```r
private_fr = private %>%
    select(Year, fr_private=France)
public_fr = public %>%
    select(Year, fr_public=France)
capital_fr = full_join(private_fr, public_fr
# Data for Figure 3.6 (Piketty, 2014, p 128)
head(capital_fr)
```

R Output

Year	fr_private	fr_public
<dbl>	<dbl>	<dbl>
1970	3.10	0.41
1971	3.04	0.43
1972	3.07	0.45
1973	3.05	0.46
1974	3.03	0.48
1975	3.17	0.53

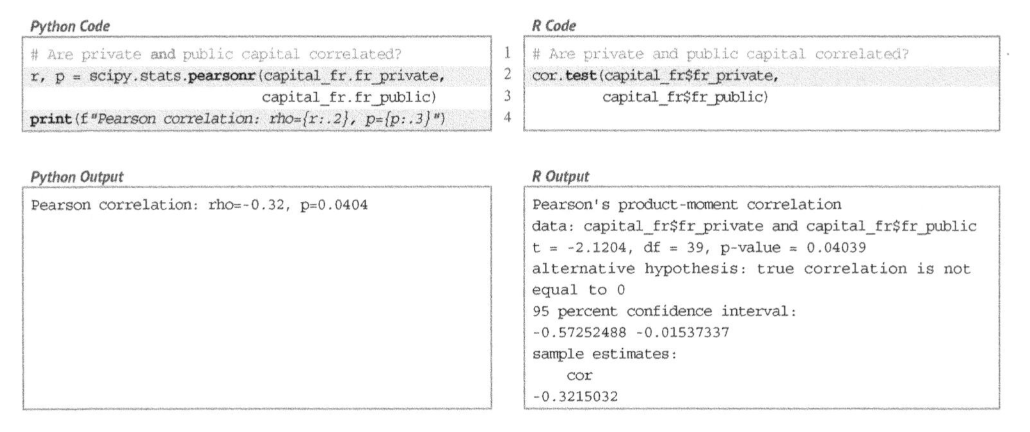

Python Code

```python
# Are private and public capital correlated?
r, p = scipy.stats.pearsonr(capital_fr.fr_private,
                            capital_fr.fr_public)
print(f"Pearson correlation: rho={r:.2}, p={p:.3}")
```

R Code

```r
# Are private and public capital correlated?
cor.test(capital_fr$fr_private,
         capital_fr$fr_public)
```

Python Output

```
Pearson correlation: rho=-0.32, p=0.0404
```

R Output

```
Pearson's product-moment correlation
data: capital_fr$fr_private and capital_fr$fr_public
t = -2.1204, df = 39, p-value = 0.04039
alternative hypothesis: true correlation is not
equal to 0
95 percent confidence interval:
-0.57252488 -0.01537337
sample estimates:
    cor
-0.3215032
```

Example 6.7 Merging private and public data for France.

This is shown in Example 6.7. The first two lines select only the *Year* and *France* columns, and rename the *France* column to indicate whether it is the private or public data. Lines 3 does the actual join, with and without the explicit selection of key column, respectively. This is then used to compute the correlation between private and public capital, which shows that there is a weak but (just) significant negative correlation ($\rho = -.32, p = .04$)[6].

Note: Next to `merge`, *pandas* data frames also have a method called `join`. It is a simplified version for joining on indices (i.e., the row labels). If you have two data frames in which corresponding rows have the same row number, you can simply write `df1.join(df2)`. In short: both methods do the same, but merge provides more options, and join is easier if you want to join on the indices.

6.4.2 Inner and Outer Joins

In the example above, both datasets had exactly one entry for each unit (year), making it the most straightforward case. If either (or both) of the datasets have missing units, however, you need to specify how to deal with this.

Table 6.1 list the four possible ways of joining, keeping all rows (*outer join*), only rows present in both (*inner join*), or all rows from one of the sets and matching rows from the other (*left* or *right join*). Left and right here literally refer to the order in which you type the data frame names. Figure 6.1 and Table 6.1 give an overview. In all cases except inner joins, this can create units where information from one of the datasets is missing. This will be lead to missing values (`NA/NaN`) being inserted in the columns of the datasets with missing units.

In most cases, you will either use inner join or left join. Inner join is useful when information should be complete, or where you are only interested in units with information in both datasets. In general, when joining sets with the same units, it is smart to check the number of rows before and after the operation. If it decreases, this shows that there are units where information is missing in either set. If it increases, it shows that apparently the sets are not at the same level of analysis, or there are duplicate units in the data. In either case, an unexpected change in the number of rows is a good indicator that something is wrong.

Left joins are useful when you are adding extra information to a "primary" dataset. For example, you might have your main survey results in a dataset, to which you want to add metadata or extra information about your respondents. If this data is not available for all respondents, you can use a left join to add the information where it is available, and simply leave the other respondents with missing values.

A similar use case is when you have a list of news items, and a separate list of items that were coded or found with some search term. Using a left join will let you keep all news items, and add the coding where it is available. Especially if items that had zero hits of a search term are excluded from the search results, you might use a left join followed by a calculation to replace missing values by zeros to indicate that the counts for items aren't actually missing, but were zero.

---

6 Of course, the fact that this is time series data means that the independence assumption of regular correlation is violated badly, so this should be interpreted as a descriptive statistic, e.g. in the years with high private capital there is low public capital and the other way around.

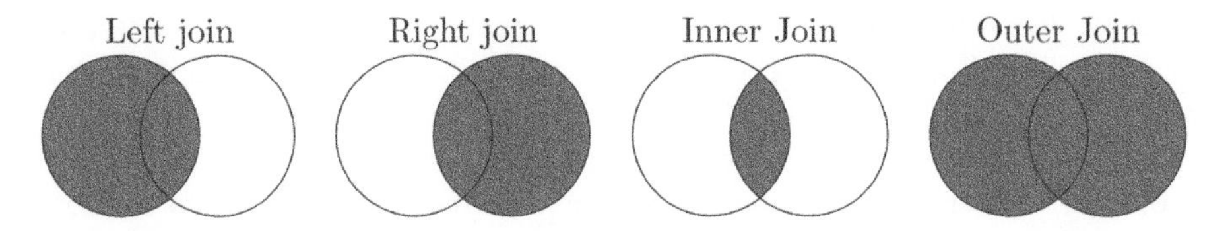

Figure 6.1 The solid area indicates whether the cases in the resulting datasets need to appear in one, both, or any of the datasets.

Table 6.1 Different types of joins between datasets d1 and d2.

Type	Description	R	Python
Outer	All units from both sets	`full_join(d1,d2)`	`d1.merge(d2, how='outer')`
Inner	Only units that are in both sets	`inner_join(d1,d2)`	`d1.merge(d2, how='inner')`
Left	All units from lefthand set	`left_join(d1,d2)`	`d1.merge(d2, how='left')`
Right	All units from righthand set	`right_join(d1,d2)`	`d1.merge(d2, how='right')`

Of course, you could also use a right join to achieve the same effect. It is more natural, however, to work from your primary dataset and add the secondary data, so you will generally use left joins rather than right joins.

Outer (or full) joins can be useful when you are adding information from e.g. multiple survey waves, and you want to include any respondent that answered any of the waves. Of course, you will have to carefully think about how to deal with the resulting missing values in the substantive analysis.

6.4.3 Nested Data

The sections above discuss merging two datasets at the same level of analysis, i.e. with rows representing the same units (respondents, items, years) in both sets. It is also possible, however, to join a more aggregate (high level) set with a more detailed dataset. For example, you might have respondents that are part of a school or organizational unit. It can be desirable to join the respondent level information with the school level information, for example to then explore differences between schools or do multilevel modeling.

For this use the same commands as for equal joins. In the resulting merged dataset, information from the group level will be duplicated for all individuals in that group.

For example, take the two datasets shown in Example 6.8. The `results` dataset shows how many votes each US 2016 presidential primary candidate received in each county: Bernie Sanders got 544 votes in Autauga County in the US state of Alabama, which was 18.2% of all votes cast in the Democratic primary. Conversely, the `counties` dataset shows a large number of facts about these counties, such as population, change in population, gender and education distribution, etc.

Suppose we hypothesize that Hillary Clinton would do relatively well in areas with more black voters. We would then need to combine the county level data about ethnic composition with the county candidate level data on vote outcomes.

Python Code

```
r="https://cssbook.net/d/2016_primary_results.csv"
results = pd.read_csv(r)
c="https://cssbook.net/d/2016_primary_county.csv")
counties = pd.read_csv(c)
counties.head()
```

R Code

```
r="https://cssbook.net/d/2016_primary_results.csv")
results = read_csv(r)
c="https://cssbook.net/d/2016_primary_county.csv")
counties = read_csv(c)
head(results)
```

Python Output

```
   fips   area_name state_abbreviation Pop_2014_count \
0     0 United States              NaN      318857056
1  1000 Alabama                    NaN        4849377
2  1001 Autauga County              AL          55395
3  1003 Baldwin County              AL         200111
4  1005 Barbour County              AL          26887

   Pop_2010_base_count Pop_change_pct Pop_2010_count
       Age_under_5_pct \
0          308758105       3.3       308745538       6.2
1            4780127       1.4         4779736       6.1
2              54571       1.5           54571       6.0
3             182265       9.8          182265       5.6
4              27457      -2.1           27457       5.7

   Age_under_18_pct Age_over_65_pct ...
     Business_hispanic_owned_pct \
0          23.1            14.5 ...           8.3
1          22.8            15.3 ...           1.2
2          25.2            13.8 ...           0.7
3          22.2            18.7 ...           1.3
4          21.2            16.5 ...           0.0

   Business_female_owned_pct Revenue_manufacture
       Revenue_wholesaler \
0              28.8     5319456312     4174286516
1              28.1      112858843       52252752
2              31.7              0              0
3              27.3        1410273              0
4              27.0              0              0

   Revenue_retail Revenue_retail_per_capita
       Revenue_food_and_hospitality \
0     3917663456          12990
      613795732
1       57344851          12364
        6426342
2         598175          12003
          88157
3        2966489          17166
         436955
4         188337           6334
             0

   Building_permits Land_area Pop_density
0       1046363    3531905.43      87.4
1         13369      50645.33      94.4
2           131        594.44      91.8
3          1384       1589.78     114.6
4             8        884.88      31.0
[5 rows x 54 columns]
```

R Output

```
  state state_abbreviation county fips party candidate
      votes
1 Alabama AL      Autauga 1001 Democrat Bernie Sanders
    544
2 Alabama AL      Autauga 1001 Democrat Hillary Clinton
   2387
3 Alabama AL      Baldwin 1003 Democrat Bernie Sanders
   2694
4 Alabama AL      Baldwin 1003 Democrat Hillary Clinton
   5290
5 Alabama AL      Barbour 1005 Democrat Bernie Sanders
    222
6 Alabama AL      Barbour 1005 Democrat Hillary Clinton
   2567
  fraction_votes
1 0.182
2 0.800
3 0.329
4 0.647
5 0.078
6 0.906
```

Example 6.8 2016 Primary results and county-level metadata. Note that to avoid duplicate output, we display the counties data in the Python example and the results data in the R example.

This is achieved in Example 6.9 in two steps. First, both datasets are cleaned to only contain the relevant data: for the `results` dataset only the Democrat rows are kept, and only the *fips* (county code), *candidate*, *votes*, and *fraction* columns. For the `counties` dataset, all rows are kept but only the *county code*, *name*, and *Race_white_pct* columns are kept.

In the next step, both sets are joined using an inner join from the results dataset. Note that we could also have used a left join here, but with an inner join it will be

Python Code

```
c=counties[["fips", "area_name", "Race_black_pct"]]

r = results.loc[
    results.candidate == "Hillary Clinton"]
r = r[["fips", "votes", "fraction_votes"]]
r = r.merge(c)
r.head()
```

R Code

```
c = counties %>%
  select("fips", "area_name", "Race_black_pct")
r = results %>%
  filter(candidate == "Hillary Clinton") %>%
  select(fips, votes, fraction_votes)
r = inner_join(r, c)
cor.test(r$Race_black_pct, r$fraction_votes)
```

Python Output

```
     fips  votes  fraction_votes area_name Race_black_pct
0  1001.0   2387           0.800 Autauga County      18.7
1  1003.0   5290           0.647 Baldwin County       9.6
2  1005.0   2567           0.906 Barbour County      47.6
3  1007.0    942           0.755    Bibb County      22.1
4  1009.0    564           0.551 Blount  County       1.8
```

R Output

```
        Pearson's product-moment correlation

data:  r$Race_black_pct and r$fraction_votes
t = 50.944, df = 2806, p-value < 2.2e-16
alternative hypothesis: true correlation is not equal
    to 0
95 percent confidence interval:
  0.6734586 0.7119165
sample estimates:
      cor
0.6931806
```

Example 6.9 Joining data at the result and the county level.

immediately obvious if county level data is missing, as the number of rows will then decrease. In fact, in this case the number of rows does decrease, because some results do not have corresponding county data. As a puzzle, can you use the dataset filtering commands discussed above to find out which results these are?

Note also that the county level data contains units that are not used, particularly the national and state level statistics. These, and the results that do not correspond to counties, are automatically filtered out by using an inner join.

Finally, we can create a scatter plot or correlation analysis of the relation between ethnic composition and electoral success (see how to create the scatter plot in Section 7.2). In this case, it turns out that Hillary Clinton did indeed do much better in counties with a high percentage of black residents. Note that we cannot take this to mean there is a direct causal relation, there could be any number of underlying factors, including the date of the election which is very important in primary races. Statistically, since observations within a state are not independent, we should really control for the state-level vote here. For example, we could use a partial correlation, but we would still be violating the independence assumption of the errors, so it would be better to take a more sophisticated (e.g. multilevel) modeling approach. This, however, is well beyond the scope of this chapter.

6.5 Reshaping Data: Wide To Long And Long To Wide

Data that you find or create does not always have the shape that you need it to be for your analysis. In many cases, for further data wrangling or for analyses you want each observation to be in its own row. However, many data sources list multiple observations in columns. For example, data from panel surveys asking the same question every week will often have one row per respondent, and one column for each weekly measurement. For a time-series analysis, however, each row should be a single measurement, i.e. the unit of analysis is a respondent per week.

Generally, data with multiple observations of the same unit is called *wide data* (as there are many columns), while a dataset with one row for each observation is called *long data* (as there are many rows). In most cases, long data is easiest to work with, and in fact in *tidyverse* jargon such data is called *tidy* data.

As a first relatively simple example, consider the datasets containing public and private capital. This data is "wide" in the sense that the measurements for the different countries are contained in the columns. To make this data "long" we would have to create rows for each country–year combination. This will make it much easier to do further data wrangling or analysis, as you can now e.g. directly merge the datasets and compute the pooled correlation between these variables. In fact, when we merged these datasets earlier in Example 6.10, we selected only the measurements for France, essentially turning it into long data.

Example 6.10 shows how you can "pivot" the capital data to long format using `pivot_longer` (R) and `melt` (*Pandas*). The second part of this example then goes on to do this for both datasets, merge them, and partially reproduce Figure 4.4 from Piketty (2014).

6.6 Restructuring Messy Data

As a final example, we will look at the data on income and wage shares from Piketty (supplemental tables S8.1 and S8.2). We want to visualize the income and wage share going to the top 1% earners in France and the US. Figure 6.2 shows a screen shot of this data in Libre Office, with the US data having a similar shape. For the previous examples, we used a clean csv version of this data, but now we will tackle the additional challenge of dealing with the Excel file including extra header rows and column names aimed at human consumption rather than easy computing.

In order to perform our visualization, we want a dataset containing a single measurement column (percentage share), and a row for each year–country–type combination, i.e. one row for wage inequality in 1910 in the US. One of the most important skills in computational social science (and data-driven analysis in general) is understanding which series of generally small steps are needed to go from one data format to the other. Although there is not a fixed set of steps that are always needed, the steps to get from the raw data visualized in Figure 6.2 to a "tidy" dataset are fairly typical:

1. Input: read the data into data frames. In this case, reading from an Excel sheet and skipping the extra header rows
2. Reshape: pivoting the data into long format
3. Normalize: normalize names, value types, etc. In this case, also separate a header like "Top 1% income share" into income type (income, wage) and percentile (10%, 1%, etc)
4. Filter: filter for the desired data
5. Analyze: create the visualization

Fortunately, these steps have been discussed before: reading csv data in Section 5.2; pivot to long data in Section 6.5; add a column in Section 6.2; joining data in Section 6.4; and visualizing in Section 7.2.

Example 6.11 shows how to perform these steps for the US case. First, we use the *readxl* (R) and *xlrd* (Python) to read a sheet from an Excel file into a data frame, manually specifying the number of header and footer rows to skip. Then, we pivot the columns into a long format. In step 3, we split the header into two columns using `separate` (R) and `split` (Python). Finally, steps 4 and 5 take the desired subset and create a line plot.

The missing step, splitting a header into two columns, is done using `separate` (R) and `split` (Python).

	A	B	C	D	E	F	G
1							
2							
3	Table S8.1. Top income and wage shares in France 1900-2010						
4		France					
5		Top 10% income share	Top 1% income share	Top 0,1% income share	Top 10% wage share	Top 1% wage share	
6	1900	45.5%					
7	1901						
8	1902						
9	1903						
10	1904						
11	1905	45.5%					
12	1906						
13	1907						
14	1908	45.5%					
15	1909	46.0%					
16	1910	46.5%	20.5%	9.5%	26.4%	5.9%	
17	1911	46.8%	20.8%	9.8%	26.6%	6.0%	

Figure 6.2 Data on top incomes as provided in Piketty (2014; digital appendix). *Source:* Used with permission from Microsoft

Python Code

```python
url="https://cssbook.net/d/private_capital.csv"
private = pd.read_csv(url)
private = private.melt(id_vars="Year",
                       var_name="country",
                       value_name="capital")
private.head()
```

R Code

```r
url="http://cssbook.net/d/private_capital.csv"
private = read_csv(url)
private = private %>% pivot_longer(cols = -Year,
    names_to="country", values_to="capital")
head(private)
```

R Output

Year <dbl>	country <chr>	capital <dbl>
1970	U.S.	3.42
1970	Japan	2.99
1970	Germany	2.25
1970	France	3.10
1970	U.K.	3.06
1970	Italy	2.39

Python Code

```python
url="https://cssbook.net/d/public_capital.csv"
public = pd.read_csv(url)
public = public.melt(id_vars="Year",
    var_name="country", value_name="capital")
d = pd.concat([private.assign(type="private"),
               public.assign(type="public")])

countries = {"France", "U.K.", "Germany"}
d = d.loc[d.country.isin(countries)]
plt = sns.lineplot(data=d, x="Year", y="capital",
                   hue="country", style="type")
plt.set(ylabel="Capital (% of national income")
plt.set_title("Capital in Europe, 1970 - 2010"
    "\nPartial reproduction of Piketty fig 4.4")
```

R Code

```r
url="https://cssbook.net/d/public_capital.csv"
public = read_csv(url) %>% pivot_longer(-Year,
    names_to="country", values_to="capital")

d = bind_rows(
    private %>% add_column(type="private"),
    public %>% add_column(type="public"))
countries = c("Germany", "France", "U.K.")
d %>% filter(country %in% countries) %>%
  ggplot(aes(x=Year, y=capital,
             color=country, lty=type)) +
  geom_line()+
  ylab("Capital (% of national income)") +
  guides(colour=guide_legend("Country"),
         linetype=guide_legend("Capital")) +
  theme_classic() +
  ggtitle("Capital in Europe, 1970 - 2010",
    "Partial reproduction of Piketty fig 4.4")
```

R Output

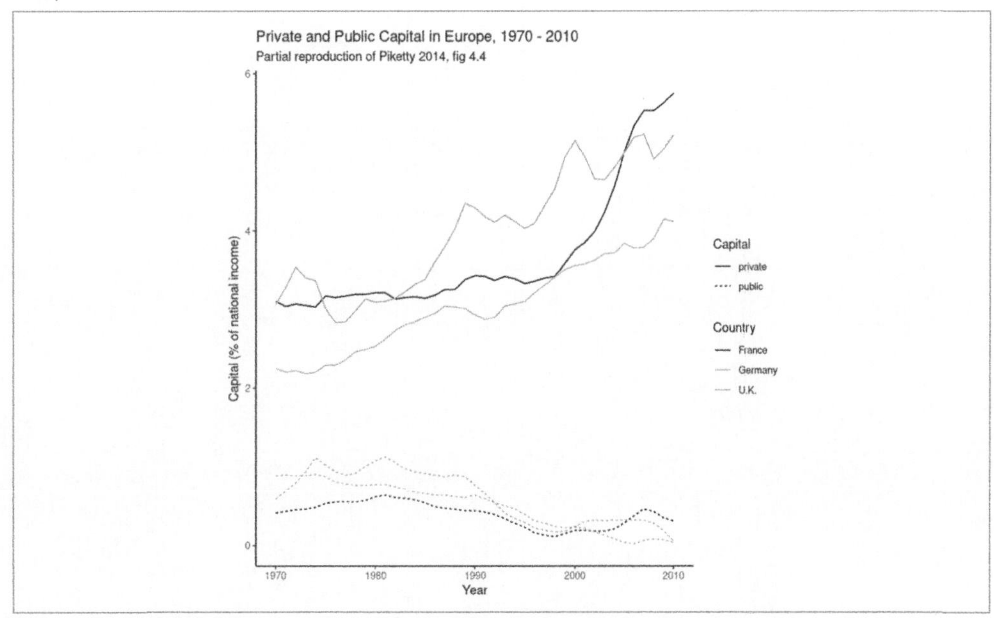

Example 6.10 Converting wide to long data to facilitate merging and visualizing.

Python Code

```python
url="https://cssbook.net/d/Chapitre8.xls"
#1 Input: Read the data into a data frame
d = pd.read_excel(url, sheet_name="TS8.2",
                  skiprows=4, skipfooter =3)

d = d.rename(columns={"Unnamed: 0": "year"})

#2 Reshape: Pivoting to long, dropping missing
d = d.melt(value_name="share", id_vars="year")

#3 Normalize
cols = ["_top", "percentile", "type",
        "_share", "capital_gains"]
d[cols] = d.variable.str.split(n=4, expand=True)
d = d.drop(columns=["variable", "_top", "_share"])
d["capital_gains"] = d["capital_gains"].notna()

d.head()
```

R Code

```r
#1 Input: Read the data into a data frame
url="https://cssbook.net/d/Chapitre8.xls"
dest = tempfile(fileext=".xls")
download.file(url, dest)
d = readxl::read_excel(dest, sheet="TS8.2", skip = 4)
d = d%>% rename("year"=1)

#2 Reshape: Pivoting to long, dropping missing
d = d%>%pivot_longer(-year, values_to="share")%>%
      na.omit()

#3 Normalize
cols = c(NA,"percent","type",NA,"capital_gains")
d = d %>% separate(name, into=cols,
     sep=" ", extra="merge", fill="right") %>%
  mutate(year=as.numeric(year),
         capital_gains=!is.na(capital_ gains))
head(d)
```

R Output

year <dbl>	percentile <chr>	type <chr>	capital_gains <lgl>	share <dbl>
1900	10%	income	FALSE	0.40500000
1900	10%	income	TRUE	0.40280054
1910	10%	income	FALSE	0.40578506
1910	10%-5%	income	FALSE	0.09886691
1910	5%-1%	income	FALSE	0.12921815
1910	1%	income	FALSE	0.17770000

Python Code

```
#1 Filter for the desired data
subset = d[(d.year >= 1910) &
           (d.percentile == "1%") &
           (d.capital_gains == False)]

#5 Analyze and/or visualize
plt = sns.lineplot(data=subset, hue="type",
                   x="year", y="share")
plt.set(xlabel="Year",
        ylabel="Share of income going to top -1%")
```

R Code

```
1  #1 Filter for the desired data
2  subset = d %>% filter(year >=1910,
3                        percentile=="1%",
4                        capital_gains==F)
5
6  #5 Analyze and/or visualization
7  ggplot(subset, aes(x=year, y=share, color=type)) +
8    geom_line() + xlab("Year") +
9  ylab("Share of income going to top-1%") +
10   theme_classic()
11
```

R Output

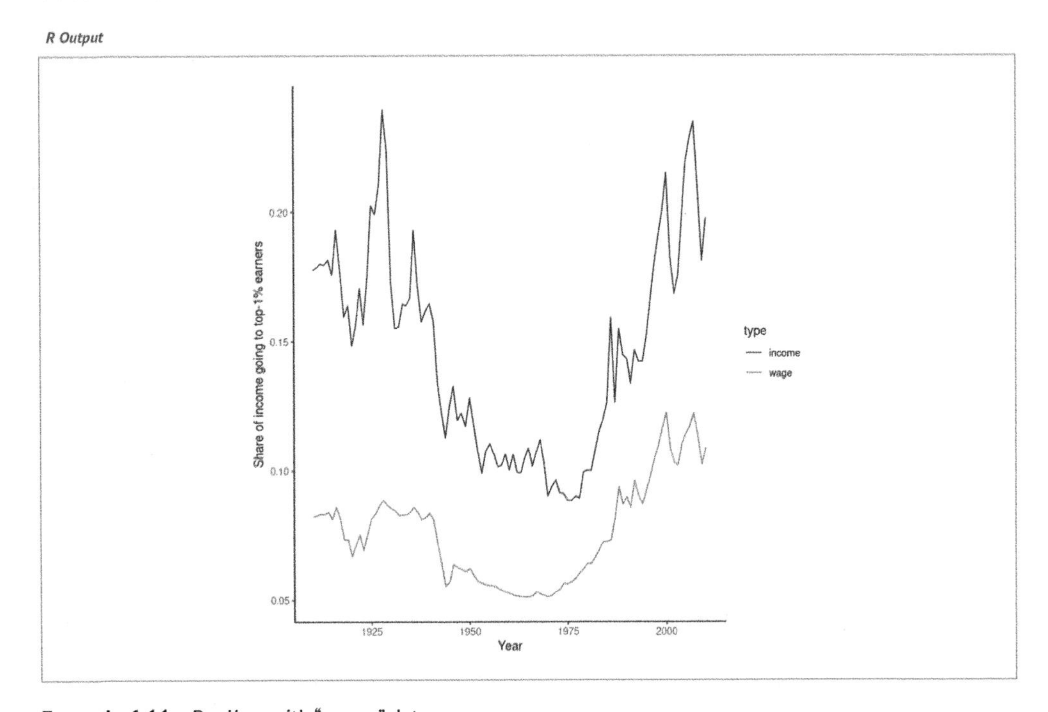

Example 6.11 Dealing with "messy" data.

7

Exploratory Data Analysis

Abstract

This chapter explains how to use data analysis and visualization techniques to understand and communicate the structure and story of our data. It first introduces the reader to exploratory statistics and data visualization in R and Python. Then, it discusses how unsupervised machine learning, in particular clustering and dimensionality reduction techniques, can be used to group similar cases or to decrease the number of features in a dataset.

Keywords descriptive statistics, visualization, unsupervised machine learning, clustering, dimensionality reduction

- Be able to conduct an exploratory data analysis
- Understand the principles of unsupervised machine learning
- Be able to conduct a cluster analysis
- Be able to apply dimension reduction techniques

In this chapter we use the R packages *tidyverse*, *maps* and *factoextra* for data analysis and visualization. For Python we use *pandas* and *numpy* for data analysis and *matplotlib*, *seaborn* and *geopandas* for visualization. Additionally, in Python we use *scikit-learn* and *scipy* for cluster analysis. You can install these packages with the code below if needed (see Section 1.4 for more details):

Python Code

```
!pip3 install pandas matplotlib seaborn geopandas
!pip3 install scikit-learn scipy bioinfokit
!pip3 install descartes
```

R Code

```
install.packages(c("tidyverse", "glue", "maps",
                   "factoextra"))
```

After installing, you need to import (activate) the packages every session:

Computational Analysis of Communication: A Practical Introduction to the Analysis of Texts, Networks, and Images with Code Examples in Python and R, First Edition. Wouter van Atteveldt, Damian Trilling & Carlos Arcila Calderón.

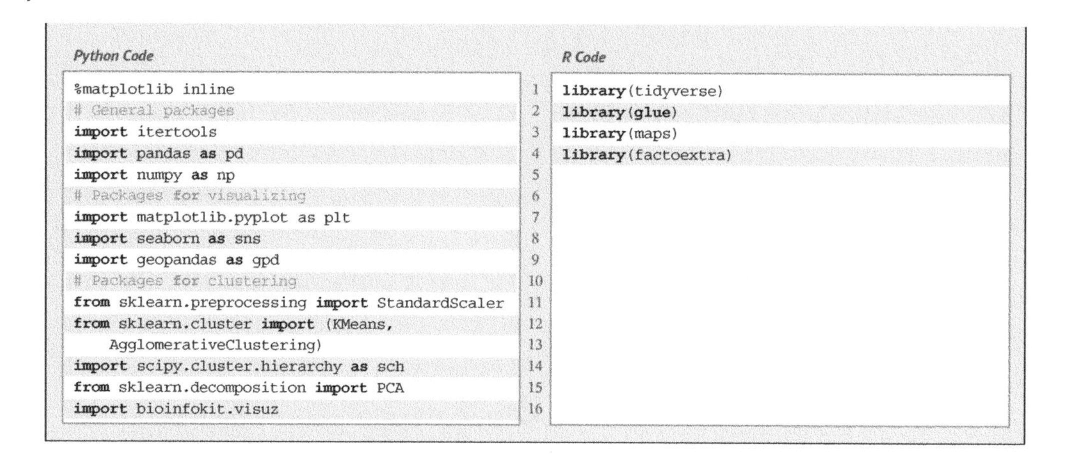

Python Code

```python
%matplotlib inline
# General packages
import itertools
import pandas as pd
import numpy as np
# Packages for visualizing
import matplotlib.pyplot as plt
import seaborn as sns
import geopandas as gpd
# Packages for clustering
from sklearn.preprocessing import StandardScaler
from sklearn.cluster import (KMeans,
    AgglomerativeClustering)
import scipy.cluster.hierarchy as sch
from sklearn.decomposition import PCA
import bioinfokit.visuz
```

R Code

```r
library(tidyverse)
library(glue)
library(maps)
library(factoextra)
```

7.1 Simple Exploratory Data Analysis

Now that you are familiar with data structures (Chapter 5) and data wrangling (Chapter 6) you are probably eager to get some real insights into your data beyond the basic techniques we briefly introduced in Chapter 2.

As we outlined in Chapter 1, the computational analysis of communication can be bottom-up or top-down, inductive or deductive. Just as in traditional research methods (for an overview, see, for example, Bryman, 2012), sometimes, an inductive bottom-up approach is a goal in itself: after all, explorative analyses are invaluable for generating hypotheses that can be tested in follow-up research. But even when you are conducting a deductive, hypothesis-testing study, it is a good idea to start by *describing* your dataset using the tools of exploratory data analysis to get a better picture of your data. In fact, we could even go as far as saying that obtaining descriptives like frequency tables, cross-tabulations, and summary statistics (mean, median, mode, etc.) is always necessary, even if your research questions or hypotheses require further complex analysis. For the computational analysis of communication, a significant amount of time may actually be invested at this stage.

Exploratory data analysis (EDA), as originally conceived by Tukey (1977), can be a very powerful framework to prepare and evaluate data, as well as to understand its properties and generate insights at any stage of your research. It is mandatory to do some EDA before any sophisticated analysis to know if the data is clean enough, if there are missing values and outliers, and how the distributions are shaped. Furthermore, before making any multivariate or inferential analysis we might want to know the specific frequencies for each variable, their measures of central tendency, their dispersion, and so on. We might also want to integrate frequencies of different variables into a single table to have an initial picture of their interrelations.

To illustrate how to do this in R and Python, we will use existing representative survey data to analyze how support for migrants or refugees in Europe changes over time and differs per country. The Eurobarometer (freely available at the Leibniz Institute for the Social Sciences – GESIS) has contained these specific questions since 2015. We might pose questions about the variation of a single variable or also describe the covariation of different variables to find patterns in our data. In this section, we will compute basic statistics to answer these questions and in the next section we will visualize them by plotting *within* and *between* variable behaviors of a selected group of features of the Eurobarometer conducted in November 2017 to 33 193 Europeans.

For most of the EDA we will use *tidyverse* in R and *pandas* as well as numpy and scipy in Python (Example 7.1). After loading a clean version of the survey data[1] stored in a csv file (using the *tidyverse* function `read_csv` in R and the *pandas* function `read_csv` in R), checking the dimensions of our data frame (33193 × 17), we probably want to get a global picture of each of our variables by getting a frequency table. This table shows the frequency of different outcomes for every case in a distribution. This means that we can know how many cases we have for each number or category in the distribution of every variable, which is useful in order to have an initial understanding of our data.

pandas versus pure _numpy/scipy_

In this book, we use *pandas* data frames a lot: they make our lives easier compared to native data types (Section 3.1), and they already integrate a lot of functionality of underlying math and statistics packages such as *numpy* and *scipy*. However, you do not have to force your data into a data frame if a different structure makes more sense in your script. *numpy* and *scipy* will happily calculate mean, media, skewness, and kurtosis of the values in a list, or the correlation between two lists. It's up to you.

Python Code

```
url="https://cssbook.net/d/eurobarom_nov_2017.csv"
d2=pd.read_csv(url)
print("Shape of my filtered data =", d2.shape)
print("Variables:", d2.columns)
```

R Code

```
url="https://cssbook.net/d/eurobarom_nov_2017.csv"
d2= read_csv(url, col_names = TRUE)
glue("{nrow(d2)} rows x {ncol(d2)} columns")
colnames(d2)
```

Python Output

```
Shape of my filtered data = (33193, 17)
Variables: Index(['survey', 'uniqid', 'date',
'country',
       'marital_status', 'educational',
       'gender', 'age', 'occupation', 'type_
       community',
       'household_composition', 'support_
             refugees', 'support_migrants',
       'date_n', 'support_refugees_n', 'support_
             migrants_n', 'educational_n'],
    dtype='object')
```

R Output

```
33193 rows x 17 columns
[1] "Variables:"
 [1] "survey"          "uniqid"        "date"
 [4] "country"         "marital_status"
     "educational"
 [7] "gender"         "age"       "occupation"
[10] "type_community"               "household_
     composition"                 "support_refugees"
[13] "support_migrants"   "date_n"
     "support_refugees_n"
[16] "support_migrants_n"
     "s"educational_n"
```

Example 7.1 Load data from Eurobarometer survey and select some variables.

[1] See https://cssbook.net/datasets for more information.

Python Code

```python
print(d2["gender"].value_counts())
print(d2["gender"].value_counts(normalize=True))
```

R Code

```r
d2 %>%
  group_by(gender) %>%
  summarise(frequency = n()) %>%
  mutate(rel_freq = frequency / sum(frequency))
```

R Output: Absolute and relative frequencies of gender:

gender <chr>	frequency <int>	rel_freq <dbl>
Man	15477	0.466273
Woman	17716	0.533727

Python Code

```python
print(d2["support_refugees"].value_counts())
print(d2["support_refugees"].value_counts(
      normalize=True, dropna=False))
```

R Code

```r
d2 %>%
  group_by(support_refugees) %>%
  summarise(frequency = n()) %>%
  mutate(rel_freq = frequency / sum(frequency))
```

R Output: Absolute and relative frequencies of support of refugees:

support_refugees <chr>	frequency <int>	rel_freq <dbl>
Tend to agree	12695	0.3824602
Tend to disagree	5391	0.1624138
Totally agree	4957	0.1493387
Totally disagree	3574	0.1076733
NA	6576	0.1981141

Example 7.2 Absolute and relative frequencies of support of refugees and gender.

Python Code

```python
n_miss = d2["support_refugees"].isna().sum()
print(f"# of missing values: {n_miss}")

d2 = d2.dropna()
print(f"Shape after dropping NAs: {d2.shape}")
```

R Code

```r
n_miss = sum(is.na(d2$support_refugees))
print(glue("# of missing values: {n_miss}"))

d2 = d2 %>% drop_na()
print(glue("Rows after dropping NAs: {nrow(d2)}"))
```

Python Output

```
Number of missing values in the variable support_
refugees: 6576
Shape of my data without missing values (23448, 17)
```

R Output

```
Number of missing values in the variable support_
refugees: 6576
Shape of my data without missing values 23448 x 17
```

Example 7.3 Drop missing values.

Let us first get the distribution of the categorical variable *gender* by creating tables that include absolute and relative frequencies. The frequency tables (using the `dplyr` functions `group_by` and `summarize` in R, and *pandas* function `value_counts` in Python) reveals that 17 716 (53.38%) women and 15 477 (46.63%) men answered this survey (Example 7.2). We can do the same with the level of support of refugees (*support_refugees*) (*To what extent do you agree or disagree with the following statement: our country should help refugees*) and obtain that 4 957 (14.93%) persons totally agreed with this statement, 12 695 (38.25%) tended to agree, 5931 (16.24%) tended to disagree and 3574 (10.77%) totally disagreed.

Before diving any further into any *between* variables analysis, you might have noticed that there might be some missing values in the data. These values represent an important amount of data in many real social and communication analysis (just remember that you cannot be forced to answer every question in a telephone or face-to-face survey!). From a statistical point of view, we can have many approaches to address missing values: For example, we can drop either the rows or columns that contain any of them, or we can impute the missing values by predicting them based on their relation with other variables – as we did in Section 6.2 by replacing the missing values with the column mean. It goes beyond the scope of this chapter to explain all the imputation methods (and, in fact, mean imputation has some serious drawbacks when used in subsequent analysis), but at least we need to know how to identify the missing values in our data and how to drop the cases that contain them from our dataset.

In the case of the variable *support_refugees* we can count its missing data (6576 cases) with the base R function `is.na` and the *pandas* method `isna`[2]. Then we may decide to drop all the records that contain these values in our dataset using the tidyr function `drop_na` in R and *pandas* function `dropna` (Example 7.3) in Python to drop the records[3]. By doing this we get a cleaner dataset and continue with a more sophisticated EDA with cross-tabulation and summary statistics for the group of cases.

Now let us cross tabulate the *gender* and *support_refugees* to have an initial idea of what the relationship between these two variables might be. With this purpose we create a contingency table or cross-tabulation to get the frequencies in each combination of categories (using *dplyr* functions `group_by`, `summarize` and `spread` in R, and *pandas* function `crosstab` in Python; Example 7.4). From this table you can easily see that 2178 women totally supported helping refugees and 1524 men totally did not. Furthermore, other interesting questions about our data might now arise if we compute summary statistics for a group of cases (using again *dplyr* functions `group_by`, `summarize` and `spread`, and base `mean` in R; and *pandas* function `groupby` and base `mean` in Python). For example, you might wonder what the average ages of the women were that totally supported (52.42) or not (53.2) to help refugees. This approach will open a huge amount of possible analysis by grouping variables and estimating different statistics beyond the mean, such as count, sum, median, mode, minimum or maximum, among others.

7.2 Visualizing Data

Data visualization is a powerful technique for both understanding data yourself and communicating the story of your data to others. Based on *ggplot2* in R and *matplotlib* and *seaborn* in Python, this section covers histograms, line and bar graphs, scatterplots and heatmaps. It touches on combining multiple graphs, communicating uncertainty with boxplots and ribbons, and plotting geospatial data. In fact, visualizing data is an

---

[2] If missing values are not correctly declared (e.g. using strings or numbers such as 999) we should first transform the initial values into proper missing values using the *tidyverse* function `na_if` in R and the *numpy* object `nan` in Python. We did this when cleaning the original Eurobarometer dataset for you.

[3] We may also use: dropna(axis='columns') if you want to drop columns instead of rows.

important stage in both EDA and advanced analytics, and we can use graphs to obtain important insights into our data. For example, if we want to visualize the age and the support for refugees of European citizens, we can plot a histogram and a bar graph, respectively.

7.2.1 Plotting Frequencies and Distributions

In the case of nominal data, the most straightforward way to visualize them is to simply count the frequency of value and then plot them as a bar chart. For instance, when we depict the support to help refugees (Example 7.5) you can quickly get that the option "tend to agree" is the most frequently voiced answer.

If we have continuous variables, however, having such a bar chart would lead to too many bars: we may lose oversight (and creating the graph may be resource-intensive). Instead, we want to group the data into *bins*, such as age groups. Hence, a histogram is used to examine the distribution of a continuous variable (*ggplot2* function geom_histogram in R and *pandas* function `hist` in Python) and a bar graph to inspect the distribution of a categorical one (*ggplot2* function geom_bar() in R and *matplotlib* function `plot` in Python). In Example 7.6 you can easily see the

Python Code

```python
print("Crosstab gender and support_refugees:")
print(pd.crosstab(d2["support_refugees"],
                  d2["gender"]))

print("Summary statistics for group of cases:")
print(d2.groupby(["support_refugees", "gender"])
      ["age"].mean())
```

R Code

```r
print("Crosstab gender and support_refugees:")
d2 %>%
    group_by(gender, support_refugees)%>%
    summarise(n=n())%>%
    pivot_wider(values_from="n",names_from="gender")

print("Summary statistics for group of cases:")
d2 %>%
     group_by(support_refugees, gender)%>%
     summarise(mean_age=mean(age, na.rm = TRUE))
```

Python Output

```
Crosstab gender and support_refugees:
gender                 Man        Woman
support_refugees
Tend to agree          5067       5931
Tend to disagree       2176       2692
Totally agree          2118       2178
Totally disagree       1524       1762
Summary statistics for group of cases:
support_refugees gender
Tend to agree        Man      54.073022
                     Woman    53.373799
Tend to disagree     Man      52.819853
                     Woman    52.656761
Totally agree        Man      53.738905
                     Woman    52.421947
Totally disagree     Man      52.368110
                     Woman    53.203746
Name: age, dtype:float64
```

R Output

```
[1] "Crosstab gender and support_refugees:"
  support_refugees Man Woman
1 Tend to agree 5067 5931
2 Tend to disagree 2176 2692
3 Totally agree 2118 2178
4 Totally disagree 1524 1762
[1] "Summary statistics for group of cases:"
  support_refugees Man Woman
1 Tend to agree 54.07302 53.37380
2 Tend to disagree 52.81985 52.65676
3 Totally agree 53.73890 52.42195
4 Totally disagree 52.36811 53.20375
```

Example 7.4 Cross tabulation of support of refugees and gender, and summary statistics.

R: GGPlot syntax

One of the nicest features of using R for data exploration is the *ggplot2* package for data visualization. This is a package that brings a unified method for visualizing with generally good defaults but that can be customized in every way if desired. The syntax, however, can look a little strange at first. Let's consider the command from Example 7.5: `ggplot (data=d2) + geom_bar(mapping=aes(x= support_refugees), fill="blue")`.

What you can see here is that every ggplot is composed of multiple sub-commands that are added together with the plus sign. At a minimum, every ggplot needs two sub-commands: `ggplot`, which initiates the plot and can be seen as an empty canvas, and one or more `geom` commands which add geometries to the plot, for example bars, lines, or points. Moreover, each geometry needs a *data* source, and a ggplot *aesthetic mapping* which tells ggplot how to map columns in the data (in this case the `support_refugees` column) to graphical (aesthetic) elements of the plot, in this case the *x* position of each bar. Graphical elements can also be set to a constant value rather than mapped to a column, in which case the argument is placed outside the `aes` function, as in the `fill="blue"` above.

Each aesthetic mapping is assigned a scale. This scale is initialized with a sensible default which depends on the data type. For example, the color of the lines in Example 7.9 are mapped to the `group` column. Since that is a nominal value (character column), ggplot automatically assigns colors to each group, in this case blue and red. In Example 7.15, on the other hand, the fill color is mapped to the `score` column, which is numerical (interval) data, to which ggplot by default assigns a color range of white to blue.

Almost every aspect of ggplot can be customized by adding more subcommands. For example, you can specify the title and axis labels by adding `+ labs(title="Title", x="Axis Label")` to the plot, and you can completely alter the look of the graph by applying a theme. For example, the *ggthemes* package defines an Economist theme, so by simply adding `+ theme_economist()` to your plot you get the characteristic layout of plots from that magazine. You can also customize the way scales are mapped using the various `scale_variable_mapping` functions. For example, Example 7.19 uses `scale_fill_viridis_c(option = "B")` to use the *viridis* scale for the *fill* aesthetic, specifying that scale B should be used. Similar commands can be used to e.g. change the colors of color ranges, the size of points, etc.

Because all geometries start with `geom_`, all scales start with `scale_`, all themes start with `theme_`, etc., you can use the RStudio autocompletion to browse through the complete list of options: simply type `geom_`, press tab or control+space, and you get a list of the options with a short description, and you can press F1 to get help on each option. The help for every geometry also lists all aesthetic elements that can or must be supplied.

Besides the built-in help, there are a number of great (online) resources to learn more. Specifically, we recommend the book *Data Visualization: A practical introduction* by Kieran Healy[4]. Another great resource is the R Graph Gallery[5], which has an enormous list of possible visualizations, all with R code included and most of them based on *ggplot*. Finally, we recommanded the Data-to-Viz[6] website, which allows you to explore a number of graph types depending on your data, lists the do's and don'ts for each graph, and links to the Graph Gallery for concrete examples.

shape of the distribution of the variable age, with many values close to the average and a slightly bigger tail to the right (not that far from the normal distribution!).

Another way to show distributions is using bloxplots, which are powerful representations of the distribution of our variables through the use of quartiles that are marked with the 25th, 50th (median) and 75th percentiles of any given variable. By examining the lower and upper levels of two or more distributions you can compare their variability and even detect possible outliers. You can generate multiple boxplots to compare the ages of the surveyed citizens by country and quickly see that in terms of age the distributions of Spain and Greece are quite similar, but we can identify some differences between Croatia and the Netherlands. In R we use the base function `geom_boxplot`, while in Python we use the *seaborn* function `boxplot`.

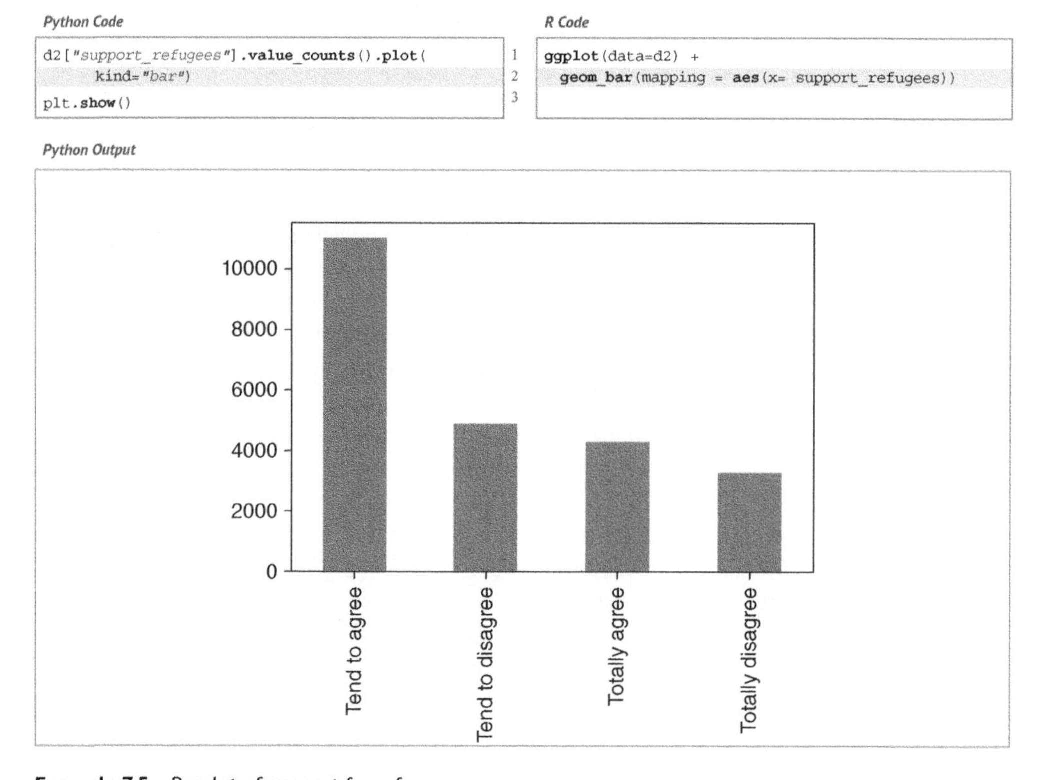
Python Code
d2["support_refugees"].value_counts().plot(
    kind="bar")
plt.show()
R Code
ggplot(data=d2) +
  geom_bar(mapping = aes(x= support_refugees))
Python Output
10000
8000
6000
4000
2000
0
Tend to agree
Tend to disagree
Totally agree
Totally disagree

Example 7.5 Barplot of support for refugees.

---

[4] Freely available at https://socviz.co/

[5] https://www.r-graph-gallery.com/

[6] https://www.data-to-viz.com/

Python Code

```
d2.hist(column="age", bins=15)
plt.show()
```

R Code

```
1  ggplot(data=d2) +
2    geom_histogram(mapping = aes(x= age), bins = 15)
```

Python Output

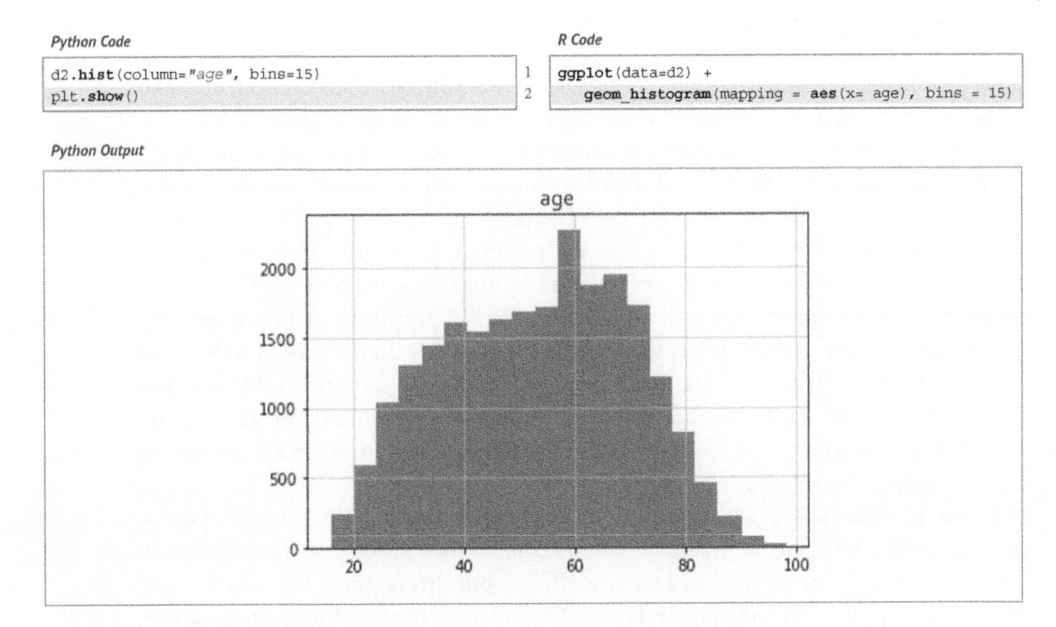

Example 7.6 Histogram of Age.

Python Code

```
d2 = d2.sort_values(by ="country")
plt.figure(figsize=(8,8))
sns.boxplot(x="age", y="country", data=d2)
plt.show()
```

R Code

```
1  ggplot(d2, aes(y=fct_rev(country), x=age))+
2    geom_boxplot()
3
4
```

R Output

Example 7.7 Bloxplots of age by country.

7.2.2 Plotting Relationships

After having inspected distributions of single variables, you may want to check how two variables are related. We are going to discuss two ways of doing so: plotting data over time, and scatterplots to illustrate the relationship between two continuous variables.

The Eurobarometer collects data for 15 days (in the example from November 5 to 19, 2017) and you may wonder if the level of support to refugees or even to general migrants changes over the time. This is actually a simple time series and you can use a line graph to represent it. Firstly you must use a numerical variable for the level of support (*support_refugees_n*, which ranges from 1 to 4, 4 being the maximum support) and group it by day in order to get the average for each day. In the case of R, you can plot the two series using the base function `plot`, or you can use the *ggplot2* function `geom_line`. In the case of Python you can use the *matplotlib* function `plot` or the *seaborn* function `lineplot`. To start, Example 7.8 shows how to create a graph for the average support for refugees by day.

To also plot the support for migrants, you can combine multiple subgraphs in a single plot, giving the reader a broader and more comparative perspective (Example 7.9). In R, the `geom_line` also takes a color aesthetic, but this requires the data to be in long format. So, we first reshape the data and also change the factor labels to get a better legend (see Section 6.5). In Python, you can plot the two lines as separate figures and add the *pyplot* function `show` to display an integrated figure.

Python Code

```
support_refugees = (d2.groupby(["date_n"])
                   ["support_refugees_n"].mean())
support_refugees = support_refugees.to_frame()

plt.plot(support_refugees.index,
         support_refugees["support_refugees_n"])
plt.xlabel("Day")
plt.ylabel("Support for refugees")
plt.show()
```

R Code

```
1  support_refugees = d2 %>%
2    group_by(date_n) %>%
3    summarise(support=mean(support_refugees_n,
4                            na.rm = TRUE))
5  ggplot(support_refugees, aes(x=date_n, y=support))+
6    geom_line() +
7    xlab("Day") +
8    ylab("Support for refugees")
9
```

Python Output

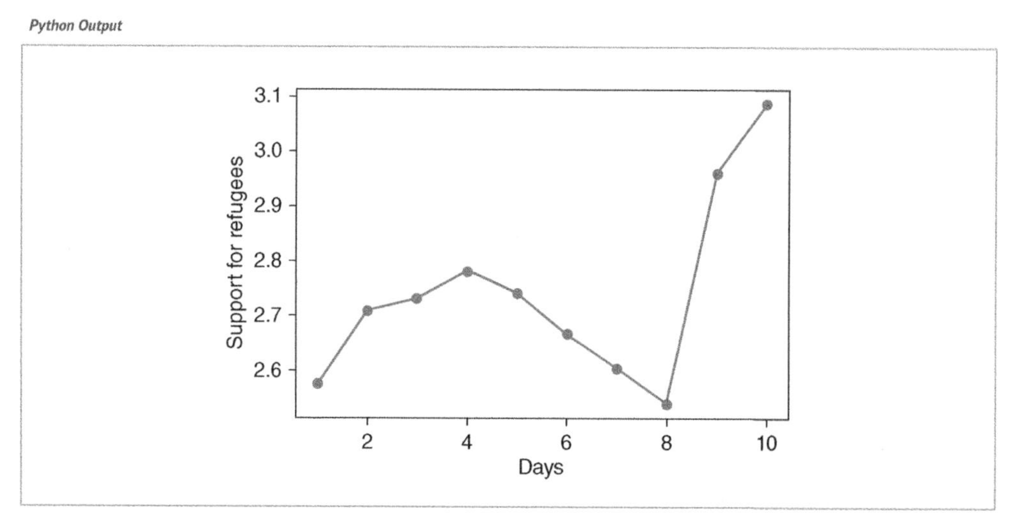

Example 7.8 Line graph of average support for refugees by day.

Python Code

```
# Combine data
support_combined = d2.groupby(["date_n"]).agg(
      refugees = ("support_refugees_n", "mean"),
      migrants = ("support_migrants_n", "mean"))


#plot
sns.lineplot(x="date_n", y="refugees",
             data=support_combined, color="blue")
sns.lineplot(x="date_n", y="migrants",
             data=support_combined, color="red")
plt.xlabel("Day")
plt.ylabel("Level of support")
plt.title("Support of refugees and migrants")
plt.show()
```

R Code

```
# Combine data
support_combined = d2 %>% group_by(date_n) %>%
  summarise(
    refugees=mean(support_refugees_n, na.rm = TRUE),
    migrants=mean(support_migrants_n, na.rm = TRUE))

# Pivot to long format and plot
support_long = support_combined %>%
  pivot_longer(-date_n, names_to="group",
               values_to="support")
ggplot(support_long,
       aes(x=date_n, y=support, colour=group)) +
  geom_line(size = 1.5) +
  labs(title="Support for refugees and migrants",
       x="Day", y="Level of Support")
```

Python Output

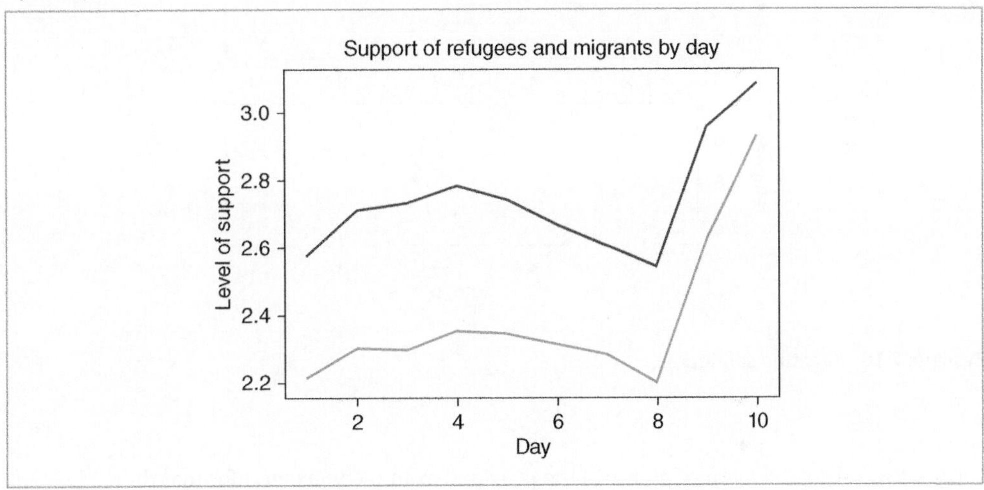

Example 7.9 Plotting multiple lines in one graph.

Alternatively, you can create multiple subplots, one for each group that you want to show (Example 7.10). In *ggplot* (R), you can use the `facet_grid` function to automatically create subplots that each show one of the groups. In the case of Python you can use the *matplotlib* function `subplots` that allows you to configure multiple plots in a single one.

Now if you want to explore the possible correlation between the average support for refugees (`mean_support_(refugees)_by_day`) and the average support to migrants by year (`mean_support_(migrants)_by_day`), you might need a scatterplot, which is a better way to visualize the type and strength of this relationship *scatter*.

A scatterplot uses dots to depict the values of two variables in a Cartesian plane (with coordinates for the axes x and y). You can easily plot this figure in R using the *ggplot2* function `geom_point` (and `geom_smooth` to display a regression line!), or in Python using *seaborn* function `scatterplot` (`lmplot` to include the regression line as shown in Example 7.12).

Python Code

```python
f, axes = plt.subplots(2,1)
sns.lineplot(x="date_n", y="refugees",
             data=support_combined, ax=axes[0])
sns.lineplot(x="date_n", y="migrants",
             data=support_combined, ax=axes[1])

sns.lineplot(x="date_n", y="support_refugees_n",
             data=d2, ci=0, ax=axes[0])
sns.lineplot(x="date_n", y="support_migrants_n",
             data=d2, ci=0, ax=axes[1])
plt.show()
```

R Code

```r
ggplot(support_long, aes(x=date_n, y=support)) +
    geom_line() + facet_grid(rows=vars(group)) +
    xlab("Day") + ylab("Support")
```

Python Output

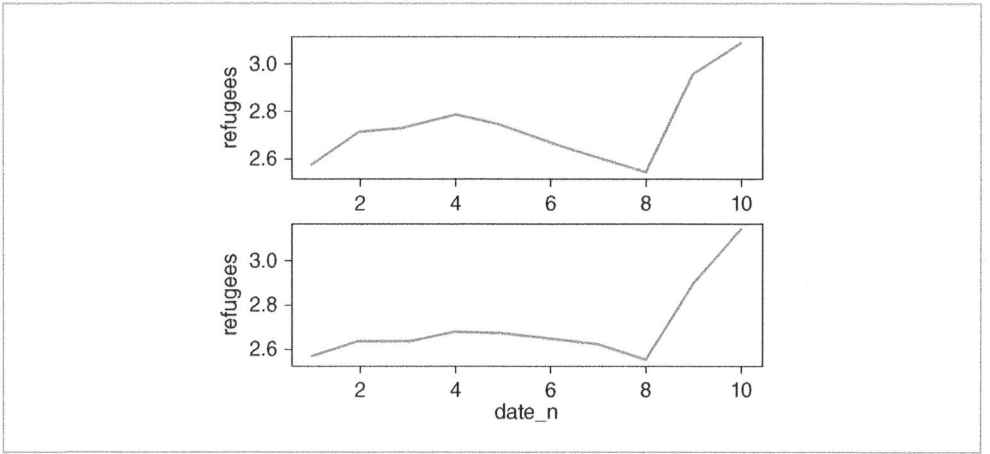

Example 7.10 Creating subfigures.

Looking at the dispersion of points in the provided example you can infer that there might be a positive correlation between the two variables, or in other words, the more the average support to refugees the more the average support to migrants over time.

We can check and measure the existence of this correlation by computing the Pearson correlation coefficient or Pearson's r, which is the most well known estimator for a correlation function. As you probably remember from your statistics class, a correlation refers to a relationship between two continuous variables and is usually applied to measure linear relationships (although there also exist nonlinear correlation coefficients, such as Spearman's ρ). Specifically, Pearson's r measures the linear correlation between two variables (X and Y) producing a value between -1 and $+1$, where 0 depicts the absence of correlation and values near to 1 a strong correlation. The signs ($+$ or $-$) represent the direction of the relationship (being positive if two variables variate in the same direction, and negative if they vary in the opposite direction). The correlation coefficient is usually represented with r or the Greek letter ρ and mathematically expressed as:

$$r = \frac{\sum_{i=1}^{n}(x_i - \bar{x})(y_i - \bar{y})}{\sqrt{\sum_{i=1}^{n}(x_i - \bar{x})^2}\sqrt{\sum_{i=1}^{n}(y_i - \bar{y})^2}}$$

Python Code

```
sns.scatterplot(data=support_combined,
                x="refugees", y="migrants")
```

R Code

```
ggplot(support_combined,
       aes(x=refugees, y=migrants))+
  geom_point()
```

Python Output.

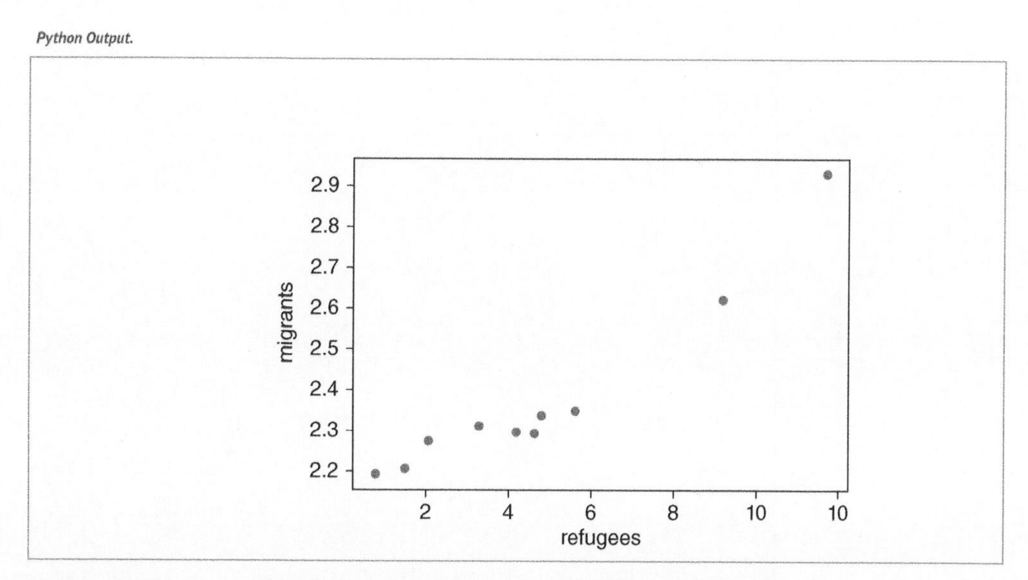

Example 7.11 Scatterplot of average support for refugees and migrants by year.

You can estimate this correlation coefficient with the *pandas* function `corr` in Python and the base R function `cor` in R. As shown in Example 7.13 the two variables plotted above are highly correlated with a coefficient of 0.95.

Another useful representation is the heatmap. This figure can help you plot a continuous variable using a color scale and shows its relation with another two variables. This means that you represent your data as colors, which might be useful for understanding patterns. For example, we may wonder what the level of support for refugees is given the nationality and the gender of the individuals. For this visualization, it is necessary to create a proper data frame (Example 7.14) to plot the heatmap, in which each number of your continuous variable *_refugees_n* is included in a table where each axis (x= gender, y=country) represents the categorical variables. This pivoted table stored in an object called `pivot_data` can be generated using some of the already explained commands.

In the first resulting figure proposed in Example 7.15, the lighter the blue the greater the support in each combination of country × gender. You can see that level of support is similar in countries such as Slovenia or Spain, and is different in the Czech Republic or Austria. It also seems that women have a higher level of support. For this default heatmap we can use the *ggplot2* function `geom_tile` in R and *seaborn* function `heatmap` in Python. To personalize the scale colors (e.g., a scale of blues) we can use

Python Code

```python
sns.lmplot(data=support_combined,
           x="refugees", y="migrants")
plt.show()
```

R Code

```r
ggplot(support_combined,
       aes(x=refugees, y= migrants))+
  geom_point()+
  geom_smooth(method = lm)
```

Python Output.

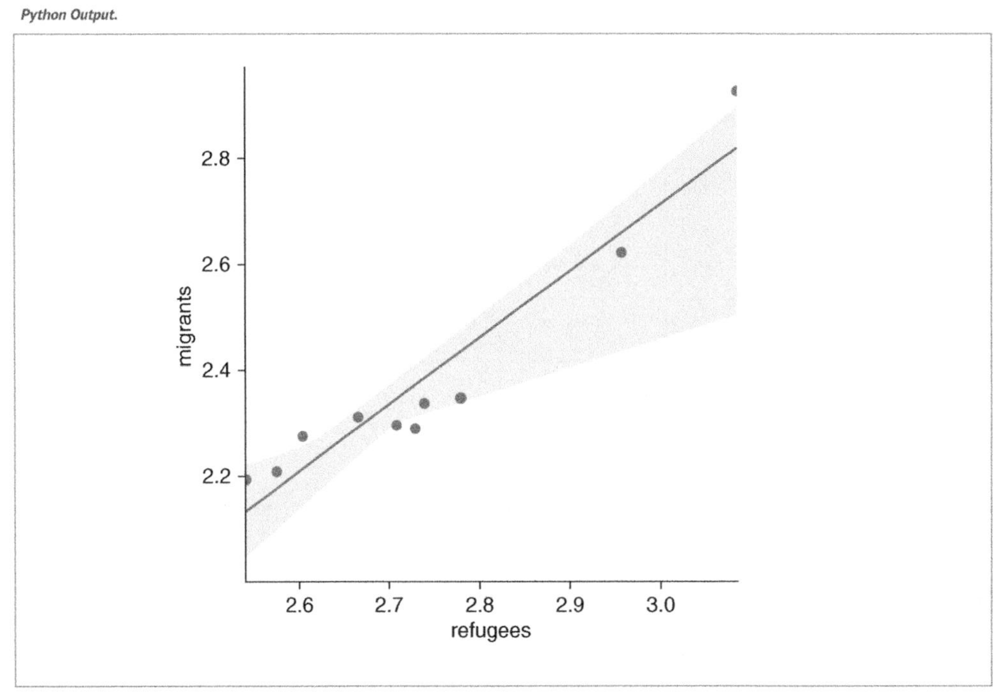

Example 7.12 Scatterplot with regression line.

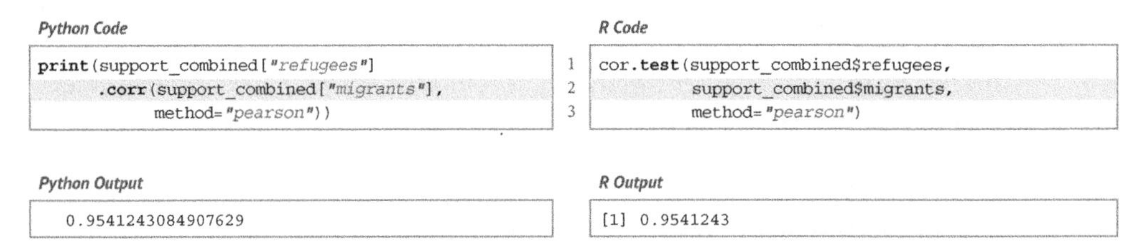

Python Code

```python
print(support_combined["refugees"]
      .corr(support_combined["migrants"],
            method="pearson"))
```

R Code

```r
cor.test(support_combined$refugees,
         support_combined$migrants,
         method="pearson")
```

Python Output

```
0.9541243084907629
```

R Output

```
[1] 0.9541243
```

Example 7.13 Pearson correlation coefficient.

Python Code

```python
pivot_data = pd.pivot_table(d2,
  values="support_refugees_n",
  index=["country"], columns="gender")
```

R Code

```r
pivot_data= d2 %>%
  select(gender, country, support_refugees_n) %>%
  group_by(country, gender) %>%
  summarise(score = mean(support_refugees_n))
```

Example 7.14 Create a data frame to plot the heatmap.

Python Code

```python
plt.figure(figsize=(10,6))
sns.heatmap(pivot_data, cmap="Blues",
      cbar_kws={"label": "support_refugees_n"})
plt.show()
```

R Code

```r
ggplot(pivot_data, aes(x = gender,
    y = fct_rev(country), fill = score)) +
  geom_tile()+
  scale_fill_gradient2(low="white", high="blue")
```

Python Output.

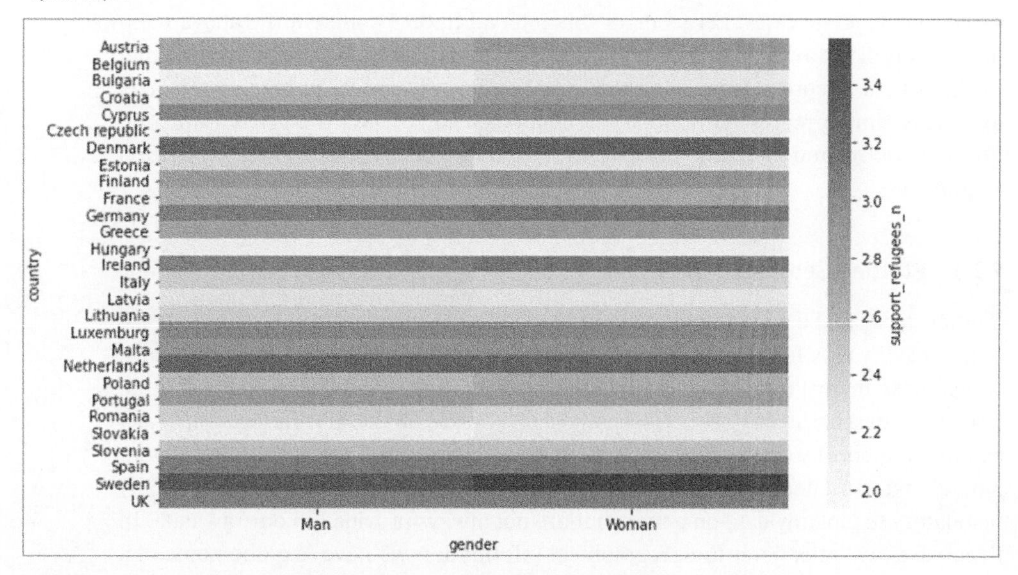

Example 7.15 Heatmap of country gender and support for refugees.

Python Code

```python
sns.lineplot(x="date_n", y="support_refugees_n",
  data=d2, color="blue", ci=100, label="Refugees")
sns.lineplot(x="date_n", y="support_migrants_n",
  data=d2, color="red", ci=100, label="Migrants")
plt.xlabel("Day")
plt.ylabel("Level of support")
plt.title("Support for refugees and migrants")
plt.show()
```

R Code

```r
ggplot(support_long,
       aes(x=date_n, y=support, color=group)) +
  geom_line(size=1.5) +
  geom_ribbon(aes(fill=group, ymin=support-0.15,
                  ymax=support+0.15),
              alpha=.1, lty=0)+
  ggtitle("Support for refugees and migrants")
```

Python Output. R Output will be slightly different

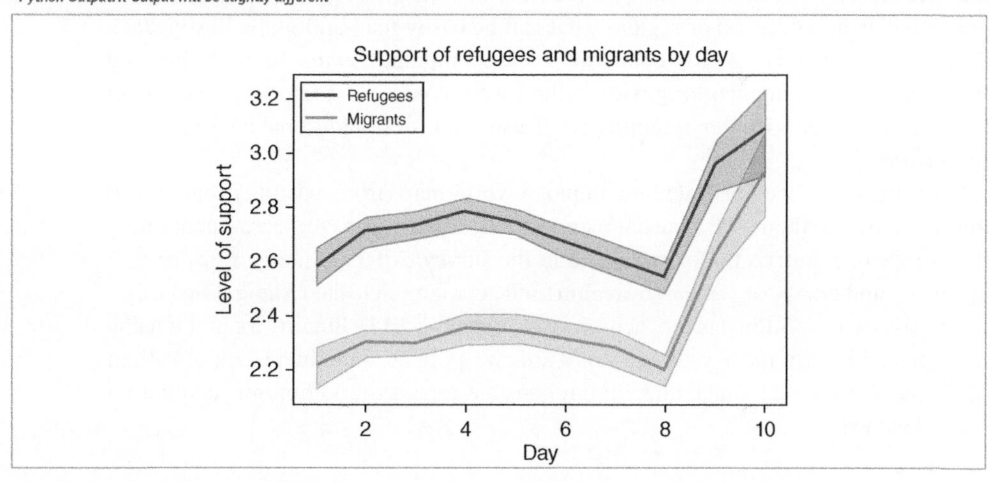

Example 7.16 Add ribbons to the line graph of support to refugees and migrants.

the *ggplot2* function `scale_fill_gradient` in R or the parameter `cmap` of the *seaborn* function `heatmap` in Python.

As you will notice, one of the goals of EDA is exploring the variance of our variables, which includes some uncertainty about their behavior. We will introduce you to two basic plots to visually communicate this uncertainty. Firstly, ribbons and area plots can help us to clearly identify a predefined interval of a variable in order to interpret its variance over some cases. Let us mark this interval in 0.15 points in the above-mentioned plots of the average support to refugees or migrants by day, and we can see that the lines tend to converge more on the very last day and are more separated by day four. This simple representation can be conducted in R using the *ggplot2* function `geom_ribbon` and in Python using the parameter `ci` of the *seaborn* function `lineplot`.

7.2.3 Plotting Geospatial Data

Plotting geospatial data is a more powerful tool to compare countries or other regions. Maps are very easy to understand and can have greater impact to all kinds of readers, which make them a useful representation for a wide range of studies that any computational analyst has to deal with. Geospatial data is based on the specific location of any country, region, city or geographical area, marked by its coordinates, latitude and longitude, that can later build points and polygon areas. The coordinates are normally mandatory to plot any data on a map, but are not always provided in our raw data. In those cases, we must joint the geographical information we have (e.g. the name of a country) with its coordinates in order to have an accurate data frame for plotting geospatial data. Some libraries in R and Python might directly read and interpret different kinds of geospatial information by recognizing strings such as "France" or "Paris", but in the end they will be converted into coordinates.

Using the very same data as our example, we might want to plot in a map the level of support to European refugees by country. Firstly, we should create a data frame with the average level of support to refugees by country (`supports_country`). Secondly, we must install an existing library that provides you with accurate geospatial information. In the case of R, we recommend the package *maps* which contains the function `map_data` that helps you generate an object with geospatial information of specific areas, countries or regions, that can be easily read and plotted by *ggplot2*. Even if not explained in this book, we also recommend *ggmap* in R (Kahle and Wickham 2013). When working with Python we recommend *geopandas* that works very well with *pandas* and *matplotlib* (it will also need some additional packages such as *descartes*).

In Example 7.17 we illustrate how to plot a world map (from existing geographical information). We then save a partial map into the object `some_eu_maps` containing the European countries that participated in the survey. After we merge `supports_country` and `some_eu_maps` (by region) and get a complete data frame called `support_map` with coordinates for each country (Example 7.18). Finally, we plot it using the *ggplot2* function `geom_polygon` in R and the *geopandas* method `plot` in Python (Example 7.19). Voilà: a nice and comprehensible representation of our data with a scale of colors!

7.2.4 Other Possibilities

There are many other ways of visualizing data. For EDA we have covered in this chapter only some of the most used techniques but they might be still limited for your future work. There are many books that cover data visualization in detail, such as Tufte (2006), Cairo (2019), and Kirk (2016). There are also many online resources,

Python Code

```
supports_country = (d2.groupby(["country"])
    ["support_refugees_n"].mean()
    .to_frame().reset_index())

#Load a world map and plot it
wmap = gpd.read_file(
    gpd.datasets.get_path("naturalearth_lowres"))
wmap = wmap.rename(columns={"name": "country"})
wmap.plot();
```

R Code

```
supports_country = d2 %>%
  group_by(country) %>%
  summarise(m=mean(support_refugees_n,na.rm=TRUE))

#Load a world map and plot it
wmap = map_data("world")
ggplot(wmap, aes(x=long, y=lat, group=group)) +
  geom_polygon(fill="lightgray", colour = "white")
```

Python Output

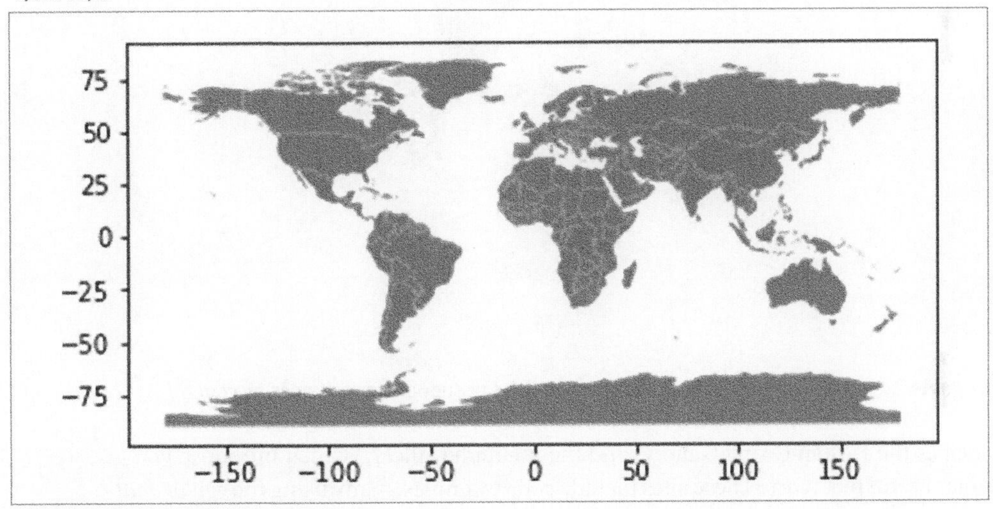

Example 7.17 Simple world map.

Python Code

```
countries = [
    "Portugal", "Spain", "France", "Germany",
    "Austria", "Belgium", "Netherlands", "Ireland",
    "Denmark", "Poland", "UK", "Latvia", "Cyprus",
    "Croatia", "Slovenia", "Hungary", "Slovakia",
    "Czech republic", "Greece", "Finland", "Italy",
    "Luxemburg", "Sweden", "Sweden", "Bulgaria",
    "Estonia", "Lithuania", "Malta", "Romania"]
m = wmap.loc[
    wmap["country"].isin(countries)]
m = pd.merge(supports_country, m, on="country")
```

R Code

```
countries = c(
  "Portugal", "Spain", "France", "Germany",
  "Austria", "Belgium", "Netherlands", "Ireland",
  "Denmark", "Poland", "UK", "Latvia", "Cyprus",
  "Croatia", "Slovenia", "Hungary", "Slovakia",
  "Czech republic", "Greece", "Finland", "Italy",
  "Luxemburg", "Sweden", "Sweden", "Bulgaria",
  "Estonia", "Lithuania", "Malta", "Romania")
m = wmap %>% rename(country=region) %>%
  filter(country %in% countries) %>%
  left_join(supports_country, by="country")
```

Example 7.18 Select EU countries and joint the map with Eurobarometer data.

Python Code

```python
m = gpd.GeoDataFrame(m, geometry=m["geometry"])
m.plot(column="support_refugees_n",
   legend=True, cmap="OrRd",
   legend_kwds={"label": "Level of suppport"}
).set_title("Support of refugees by country")
```

R Code

```r
ggplot(m, aes(long, lat, group=group))+
  geom_polygon(aes(fill = m), color="white")+
  scale_fill_viridis_c(option="B")+
  labs(title="Support of refugees by country",
       fill="Level of support")
```

R Output

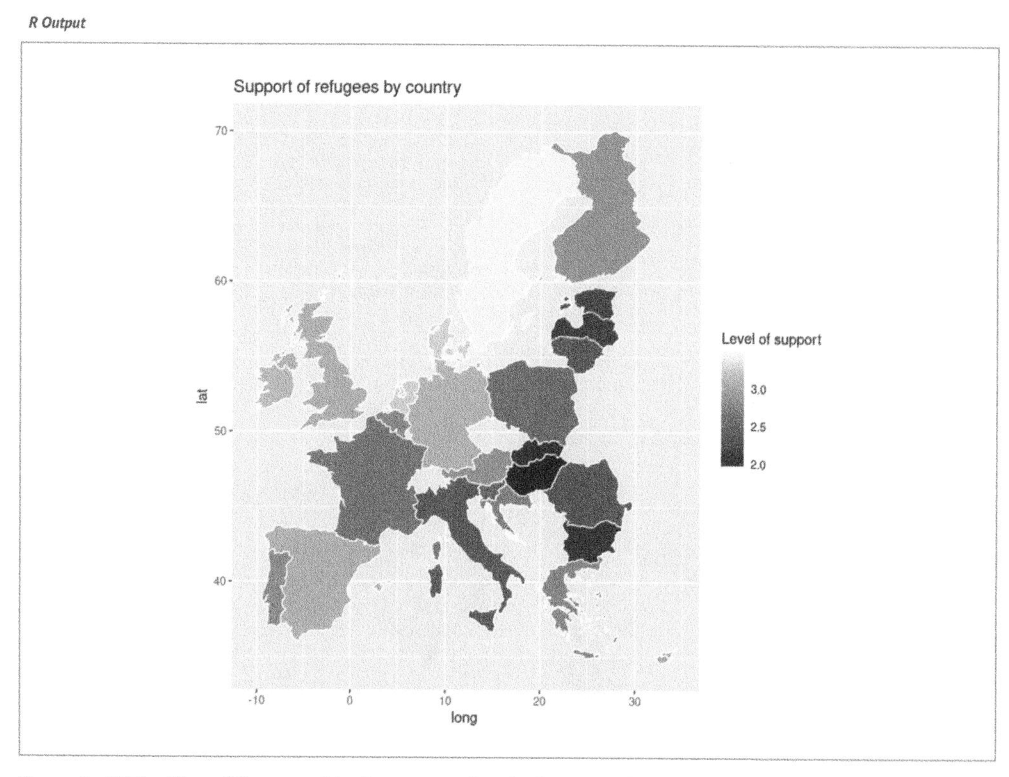

Example 7.19 Map of Europe with the average level of support for refugees by country.

such as the Python Graph Gallery[7] and the R Graph Gallery,[8] which introduce you to other useful plot types. These sites include code examples, many using the *ggplot*, *matplotlib* and *seaborn* packages introduced here, but also using other packages such as *bokeh* or plotly for interactive plots.

7.3 Clustering and Dimensionality Reduction

So far, we have reviewed traditional statistical exploratory and visualization techniques that any social scientist should be able to apply. A more computational next step in your EDA workflow is using machine learning (ML) to let your computer "learn" about our data and in turn give more initial insights. ML is a branch of artificial intelligence that uses algorithms to interact with data and obtain some patterns or rules that characterize that data. We normally distinguish between supervised machine

[7] https://www.python-graph-gallery.com

[8] https://r-graph-gallery.com/

learning (SML) and unsupervised machine learning (UML). In Chapter 8, we will come back to this distinction. For now, it may suffice to say that the main characteristic of unsupervised methods is that we do not have any measurement available for a dependent variable, label, or categorization, which to predict. Instead, we want to identify *patterns* in the data without knowing in advance what these may look like. In this, unsupervised machine learning is very much of a inductive, bottom-up technique (see Chapter 1 and Boumans and Trilling, 2016).

In this chapter, we will focus on UML as a means of finding groups and latent dimensions in our data, which can also help to reduce our number of variables. Specifically, we will use base R and Python's *scikit-learn* to conduct k-means clustering, hierarchical clustering, and principal component analysis (PCA) as well as the closely related singular value decomposition (SVD).

In data mining, we use clustering as a UML technique that aims to find the relationship between a set of descriptive variables. By doing cluster analysis we can identify underlying groups in our data that we will call *clusters*. Imagine we want to explore how European countries can be grouped based on their average support to refugees/migrants, age and educational level. We might create some *a priori* groups (such as southern versus northern countries), but cluster analysis would be a great method to let the data "talk" and then create the most appropriate groups for this specific case. As in all UML, the groups will come unlabeled and the computational analyst will be in charge of finding an appropriate and meaningful label for each cluster to better communicate the results.

7.3.1 *k*-means Clustering

k-means is a very frequently used algorithm to perform cluster analysis. Its main advantage is that, compared to the hierarchical clustering methods we will discuss later, it is very fast and does not consume much resources. This makes it especially useful for larger datasets.

k-means cluster analysis is a method that takes any number of observations (cases) and groups them into a given number of clusters based on the proximity of each observation to the mean of the formed cluster (centroid). Mathematically, we measure this proximity as the distance of any given point to its cluster center, and can be expressed as

$$J = \sum_{n=1}^{N}\sum_{k=1}^{K} r_{nk} \left\| x_n - \mu_k \right\|^2$$

where $\left\| x_n - \mu_k \right\|$ is the distance between the data point x_n and the center of the cluster μ_k.

Instead of taking the mean, some variations of this algorithm take the median (k-medians) or a representative observation, also called medoid (k-medoids or partitioning around medoids, PAM) as a way to optimize the initial method.

Because k-means clustering calculates *distances* between cases, these distances need to be meaningful – which is only the case if the scales on which the variables are measured are comparable. If all your variables are measured on the same (continuous) scale with the same endpoints, you may be fine. In most cases, you need to normalize your data by transforming them into, for instance, z-scores[9], or a scale from 0 to 1.

[9] A z-transformation means rescaling data to a mean of 0 and a standard deviation of 1.

Hence, the first thing we do in our example, is to prepare a proper dataset with only continuous variables, scaling the data (for comparability) and avoiding missing values (drop or impute). In Example 7.20, we will use the variables support to refugees (*support_refugees_n*), support to migrants (*support_migrants_n*), age (*age*) and educational level (number of years of education) (*educational_n*) and will create a data frame `d3` with the mean of all theses variables for each *country* (each observation will be a country). *k*-means requires us to specify the number of clusters, *k*, in advance. This is a tricky question, and (besides arbitrarily deciding *k*!) you essentially need to re-estimate your model multiple times with different *k*s.

The simplest method to obtain the optimal number of clusters is to estimate the variability within the groups for different runs. This means that we must run *k*-means for different number of clusters (e.g. 1 to 15 clusters) and then choose the number of clusters that decreases the variability maintaining the highest number of clusters. When you generate and plot a vector with the variability, or more technically, the within-cluster sum of squares (WSS) obtained after each execution, it is easy to identify the optimal number: just look at the bend (*knee* or *elbow*) and you will find the point where it decreases the most and then get the optimal number of clusters (three clusters in our example).

Now we can estimate our final model (Example 7.21). We generate 25 initial random centroids (the algorithm will choose the one that optimizes the cost). The default of this parameter is 1, but it is recommended to set it with a higher number (e.g. 20 to 50) to guarantee the maximum benefit of the method. The base R function `kmeans` and *scikit-learn* function `KMeans` in Python will produce the clustering. You can observe the mean (scaled) for each variable in each cluster, as well as the corresponding cluster for each observation.

Using the function `fviz_cluster` of the library *factoextra* in R, or the *pyplot* function `scatter` in Python, you can get a visualization of the clusters. In Example 7.22 you can clearly identify that the clusters correspond to Nordic countries (more support to foreigners, more education and age), Central and Southern European countries (middle support, lower education and age), and Eastern European countries (less support, lower education and age)[10].

7.3.2 Hierarchical Clustering

Another method to conduct a cluster analysis is hierarchical clustering, which builds a hierarchy of clusters that we can visualize in a dendogram. This algorithm has two versions: a bottom-up approach (observations begin in their own clusters), also called *agglomerative*, and a top-down approach (all observations begin in one cluster), also called *divisive*. We will follow the bottom-up approach in this chapter and when you look at the dendogram you will realize how this strategy repeatedly combines the two *nearest* clusters at the bottom into a larger one in the top. The distance between clusters is initially estimated for every pair of observation points and then put every point in its own cluster in order to get the closest pair of points and iteratively compute the

[10] We can re-run this cluster analysis using *k*-medoids or partitioning around medoids (PAM) and get similar results (the three medoids are: Slovakia, Belgium and Denmark), both in data and visualization. In R you must install the package *cluster* than contains the function `pam`, and in Python the package *scikit-learn-extra* with the function `Kmedoids`.

Python Code

```
# Average variables by country and scale
d3 = d2.groupby(["country"])[[
     "support_refugees_n",
     "support_migrants_n",
     "age",
     "educational_n"]].mean()

scaler = StandardScaler()
d3_s = scaler.fit_transform(d3)

# Store sum of squares for 1..15 clusters
wss = []
for i in range(1, 15):
     km_out = KMeans(n_clusters=i, n_init=20)
     km_out.fit(d3_s)
     wss.append(km_out.inertia_)

plt.plot(range(1, 15), wss, marker="o")
plt.xlabel("Number of clusters")
plt.ylabel("Within groups sum of squares")
plt.show()
```

R Code

```
# Average variables by country and scale
d3_s = d2%>%
   group_by(country)%>%
   summarise(
     m_refugees=mean(support_refugees_n, na.rm=T),
     m_migrants=mean(support_migrants_n, na.rm=T),
     m_age=mean(age, na.rm=T),
     m_edu=mean(educational_n, na.rm=T)) %>%
   column_to_rownames(var="country") %>%
   scale()
# Store sum of squares for 1..15 clusters
wss = list()
for (i in 1:15) {
   km.out = kmeans(d3_s, centers=i, nstart=25)
   wss[[i]] = tibble(k=i, ss=km.out$tot.withinss)
}
wss = bind_rows(wss)
ggplot(wss, aes(x=k, y=ss)) +
   geom_line() + geom_point() +
   xlab("Number of Clusters") +
   ylab("Within groups sum of squares")
```

Python Output

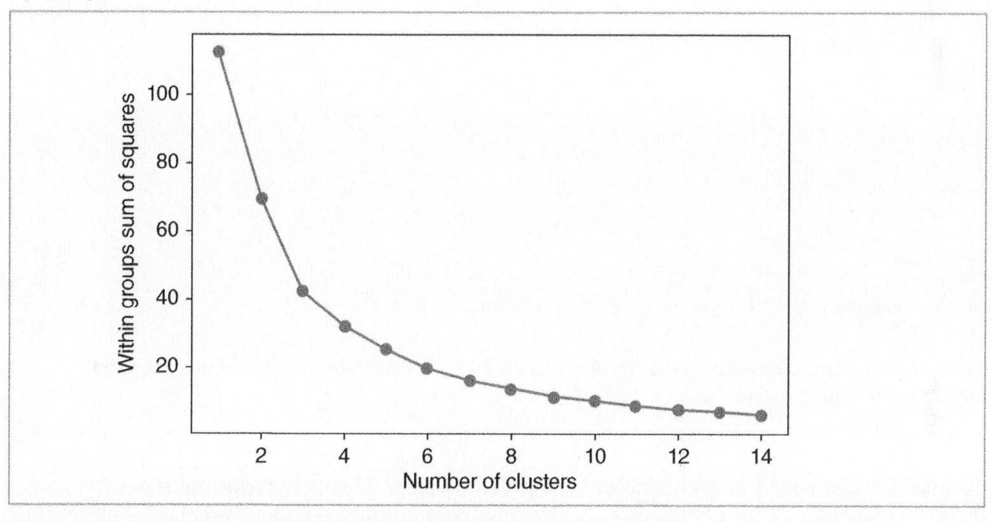

Example 7.20 Getting the optimal number of clusters.

Python Code

```
# Compute k-means with k = 3
km_res = KMeans(n_clusters=3, n_init=25).fit(d3_s)
print(km_res)
print("K-means cluste sizes:",
    np.bincount(km_res.labels_[km_res.labels_>=0]))
print(f"Cluster means: {km_res.cluster_centers_}")
print("Clustering vector:")
print(np.column_stack((d3.index, km_res.labels_)))
print("Within cluster sum of squares:")
print(km_res.inertia_)
```

R Code

```
set.seed(123)
km.res = kmeans(d3_s, 3, nstart=25)
print(km.res)
```

Python Output

```
KMeans(n_clusters=3, n_init=25)
K-means clustering with 3 clusters of sizes [13 3 12]
Cluster means: [[-0.89000978 -0.82574663 -0.3892184 -0.21560025]
 [ 1.2101425 1.01720791 1.78536032 2.49604445]
 [ 0.66164163 0.64025687 -0.02468681 -0.39044418]]
Clustering vector:
[['Austria' 2]
 ['Belgium' 2]
 ['Bulgaria' 0]
 ['Croatia' 0]
 ['Cyprus' 2]
 ['Czech republic' 0]
 ['Denmark' 1]
 ['Estonia' 0]
 ['Finland' 1]
 ['France' 2]
 ['Germany' 2]
 ['Greece' 0]
 ['Hungary' 0]
 ['Ireland' 2]
 ['Italy' 0]
 ['Latvia' 0]
 ['Lithuania' 0]
 ['Luxemburg' 2]
 ['Malta' 2]
 ['Netherlands' 2]
 ['Poland' 0]
 ['Portugal' 2]
 ['Romania' 0]
 ['Slovakia' 0]
 ['Slovenia' 0]
 ['Spain' 2]
 ['Sweden' 1]
 ['UK' 2]]
Within cluster sum of squares: 42.50488485460034
```

Example 7.21 Using Kmeans to group countries based on the average support of refugees and migrants, and educational level.

distance between each new cluster and the previous ones. This is the internal rule of the algorithm and we must choose a specific linkage method (complete, single, average or centroid, or Ward's linkage). Ward's linkage is a good default choice: it minimizes the variance of the clusters being merged. In doing so, it tends to produce roughly evenly sized clusters and is less sensitive to noise and outliers than some of the other methods. In Example 7.23 we will use the function `hcut` of the package *factoextra* in R and *scikit-learn* function `AgglomerativeClustering` in Python, to compute the hierarchical clustering.

A big advantage of hierarchical clustering is that, once estimated, you can freely choose the number of clusters in which to group your cases without re-estimating the model. If you decide, for instance, to use four instead of three clusters, then the cases in one of your three clusters are divided into two subgroups. With k-means, in contrast, a three-cluster solution can be completely different from a four-cluster solution. However, this comes at a big cost: hierarchical clustering requires a lot more computing resources and may therefore not be feasible for large datasets.

Python Code

```
for cluster in range(km_res.n_clusters):
    plt.scatter(d3_s[km_res.labels_ == cluster, 0],
                d3_s[km_res.labels_ == cluster, 1])
plt.scatter(km_res.cluster_centers_[:, 0],
            km_res.cluster_centers_[:, 1],
            s=250, marker="*")
plt.legend(scatterpoints=1)
plt.show()
```

R Code

```
fviz_cluster(km.res, d3_s, ellipse.type="norm")
```

R Output

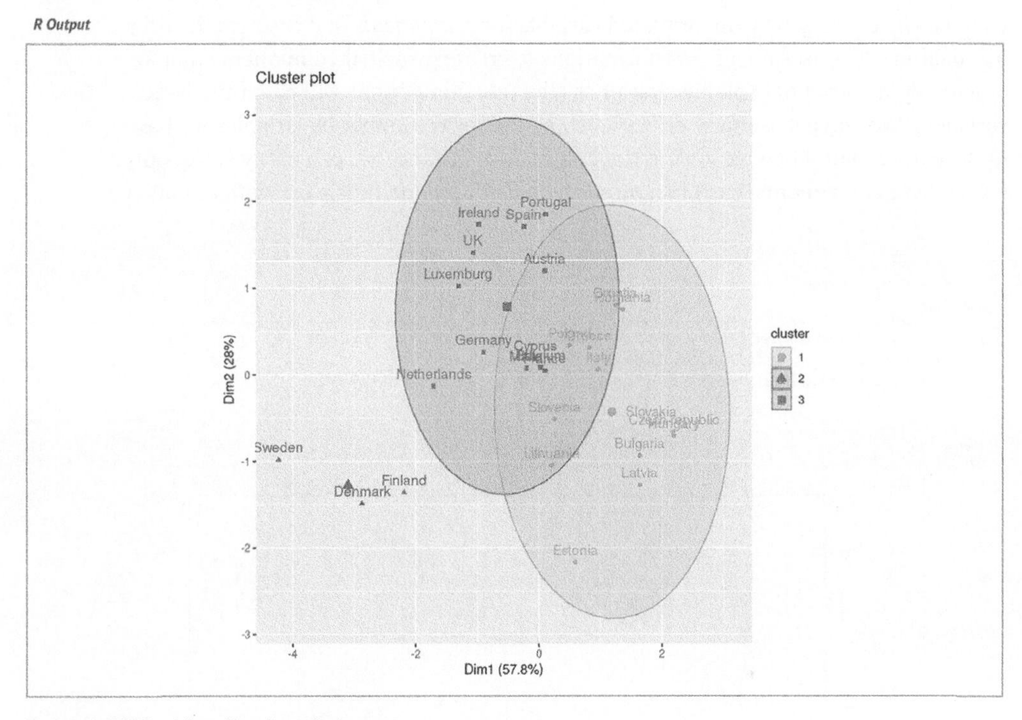

Example 7.22 Visualization of clusters.

Python Code

```
hc_res = AgglomerativeClustering(
    affinity = "euclidean", linkage = "complete")
hc_res.fit_predict(d3_s)
print(hc_res)
```

R Code

```
hc.res <- hcut(d3_s, hc_method="complete")
summary(hc.res)
```

Python Output

```
AgglomerativeClustering(linkage='complete')
```

Example 7.23 Using hierarchical clustering to group countries based on the average support of refugees and migrants, age and educational level.

We can then plot the dendogram with base R function `plot` and *scipy* (module `cluster.hierarchy`) function `dendogram` in Python. The summary of the initial model suggest two clusters (size=2) but by looking at the dendogram you can choose the number of clusters you want to work with by choosing a height (for example four to get three clusters).

If you re-run the hierarchical clustering for three clusters (Example 7.25) and visualize it (Example 7.26) you will get a graph similar to the one produced by *k*-means.

7.3.3 Principal Component Analysis and Singular Value Decomposition

Cluster analyses are in principle used to group similar cases. Sometimes, we want to group similar variables instead. A well-known method for this is principal component analysis (PCA)[11]. This unsupervised method is useful to reduce the dimensionality of your data by creating new uncorrelated variables or *components* that describe the original dataset. PCA uses linear transformations to create principal components that are ordered by the level of explained variance (the first component will catch the largest variance). We will get as many principal components as number of variables we have in the dataset, but when we look at the cumulative variance we can easily select only few of these components to explain most of the variance and thus work with a smaller

Python Code

```
dendrogram = sch.dendrogram(
    sch.linkage(d3_s, method="complete"),
    labels=list(d3.index), leaf_rotation=90)
```

R Code

```
plot(hc.res, cex=0.5)
```

Python Output

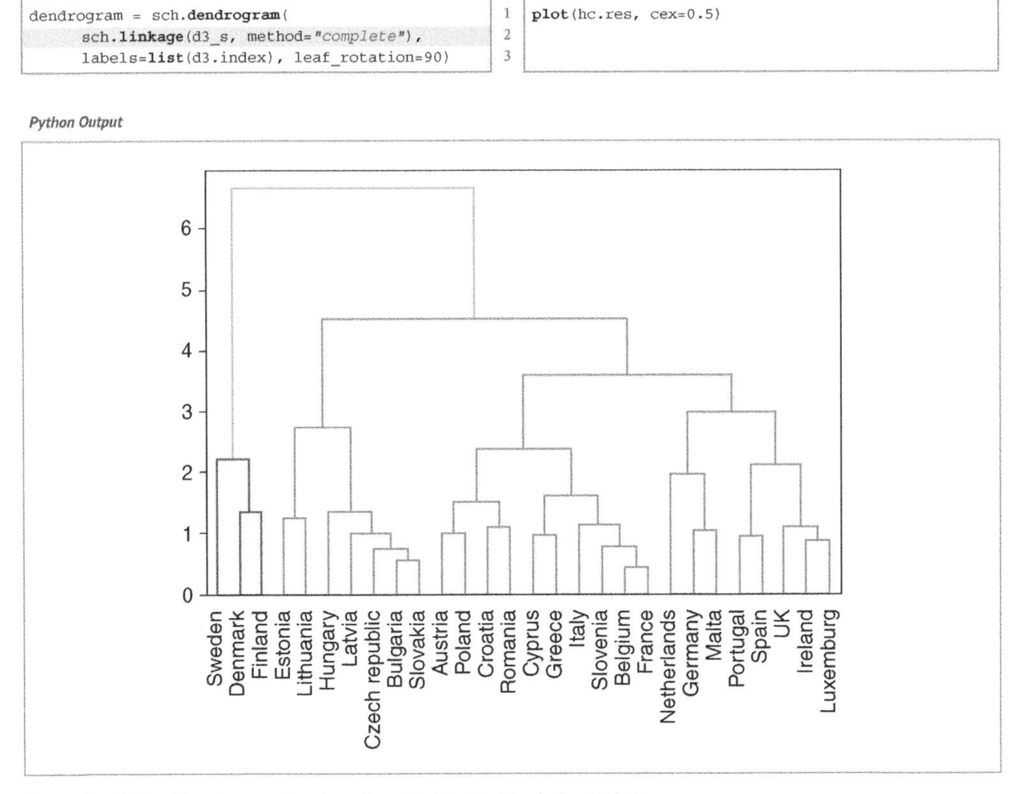

Example 7.24 Dendogram to visualize the hierarchical clustering.

[11] If your had to learn statistics using SPSS, you have almost certainly already conducted a PCA. Quite counter-intuitively, the default analysis that is run when clicking on the "Factor" menu in SPSS, is a PCA.

Python Code

```python
hc_res = AgglomerativeClustering(n_clusters=3,
  affinity = "euclidean", linkage = "ward")
hc_res.fit_predict(d3_s)
print(hc_res)
```

R Code

```r
hc.res = hcut(d3_s, k=3, hc_method="complete")
summary(hc.res)
```

Python Output

```
AgglomerativeClustering(n_clusters=3)
```

Example 7.25 Re-run hierarchical clustering with three clusters.

and summarized data frame that might be more convenient for many tasks (e.g. those that require avoiding multicollinearity or just need to be more computationally efficient). By simplifying the complexity of our data we can have a first understanding of how our variables are related and also of how our observations might be grouped. All components have specific loadings for each original variable, which can tell you how the old variables are represented in the new components. This statistical technique is especially useful in EDA when working with high dimensional datasets but it can be used in many other situations.

The mathematics behind PCA can be relatively easy to understand. However, for the sake of simplicity, we will just say that in order to obtain the principal components the algorithm firstly has to compute the mean of each variable and then compute the covariance matrix of the data. This matrix contains the covariance between the elements of a vector and the output will be a square matrix with an identical number of rows and columns, corresponding to the total number of dimensions of the original dataset. Specifically, we can calculate the covariance matrix of the variables X and y with the formula:

$$cov_{x,y} = \frac{\sum_{i=1}^{N}(x_i - \bar{x})(y_i - \bar{y})}{N-1}$$

Secondly, using the covariance matrix the algorithm computes the eigenvectors and their corresponding eigenvalues, and then drop the eigenvectors with the lowest eigenvalues. With this reduced matrix it transforms the original values to the new subspace in order to obtain the principal components that will synthesize the original dataset.

Let us now conduct a PCA over the Eurobarometer data. In Example 7.27 we will re-use the sub-data frame *d3* containing the means of 4 variables (support to refugees, support to migrants, age and educational level) for each of the 30 European countries (30×4). The question is can we have a new data frame containing less than 4 variables but that explains most of the variance, or in other words, that represents our original dataset well enough, but with fewer dimensions? As long as our features are measured on different scales, it is normally suggested to center (to mean 0) and scale (to standard deviation 1) the data. You may also know this transformation as "calculating z-scores". We can perform the PCA in R using the base function `prcomp` and in Python using the function `PCA` of the module `decomposition` of *scikit-learn*.

The generated object with the PCA contains different elements (in R`sdev`, `rotation`, `center`, `scale` and `x`) or attributes in Python (`components_`,

Python Code

```python
for cluster in range(hc_res.n_clusters):
    plt.scatter(d3_s[hc_res.labels_==cluster, 0],
                d3_s[hc_res.labels_==cluster, 1])
plt.legend(scatterpoints=1)
plt.show()
```

R Code

```r
fviz_cluster(hc.res, d3_s, ellipse.type="convex")
```

R Output

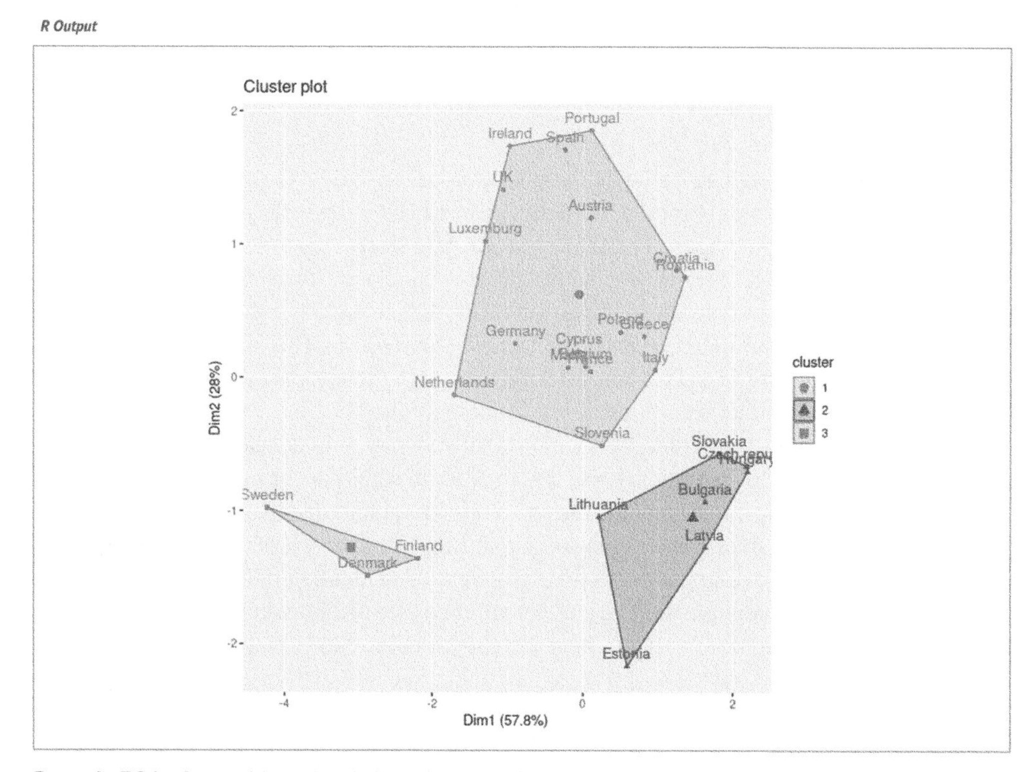

Example 7.26 Re-run hierarchical clustering with three clusters.

explained_variance_, explained_variance_ratio, singular_values_, mean_, n_components_, n_features_, n_samples_ and noise_variance_). In the resulting object we can see the values of four principal components of each country, and the values of the loadings, technically called *eigenvalues*, for the variables in each principal component. In our example we can see that support for refugees and migrants are more represented on PC1, while age and educational level are more represented on PC2. If we plot the first two principal components using base function biplot in R and and the library *bioinfokit* in Python (Example 7.28), we can clearly see how the variables are associated with either PC1 or with PC2 (we might also want to plot any pair of the four components!). But we can also get a picture of how countries are grouped based only in these two new variables.

So far we are not sure how many components are enough to accurately represent our data, so we need to know how much variance (which is the square of the standard deviation) is explained by each component. We can get the values (Example 7.29) and plot the proportion of explained variance (Example 7.30). We get that the first component explains 57.85% of the variance, the second 27.97%, the third 10.34% and the fourth just 3.83%.

Python Code

```python
pca_m = PCA()
pca = pca_m.fit(d3_s)
pca_n = PCA()
pca = pca_n.fit_transform(d3_s)
pca_df = pd.DataFrame(data=pca,
   columns=["PC1", "PC2", "PC3", "PC4"])
pca_df.index = d3.index
print(pca_df.head())

pca_df_2 = pd.DataFrame(data=pca_n.components_.T,
  columns=["PC1", "PC2", "PC3", "PC4"])
pca_df_2.index = d3.columns
print(pca_df_2)
```

R Code

```r
pca = prcomp(d3_s, scale=TRUE)
head(pca$x)
pca$rotation
```

R Output

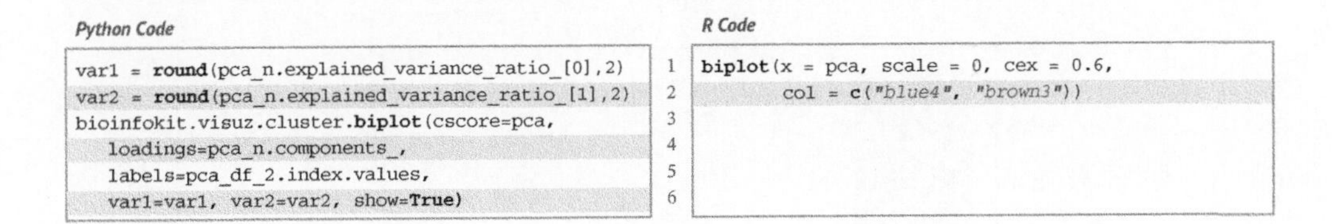

```
              PC1        PC2        PC3        PC4
country
Austria -0.103285 -1.220018 -0.535673 0.066888
Belgium -0.029355 -0.084707 0.051515 0.227609
Bulgaria -1.660518 0.949533 -0.480337 -0.151837
Croatia -1.267502 -0.819093 -0.920657 0.843682
Cyprus    0.060590 -0.195928 0.573670 0.812519
                         PC1       PC2        PC3        PC4
support_refugees_n 0.573292 -0.369010 0.139859 0.718058
support_migrants_n 0.513586 -0.533140 -0.094283 -0.665659
age                0.445117 0.558601 0.670994 -0.199005
educational_n      0.457642 0.517261 -0.722023 0.041073
```

Example 7.27 Principal component analysis (PCA) of a data frame with 30 records and 4 variables.

Python Code

```python
var1 = round(pca_n.explained_variance_ratio_[0],2)
var2 = round(pca_n.explained_variance_ratio_[1],2)
bioinfokit.visuz.cluster.biplot(cscore=pca,
   loadings=pca_n.components_,
   labels=pca_df_2.index.values,
   var1=var1, var2=var2, show=True)
```

R Code

```r
biplot(x = pca, scale = 0, cex = 0.6,
        col = c("blue4", "brown3"))
```

Python Output

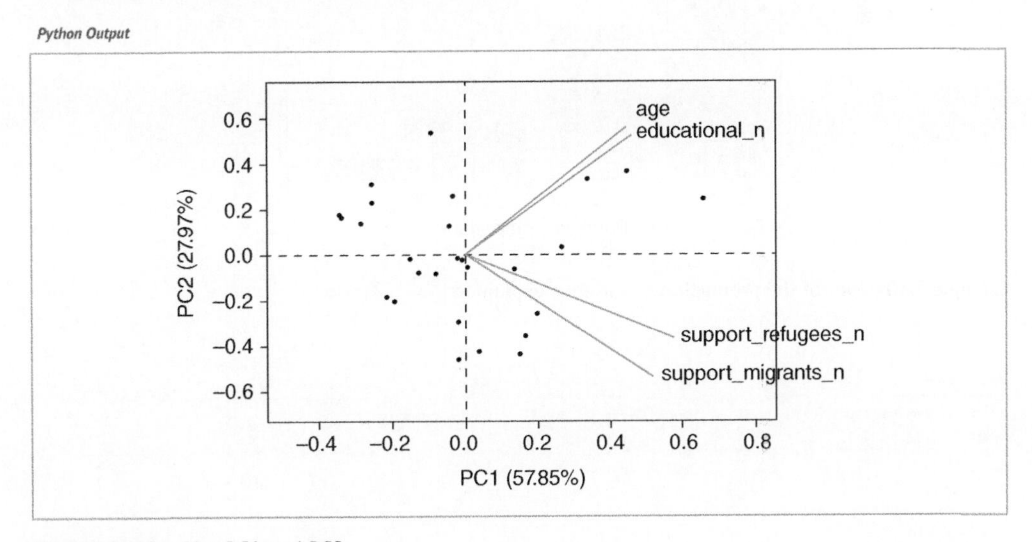

Example 7.28 Plot PC1 and PC2.

When we estimate (Example 7.31) and plot (Example 7.32) the cumulative explained variance it is easy to identify that with just the two first components we explain 88.82% of the variance. It might now seem a good deal to reduce our dataset from four to two variables, or let's say half of the data, but retaining most of the original information.

Python Code

```python
print("Proportion of variance explained:")
print(pca_n.explained_variance_ratio_)
```

R Code

```r
print("Proportion of variance explained:")
prop_var = tibble(pc=1:4,
    var = pca$sdev^2 / sum(pca$sdev^2))
prop_var
```

Python Output

```
Proportion of variance explained: [0.57848569 0.27974794 0.10344996 0.03831642]
```

Example 7.29 Proportion of variance explained.

Python Code

```python
plt.bar([1,2,3,4],pca_n.explained_variance_ratio_)
plt.ylabel("Proportion of variance explained")
plt.xlabel("Principal component")
plt.xticks([1,2,3,4])
plt.show()
```

R Code

```r
ggplot(prop_var, aes(x=pc, y=var)) +
    geom_col() +
    scale_y_continuous(limits = c(0,1)) +
    xlab("Principal component") +
    ylab("Proportion of variance explained")
```

Python Output

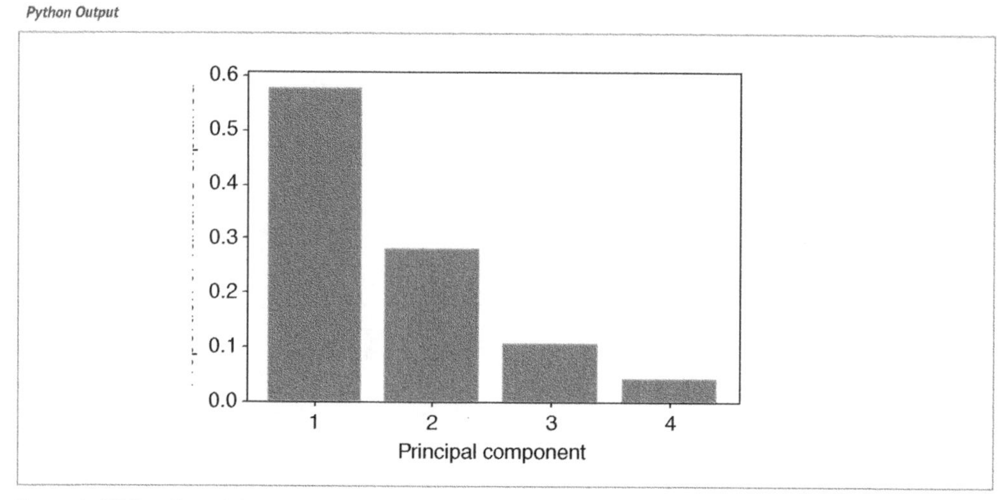

Example 7.30 Plot of the proportion of variance explained.

Python Code

```python
cvar=np.cumsum(pca_n.explained_variance_ratio_)
cvar
```

R Code

```r
cvar = cumsum(prop_var)
cvar
```

Python Output

```
Cumulative explained variance: [0.57848569 0.85823362 0.96168358 1. ]
```

Example 7.31 Cumulative explained variance.

Python Code

```python
plt.plot(cvar)
plt.xlabel("number of components")
plt.xticks(np.arange(len(cvar)),
           np.arange(1, len(cvar)+1))
plt.ylabel("cumulative explained variance")
plt.show()
```

R Code

```r
ggplot(cvar, aes(x=pc, y=var)) +
  geom_point() +
  geom_line() +
  theme_bw() +
  xlab("Principal component")+
  ylab("Cumulative explained variance")
```

Python Output

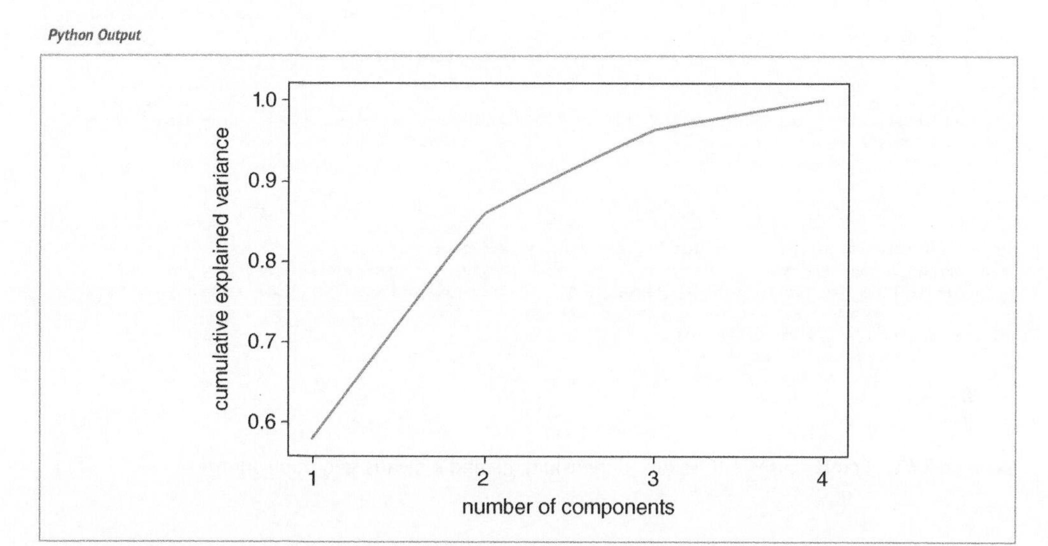

Example 7.32 Plot of the cumulative explained variance.

And what if we want to use this PCA and deploy a clustering (as explained above) with just these two new variables instead of the four original ones? Just repeat the *k*-means procedure but now using a new smaller data frame selecting PC1 and PC2 from the PCA. After estimating the optimal number of clusters (three again!) we can compute and visualize the clusters, and get a very similar picture to the one obtained in the previous examples, with little differences such as the change of cluster of the Netherlands (more similar now to the Nordic countries!). This last exercise is a good example of how to combine different techniques in EDA.

When your dataset gets bigger, though, you may actually not use PCA but the very much related singular value decomposition, SVD. They are closely interrelated, and in fact SVD can be used "under the hood" to estimate a PCA. While PCA is taught in a lot of classical textbooks for statistics in the social sciences, SVD is usually not. Yet, it has a great advantage: in the way that it is implemented in *scikit-learn*, it does not require to store the (dense) covariance matrix in memory (see the feature box on p. 192 for more information on sparse versus dense matrices). This means that once your dataset grows bigger than typical survey datasets, a PCA may qucikly become impossible to estimate, while the SVD can still be estimated without much resource required. Therefore, especially when you are working with textual data, you will see that SVD is used instead of PCA. For all practical purposes, the way that you can use and interpret the results stays the same.

Python Code

```python
#Generate a new dataset with first components
d5 = pca[:,0:2]
d5[0:5]

#Get optimal number of clusters
wss = []
for i in range(1, 15):
    km_out = KMeans(n_clusters=i, n_init=20)
    km_out.fit(d5)
    wss.append(km_out.inertia_)


# Plot sum of squares vs. number of clusters
plt.plot(range(1, 15), wss, marker="o")
plt.xlabel("Number of clusters")
plt.ylabel("Within groups sum of squares")
plt.show()

# Compute again with k = 3 and visualize
km_res_5 = KMeans(n_clusters=3, n_init=25).fit(d5)
for cluster in range(km_res_5.n_clusters):
    plt.scatter(d3_s[km_res_5.labels_== cluster, 0],
                d3_s[km_res_5.labels_== cluster, 1])
plt.scatter(km_res_5.cluster_centers_[:, 0],
            km_res_5.cluster_centers_[:, 1],
            s=250, marker="*")
plt.legend(scatterpoints=1)
plt.show()
```

R Code

```r
#Generate a new dataset with first components
d5 = pca$x[, c("PC1", "PC2")]
head(d5)

#Get optimal number of clusters
wss = list()
for (i in 1:15) {
  km.out = kmeans(d5, centers = i, nstart = 20)
  wss[[i]] = tibble(k=i, ss=km.out$tot.withinss)
}
wss = bind_rows(wss)

# Plot sum of squares vs. number of clusters
ggplot(wss, aes(x=k, y=ss)) + geom_line() +
      xlab("Number of Clusters")+
      ylab("Within groups sum of squares")


# Compute again with k = 3 and visualize
set.seed(123)
km.res_5 <- kmeans(d5, 3, nstart = 25)
fviz_cluster(km.res_5, d5, ellipse.type = "norm")
```

Example 7.33 Combining PCA to reduce dimensionality and k-means to group countries.

8

Statistical Modeling and Supervised Machine Learning

Abstract

This chapter introduces the reader to the world of supervised machine learning. It starts by outlining how classical statistical techniques such as regression models can be used for prediction. It then provides an overview of frequently-used techniques from Naïve Bayes classifiers to neural networks.

Keywords supervised machine learning

- Understand the principles of supervised machine learning
- Be able to run a predictive model
- Be able to evaluate the performance of a predictive model

In this chapter, we use the Python package *statsmodels* for classical statistical modeling, before we move on to use a dedicated machine learning package, *scikit-learn*. In R, we use base R for statistical modeling, *rsample* for splitting our dataset, *caret* for machine learning, and *pROC* for determining the Receiver Operating Characteristic (ROC) curve. Note that caret requires additional packages for the actual machine learning models: *naivebayes*, *LiblineaR*, and *randomforest*. You can install them as follows (see Section 1.4 for more details):

Python Code

```
!pip3 install pandas statsmodels sklearn
```

R Code

```
install.packages(c("randomForest", "rsample",
    "glue", "caret", "naivebayes", "LiblineaR",
    "randomForest", "pROC","e1071"))
```

After installing, you need to import (activate) the packages every session:

Computational Analysis of Communication: A Practical Introduction to the Analysis of Texts, Networks, and Images with Code Examples in Python and R, First Edition. Wouter van Atteveldt, Damian Trilling & Carlos Arcila Calderón.

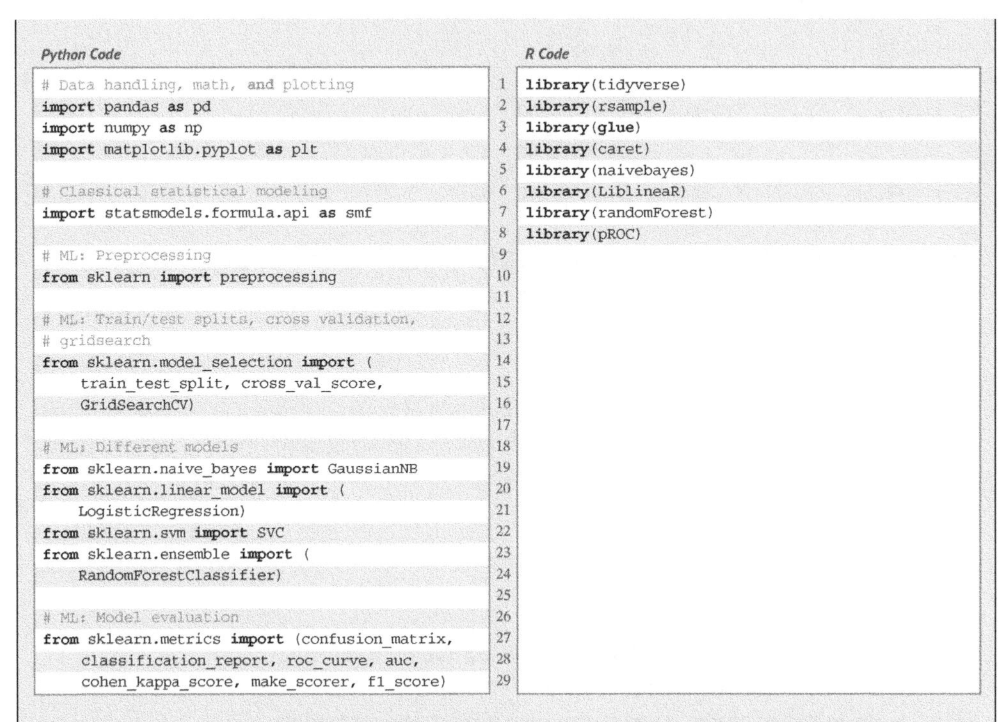

Python Code

```python
# Data handling, math, and plotting
import pandas as pd
import numpy as np
import matplotlib.pyplot as plt

# Classical statistical modeling
import statsmodels.formula.api as smf

# ML: Preprocessing
from sklearn import preprocessing

# ML: Train/test splits, cross validation,
# gridsearch
from sklearn.model_selection import (
    train_test_split, cross_val_score,
    GridSearchCV)

# ML: Different models
from sklearn.naive_bayes import GaussianNB
from sklearn.linear_model import (
    LogisticRegression)
from sklearn.svm import SVC
from sklearn.ensemble import (
    RandomForestClassifier)

# ML: Model evaluation
from sklearn.metrics import (confusion_matrix,
    classification_report, roc_curve, auc,
    cohen_kappa_score, make_scorer, f1_score)
```

R Code

```r
library(tidyverse)
library(rsample)
library(glue)
library(caret)
library(naivebayes)
library(LiblineaR)
library(randomForest)
library(pROC)
```

Note that in Python, we could also simply write `import sklearn` once instead of all the `from sklearn import` ... lines. But our approach saves a lot of typing later on, as we can simply write (`classification_report`) instead of `sklearn.metrics.classification_report`, for instance.

In this chapter, we introduce the basic concepts and ideas behind machine learning. We will outline how machine learning relates to traditional statistical approaches that you already might know (and as you will see, there is a lot of overlap), present different types of models, and discuss how to validate them. Later in this book (Section 11.4), we will specifically apply the knowledge you gain from this chapter to the analysis of textual data, arguably one of the most interesting tasks in the computational analysis of communication.

In this chapter, we focus on *supervised* machine learning (SML) – a form of machine learning, where we aim to predict a variable that, for at least a part of our data, is known. SML is usually applied to *classification* and *regression* problems. To illustrate the idea imagine that you are interested in predicting gender, based on Twitter biographies. You determine the gender for some of the biographies yourself and hand these examples over to the computer. The computer "learns" this *classification* from your examples, and can then be used to predict the gender for other Twitter biographies for which you do not know the gender.

In unsupervised machine learning (UML), in contrast, you do not have such examples. Therefore, UML is usually applied to *clustering* and *associations* problems. We have discussed some of such techniques in Section 7.3, in particular cluster analysis and principal component analysis (PCA). Later, in Section 11.5, we will discuss topic modeling, an unsupervised method to extract so-called topics from textual data.

Even though both approaches can be combined (for instance, one could first reduce the amount of data using PCA or SVD, and then predict some outcome), they can be seen as fundamentally different, from both theoretical and conceptual points of view. Unsupervised machine learning is a bottom-up approach and corresponds to an inductive reasoning: you do not have a hypothesis of, for instance, which topics are present in a corpus of text; you rather let the topics emerge from the data. Supervised machine learning, in contrast, is a top-down approach and can be seen as more deductive: you define *a priori* which topics to predict.

8.1 Statistical Modeling and Prediction

Machine learning, many people joke, is nothing other than a fancy name for statistics. And, in fact, there is some truth to this: if you say "logistic regression", this will sound familiar to both statisticians and machine learning practitioners. Hence, it does not make much sense to distinguish between statistics on the one hand and machine learning on the other hand. Still, there are some differences between traditional statistical approaches that you may have learned about in your statistics classes and the machine learning approach, even if some of the same mathematical tools are used. One may say that the focus is a different one, and the objective we want to achieve may differ.

Let us illustrate this with an example: media.csv[1] contains a few columns from survey data on how many days per week respondents turn to different media types (*radio, newspaper, tv* and *Internet*) in order to follow the news[2]. It also contains their *age* (in years), their *gender* (coded as female = 0, male = 1), and their *education* (on a 5-point scale).

A straightforward question to ask is how far the sociodemographic characteristics of the respondents explain their media use. Social scientists would typically approach this question by running a regression analysis. Such an analysis tells us how some independent variables $x_1, x_2, \ldots, x_n$ can explain y. In an ordinary least square regression (OLS), we would estimate $y = \beta_0 + \beta_1 x_1 + \beta_2 x_2 + \ldots + \beta_n x_n$.

In a typical social-science paper, we would then interpret the coefficients that we estimated, and say something like: when x_1 increases by one unit, y increases by β_1. We sometimes call this "the effect of x_1 on y" (even though, of course, it depends on the study design whether the relationship can really be interpreted as a causal effect). Additionally, we might look at the explained variance R^2, to assess how well the model fits our data. In Example 8.1 we use this regression approach to model the relationship of *age* and *gender* over the number of days per week a person reads a *newspaper*. We fit the linear model using the *stats* function `lm` in R and the *statsmodels* function `ols` (imported from the module statsmodels.formula.api) in Python.

[1] You can download the file from https://cssbook.nl/d/media.csv

[2] For a detailed description of the dataset, see Trilling (2013).

Python Code

```python
df = pd.read_csv("https://cssbook.net/d/media.csv")
mod = smf.ols(formula="newspaper ~ age + gender",
              data=df).fit()
# mod.summary() would give a lot more info,
# but we only care about the coefficients:
mod.params
```

R Code

```r
df = read.csv("https://cssbook.net/d/media.csv")
mod = lm(formula = "newspaper ~ age + gender",
         data = df)
# summary(mod) would give a lot more info,
# but we only care about the coefficients:
mod
```

Python Output

```
Intercept -0.089560
agte       0.067620
gender     0.176665
dtype: float64
```

Example 8.1 Obtaining a model through estimating an OLS regression.

Most traditional social-scientific analyses stop after reporting and interpreting the coefficients of *age* ($\beta = 0.0676$) and *gender* ($\beta = -0.0896$), as well as their standard errors, confidence intervals, p-values, and the total explained variance (19%). But we can go a step further. Given that we have already estimated our regression equation, why not use it to do some *prediction*?

We have just estimated that

newspaperreading = –0.0896 + 0.0676 · age + 0.1767 · gender

By just filling in the values for a 20 year old man, or a 40 year old woman, we can easily calculate the expected number of days such a person reads the newspaper per week, *even if no such person exists in the original dataset.*

We learn that

$$\hat{y}_{man20} = -0.0896 + 0.676 \cdot 20 + 1 \cdot 0.1767 = 1.4391$$

$$\hat{y}_{woman40} = -0.8096 + 0.0676 \cdot 40 + 0 \cdot 0.1767 = 2.6144$$

This was easy to do by hand, but of course, we could do this automatically for a large and essentially unlimited number of cases. This could be as simple as shown in Example 8.2.

Python Code

```python
newdata = pd.DataFrame([{"gender":1, "age":20},
                        {"gender": 0, "age":40} ])
mod.predict(newdata)
```

R Code

```r
gender = c(1,0)
age = c(20,40)
newdata = data.frame(age, gender)
predict(mod, newdata)
```

Python Output

```
0   1.439508
1   2.615248
dtype: float64
```

Example 8.2 Using the OLS model we estimated before to predict the dependent variable for new data where the dependent variable is unknown.

In doing so, we shift our attention from the interpretation of coefficients to the prediction of the dependent variable for new, unknown cases. We do not care about the actual values of the coefficients, we just need them for our prediction. In fact, in many machine learning models, we will have so many of them that we do not even bother to report them.

As you see, this implies that we proceed in two steps: first, we use some data to estimate our model. Second, we use that model to make predictions.

We used an OLS regression for our first example, because it is very straightforward to interpret and most of our readers will be familiar with it. However, a model can take the form of *any* function, as long as it takes some characteristics (or "features") of the cases (in this case, people) as input and returns a prediction.

Using such a simple OLS regression approach for prediction, as we did in our example, can come with a couple of problems, though. One problem is that in some cases, such predictions do not make much sense. For instance, even though we know that the output should be something between 0 and 7 (as that is the number of days in a week), our model will happily predict that once a man reaches the age of 105 (rare, but not impossible), he will read a newspaper on 7.185 out of 7 days. Similarly, a one year old girl will even have a negative amount of newspaper reading. A second problem relates to the models' inherent assumptions. For instance, in our example it is quite an assumption to make that the relationships between these variables are linear — we will therefore discuss multiple models that do not make such assumptions later in this chapter. And, finally, in many cases, we are actually not interested in getting an accurate prediction of a continuous number (a *regression* task), but rather in predicting a category. We may want to predict whether a tweet goes viral or not, whether a user comment is likely to contain offensive language or not, whether an article is more likely to be about politics, sports, economy, or lifestyle. In machine learning terms, these tasks are known as *classification*.

In the next section, we will outline key terms and concepts in machine learning. After that, we will discuss specific models that you can use for different use applications.

8.2 Concepts and Principles

The goal of Supervised Machine Learning can be summarized in one sentence: estimate a model based on some data, and then use the model to predict the expected outcome for some new cases, for which we do not know the outcome yet. This is exactly what we have done in the introductory example in Section 8.1.

But when do we need it?

In short, in any scenario where the following two preconditions are fulfilled. First, we have a large dataset (say, 100 000 headlines) for which we want to predict to which class they belong to (say, whether they are clickbait or not). Second, for a random subset of the data (say, 2000 of the headlines), we already know the class. For example because we have manually coded ("annotated") them.

Before we start using SML, though, we first need to have a common terminology. At the risk of oversimplifying matters, Table 8.1 provides a rough guideline of how some typical machine learning terms translate to statistical terms that you may be familiar with.

Let us explain them more in detail by walking through a typical SML workflow.

Before we start, we need to get a *labeled dataset*. It may be given to us, or we may need to create it ourselves. For instance, often we can draw a random sample of our data and use techniques of manual content analysis (e.g., Riffe et al., 2019) to *annotate* (i.e., to manually code) the data. You can download an example for this process (annotating the topic of news articles) from http://dx.doi.org/10.6084/m9.figshare.7314896.v1 (Vermeer, 2018).

Table 8.1 Some common machine learning terms explained.

Machine learning lingo	Statistics lingo
feature	independent variable
label	dependent variable
labeled dataset	dataset with both independent and dependent variables
to train a model	to estimate
classifier (classification)	model to predict nominal outcomes
to annotate	to (manually) code (content analysis)

It is hard to give a rule of thumb for how much labeled data you need. It depends heavily on the type of data you have (for instance, if it is a *binary* or a *multi-class* classification problem), and on how evenly distributed (*class balance*) they are (after all, having 10 000 annotated headlines doesn't help you if 9990 are not clickbait and only 10 are). These reservations notwithstanding, it is fair to say that typical sizes in our field are (very roughly) speaking often in the order of 1000 to 10 000 when classifying longer texts (see Burscher et al., 2014), even though researchers studying less rich data sometimes annotate larger datasets (e.g., 60 000 social media messages in Vermeer et al., 2019).

Once we have established that this labeled dataset is available and have ensured that it is of good quality, we randomly split it into two datasets: a *training dataset* and a *test dataset.* [3] We will use the first one to train our model, and the second to test how well our model performs. Common ratios range from 50:50 to 80:20; and especially if the size of your labeled dataset is rather limited, you may want to have a slightly larger training dataset at the expense of a slightly smaller test dataset.

In Example 8.3, we prepare the dataset we already used in Section 8.1 for classification by creating a dichotomous variable (the label) and splitting it into a training and a test dataset. We use `y_train` to denote the training labels and `X_train` to denote the feature matrix of the training dataset; `y_test` and `X_test` is the corresponding test dataset. We set a so-called random-state seed to make sure that the random splitting will be the same when re-running the code. We can easily split these datasets using the *rsample* function `initial_split` in R and the *sklearn* function `train_test_split` in Python.

We now can *train our classifier* (i.e., estimate our model using the training dataset contained in the objects `X_train` and `y_train`). This can be as straightforward as estimating a logistic regression equation (we will discuss different classifiers in Section 8.3). It may be that we first need to create new independent variables, so-called features, a step known as *feature engineering*, for example by transforming existing variables, combining them, or by converting text to numerical word frequencies. Example 8.4 shows how easy it is to train a classifier using the Naïve Bayes

[3] In Section 8.5, we discuss more advanced approaches, such as splitting or cross-validation.

Python Code

```python
df=pd.read_csv("https://cssbook.net/d/media.csv")

df["uses-internet"] = (df["internet"]>0).replace(
    {True:"user", False:"non-user"})
df.dropna(inplace=True)
print("How many people used online news at all?")
print(df["uses-internet"].value_counts())

X_train, X_test, y_train, y_test = \
train_test_split(df[["age","education", "gender"]],
        df["uses-internet"], test_size=0.2,
        random_state=42)

print(f"We have {len(X_train)} training and "\
      f"{len(X_test)} test cases.")
```

R Code

```r
df = read.csv("https://cssbook.net/d/media.csv")
df = na.omit(df %>% mutate(
    usesinternet=recode(internet,
            .default="user", '0'="non-user")))

set.seed(42)
df$usesinternet = as.factor(df$usesinternet)
print("How many people used online news at all?")
print(table(df$usesinternet))


split = initial_split(df, prop = .8)
traindata = training(split)
testdata = testing(split)

X_train = select(traindata,
                 c("age", "gender", "education"))
y_train = traindata$usesinternet
X_test = select(testdata,
                 c("age", "gender", "education"))
y_test = testdata$usesinternet

glue("We have {nrow(X_train)} training and ",
     "{nrow(X_test)} test cases.")
```

Python Output

```
How many people used online news at all?
user        1262
non-user     803
Name: uses-internet, dtype: int64
We have 1652 training and 413 test cases.
```

Example 8.3 Preparing a dataset for supervised machine learning.

Python Code

```python
myclassifier = GaussianNB()
myclassifier.fit(X_train, y_train)
y_pred = myclassifier.predict(X_test)
```

R Code

```r
myclassifier = train(x = X_train, y = y_train,
                     method = "naive_bayes")
y_pred = predict(myclassifier, newdata = X_test)
```

Example 8.4 A simple Naïve Bayes classifier.

algorithm with packages *caret*/*naivebayes* in R and *sklearn* in Python (this approach will be better explained in Subsection 8.3.1).

But before we can actually use this classifier to do some useful work, we need to test how capable it is to predict the correct labels, given a set of features. One might think that we could just feed it the same input data (i.e., the same features) again and see whether the predicted labels match the actual labels of the test dataset. In fact, we could do that. But this test would not be strict enough: after all, the classifier has been trained on exactly these data, and therefore one would expect it to perform pretty well. In particular, it may be that the classifier is very good in predicting its own training data, but fails at predicting other data, because it overgeneralizes some idiosyncrasy in the data, a phenomenon known as overfitting (see Figure 8.1).

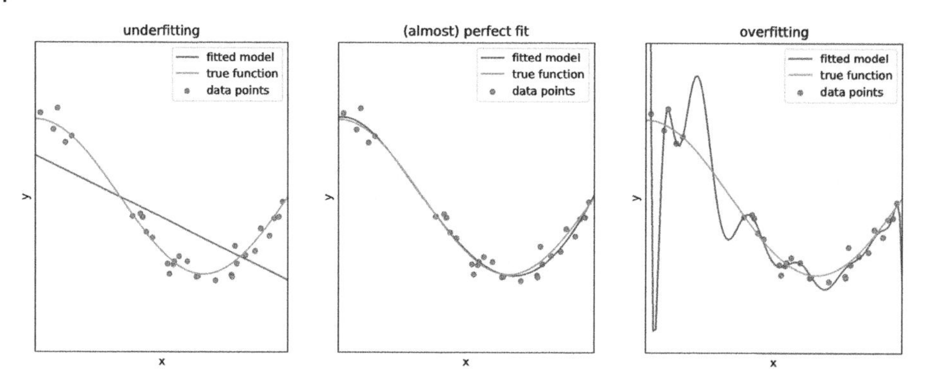

Figure 8.1 Underfitting and overfitting. Example adapted from https://scikit-learn.org/stable/auto_examples/model_selection/plot_underfitting_overfitting.html

Instead, we use the features of the *test dataset* (stored in the objects `X_test` and `y_test`) as input for our classifier, and evaluate how far the predicted labels match the actual labels. Remember: the classifier has at no point in time seen the actual labels. Therefore, we can in fact calculate how often the prediction is right.[4]

As shown in Example 8.5, we can create a *confusion matrix* (generated with *caret* function `confusionMatrix` in R and *sklearn* function `confusion_matrix` in Python), and then estimate two measures: *precision* and *recall* (using base R calculations in R and *sklearn* function `classification_report` in Python). In a binary classification, the *confusion matrix* is a useful table in which each column usually represents the number of cases in a predicted class, and each row the number of cases in the real or actual class. With this matrix (see Figure 8.2) we can then estimate the number of *true positives* (TP) (correct prediction), *false positives* (FP) (incorrect prediction), *true negatives* (TN) (correct prediction) and *false negatives* (FN) (incorrect prediction).

For a better understanding of these concepts, imagine that we build a sentiment classifier, that predicts – based on the text of a movie review – whether it is a positive review or a negative review. Let us assume that the goal of training this classifier is to build an app that recommends only good movies to the user. There are two things that we want to achieve: we want to find as many positive films as possible (recall), but we also want that the selection we found *only* contains positive films (precision).

Precision is calculated as $\frac{\mathrm{TP}}{\mathrm{TP}+\mathrm{FP}}$, where TP are true positives and FP are false positives. For example, if our classifier retrieves 200 articles that it classifies as positive films, but only 150 of them indeed are positive films, then the precision is $\frac{150}{150+50}=\frac{150}{200}=0.75$.

[4] We assume here that the manual annotation is always right; an assumption that one may, of course, challenge. However, in the absence of any better proxy for reality, we assume that this manual annotation is the so-called *gold standard* that reflects the *ground truth* as closely as possible, and that it by definition cannot be outperformed. When creating the manual annotations, it is therefore important to safeguard their quality. In particular, one should calculate and report some reliability measures, such as the *intercoder reliability* which tests the degree of agreement between two or more annotators in order to check if our classes are well defined and the coders are doing their work correctly.

Python Code

```python
print("Confusion matrix:")
print(confusion_matrix(y_test, y_pred))
print(classification_report(y_test, y_pred))
```

R Code

```r
1  print(confusionMatrix(y_pred, y_test))
2
3  print("Confusion matrix:")
4  confmat = table(testdata$usesinternet, y_pred)
5  print(confmat)
6
7  print("Precision for predicting True internet")
8  print("users and non-internet-users:")
9  precision = diag(confmat) / colSums(confmat)
10 print(precision)
11
12 print("Recall for predicting True internet")
13 print("users and non-internet-users:")
14 recall = (diag(confmat) / rowSums(confmat))
15 print(recall)
16
```

Python Output

```
Confusion matrix:
[[ 55 106]
 [ 40 212]]
              precision    recall  f1-score   support
    non-user       0.58      0.34      0.43       161
        user       0.67      0.84      0.74       252

    accuracy                           0.65       413
   macro avg       0.62      0.59      0.59       413
weighted avg       0.63      0.65      0.62       413
```

Example 8.5 Calculating precision and recall.

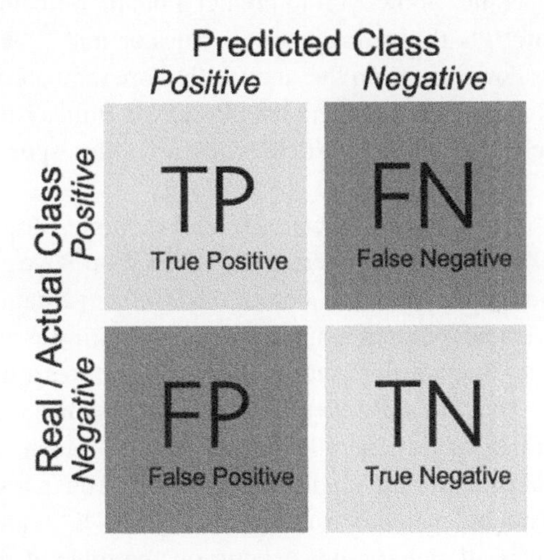

Figure 8.2 Visual representation of a confusion matrix.

Recall is calculated as $\frac{TP}{TP+FN}$, where TP are true positives and FN are false negatives. If we know that the classifier from the previous paragraph missed 20 positive films, then the recall is $\frac{150}{150+20}=\frac{150}{170}=0.88$.

In other words: recall measures how many of the cases we wanted to find we actually found. Precision measures how much of what we have found is actually correct.

Often, we have to make a trade-off between precision and recall. For example, just retrieving *every* film would give us a recall of 1.0 (after all, we didn't miss a single positive film). But on the other hand, we retrieved all the negative films as well, so precision will be extremely low. It can depend on the task at hand whether precision or recall is more important. In Section 8.5, we discuss this trade-off in detail, as well as other metrics such as *accuracy*, F_1*-score* or the *area under the curve* (AUC).

8.3 Classical Machine Learning: From Naïve Bayes to Neural Networks

To do supervised machine learning, we can use several models, all of which have different advantages and disadvantages, and are more useful for some use cases than for others. We limit ourselves to the most common ones in this chapter. The website of scikit-learn (http://www.scikit-learn.org) gives a good overview of more alternatives.

8.3.1 Naïve Bayes

The Naïve Bayes classifier is a very simple classifier that is often used as a "baseline". Before estimating more complicated and resource-intensive models, it is a good idea to estimate a simpler model first, to assess how much better the other model actually is. Sometimes, the simple model might even be just fine.

The Naïve Bayes classifier allows you to predict a binary outcome, such as: "Is this message spam or not?", "Is this article about politics or not?", "Will this go viral or not?". It, in fact, also allows you to do the same with more than one category, and both the Python and the R implementation will happily let you train a Naïve Bayes classifier on nominal data, such as whether an article is about politics, sports, the economy, or something different.

For the sake of simplicity, we will discuss a binary example, though.

As its name suggests, a Naïve Bayes classifier is based on Bayes' theorem, and it is "Naïve". It may sound a bit weird to call a model "Naïve", but what it actually means is not so much that it is stupid, but that it makes very far-reaching assumptions about the data (hence, it is Naïve). Specifically, it assumes that all features are independent from each other. Of course, that is hardly ever the case – for instance, in a survey data set, while age and gender indeed are generally independent from each other, this is not the case for education, political interest, media use, and so on. And in textual data, whether a word W_1 is used is not independent from the use of word W_2 – after all, both are not randomly drawn from a dictionary, but depend on the topic of the text (and other things). Astonishingly, even though these assumptions are regularly violated, the Naïve Bayes classifier works reasonably well in practice.

The Bayes part of the Naïve Bayes classifier comes from the fact that it uses Bayes' formula,

$$P(\mathrm{A} \mid \mathrm{B}) = \frac{\mathrm{P}(\mathrm{B} \mid \mathrm{A}) \cdot \mathrm{P}(\mathrm{A})}{\mathrm{P}(\mathrm{B})}.$$

As a short refresher: The $P(A \mid B)$ can be read as: the probability of A, given B. Or: the probability of A if B is the case/present/true. Applied to our problem, this means that we are interested in estimating the probability of an item having a label, given a set of features:

$$P(\text{label} \mid \text{features}) = \frac{P(\text{features} \mid \text{label}) \cdot P(\text{label})}{P(\text{features})}.$$

P(label) can be easily calculated: it's just the fraction of all cases with the label we are interested in. Because we assume that our features are independent (remember, the "Naïve" part), we can calculate P(features) and P(features | label) by just multiplying the probabilities of each individual feature. Let's assume we have three features, x_1, x_2, x_3. We now simply calculate the percentage of *all* cases that contain these features, $P(x_1), P(x_2)$ and $P(x_3)$. Then we do the same for the conditional probabilities and calculate the percentage of cases *with our label* that contain these features, $P(x_1 \mid \text{label}), P(x_2 \mid \text{label})$ and $P(x_3 \mid \text{label})$.

If we fill this in our formula, we get:

$$P(\text{label} \mid \text{features}) = \frac{P(x_1 \mid \text{label}) \cdot P(x_2 \mid \text{label}) \cdot P(x_3 \mid \text{label}) \cdot P(\text{label})}{P(x_1) \cdot P(x_2) \cdot P(x_3)}$$

Remember that all we need to do to calculate this formula is: (1) count how many cases we have in total; (2) count how many cases have our label; (3) count how many cases in (1) have feature x; (4) count how many cases in (2) have feature x. As you can imagine, doing this does not take much time to do, which is what makes the Naïve Bayes classifier such a fast and efficient choice. This may in particular be true if you have a lot of features (i.e., high-dimensional data).

Counting whether a feature is present or not, of course, is only possible for binary data. We could for example simply check whether a given word is present in a text or not. But what if our features are continuous data, such as the number of times the word is present? We could dichotomize it, but that would discard information. So, what we do instead, is that we estimate $P(x_i)$ using a distribution, for example a Gaussian, Bernoulli, or multinomial distribution. The core idea, though, stays the same.

Our examples in Section 8.2 illustrate how to train a Naïve Bayes classifier. We first create the labels (whether someone uses online news at all or not), split our data into a training and a test dataset (here, we use 80% for training and 20% for testing) (Example 8.3), then fit (train) a classifier (Example 8.4), before we assess how well it predicts our training data (Example 8.5).

In Section 8.5, we discuss in more detail how to evaluate different classifiers, but let's have a sneak preview at the most used measures of how well our classifier performs. The confusion matrix from Example 8.5 tells us how many users were indeed classified as users (55), and how many (wrongly) as non-users (106).[5] That doesn't look very good; but on the other hand, 212 of the non-users were correctly classified as such, and only 40 were not.

---

[5] These are the values from the Python example, the R example slightly differs, amongst other things due to different sampling.

More formally, we can express this using precision and recall. When we are interested in finding true users, we get a precision of $\frac{212}{212+106}=0.67$ and a recall of $\frac{212}{212+40}=0.84$. However, if we want to know how good we are in identifying those who do *not* use online news, we do – as we saw in the confusion matrix – considerably worse: precision and recall are 0.58 and 0.34, respectively.

8.3.2 Logistic Regression

Regression analysis does not make as strong an assumption about the independence of features as the Naïve Bayes classifier does. Sure, we have been warned about the dangers of multicollinearity in statistics classes, but correlation between features (for which multicollinearity is a fancy term) affects the coefficients and their p values, but not the predictions of the model as a whole. To put it differently, in regression models, we do not estimate the probability of a label given a feature, independent of all the other features, but are able to "control for" their influence. In theory, this should make our models better, and also in practice, this regularly is the case. However, ultimately, it is an empirical question which model performs best.

While we started this chapter with an example of an OLS regression to estimate a continuous outcome (well, by approximation, as for "days per week" not all values make sense), we will now use a regression approach to predict nominal outcomes, just as in the Naïve Bayes example. The type of regression analysis to use for this is called *logistic regression*.

In a normal OLS regression, we estimate

$$y=\beta_o+\beta_1x_1+\beta_2x_2+\ldots+\beta_nx_n$$

But this gives us a continuous outcome, which we do not want. In a logistic regression, we therefore use the sigmoid function to map this continuous outcome to a value between 0 and 1. The sigmoid function is defined as $sigmoid(x)=\frac{1}{1+e^{-x}}$ and depicted in Figure 8.3.

Combining these formulas gives us:

$$P=\frac{1}{1+e^{-(\beta_o+\beta_1x_1+\beta_2x_2=\ldots+\beta_nx_n)}}$$

Wait, you might say. Isn't P still continuous, even though it is now bounded between 0 and 1? Yes, it is. Therefore, after having estimated the model, we use a threshold value (typically, 0.5, but we will discuss in Section 8.5.1 how to select different ones) to predict the label. If $P>0.5$, we predict that the case is spam/about politics/will go viral, if not, we predict it's not. A nice side effect of this is that we still can use the probabilities in case we are interested in them, for example to figure out for which cases we are more sure of in our prediction.

Just as with the Naïve Bayes classifier, also for logistic regression classifiers, Python and R will happily allow us to estimate models with multiple nominal outcomes instead of a binary outcome. In Example 8.6 we fit the logistic regression using the

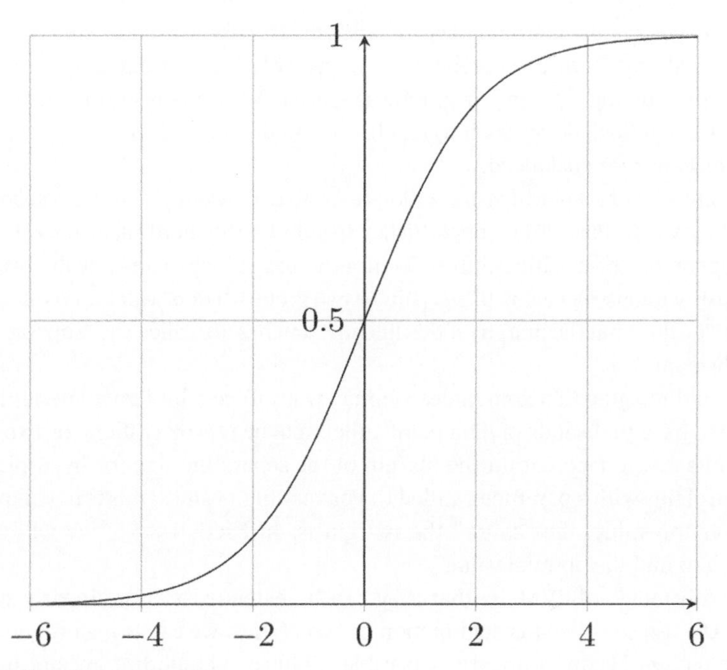

Figure 8.3 The sigmoid function.

Python Code

```
myclassifier = LogisticRegression(solver="lbfgs")
myclassifier.fit(X_train, y_train)

y_pred = myclassifier.predict(X_test)
```

R Code

```
myclassifier = train(x = X_train, y = y_train,
    method = "glm", family = "binomial")
y_pred = predict(myclassifier, newdata = X_test)
```

Example 8.6 A simple logistic regression classifier.

caret method `logreg` in R and the *sklearn* (module linear_model) function `LogisticRegression` in Python.

And, of course, you actually can do OLS regression (or more advanced regression models) if you want to estimate a continuous outcome.

8.3.3 Support Vector Machines

Support Vector Machines (SVM) are another very popular and versatile approach to supervised machine learning. In fact, they are quite similar to logistic regression, but try to optimize a different function. In technical terms, SVM minimizes *hinge loss* instead of logistic loss.

What does that mean to us? When estimating logistic regressions, we are interested in estimating probabilities, while when training a SVM, we are interested in finding a plane (more specifically, a hyperplane) that best separates the data points of the two classes (e.g., spam versus non-spam messages) that we want to distinguish. This also means that a SVM does not give you probabilities associated with your prediction, but just the label. But usually, that's all that you want anyway.

Without going into mathematical detail here (for that, a good source would be Kelleher et al., 2015), we can say that finding the widest separating margin that we can achieve constructing a plane in a graphical space (SVM) versus optimizing a log-likelihood function (logistic regression) results in a model that is less sensitive to outliers, and tends to be more balanced.

There are a lot of graphical visualizations available, for example in the notebooks supplementing VanderPlas (2016) (https://jakevdp.github.io/PythonDataScienceHandbook/05.07-support-vector-machines.html). For now, it may suffice to imagine the two-dimensional case: we construct a line that separates two groups of dots *with the broadest possible margin*. The dots that the margin of this line just touches are called the "support vectors", hence the name.

You could imagine that sometimes we may want to be a bit lenient about the margins. If we have thousands of data points, then maybe it is okay if one or two of these data points are, in fact, within the margin of the separating line (or hyperplane). We can control this with a parameter called C: For very high values, this is not allowed, but the lower the value, the "softer" the margin is. In Section 8.5.3, we will show an approach to find the optimal value.

A big advantage of SVMs is that they can be extended to non-linearly separable classes. Using a so-called kernel function or *kernel trick*, we can transform our data so that the dataset becomes linearly separable. Choices include but are not limited to multinomial kernels, the radial basis function (RBF), or Gaussian kernels. If we, for example, have two concentric rings of data points (like a donut), then we cannot find a straight line separating them. But a RBF kernel can transfer them into a linearly separable space. The aforementioned online visualizations can be very instructive here.

Example 8.7 shows how we implement standard SVM to our data using the *caret* method `svmLinear3` in R and the *sklearn* (module svm) function `SVC` in Python. You can see in the code that feature data is standardized or normalized (with $m = 0$ and $\text{std} = 1$) before model training in order to have all the features measured at the same scale, as required by SMV.

Python Code

```python
# !!! We normalize our features to have M=0 and
# SD=1. This is necessary as our features are not
# measured on the same scale, which SVM requires.
# Alternatively, rescale to [0:1] or [-1:1]

scaler=preprocessing.StandardScaler().fit(X_train)

X_train_scaled = scaler.transform(X_train)
X_test_scaled = scaler.transform(X_test)

myclassifier = SVC(gamma="scale")
myclassifier.fit(X_train_scaled, y_train)

y_pred = myclassifier.predict(X_test_scaled)
```

R Code

```r
# !!! We normalize our features to have M=0 and
# SD=1. This is necessary as our features are not
# measured on the same scale, which SVM requires.
# Alternatively, rescale to [0:1] or [-1:1]

myclassifier = train(x = X_train, y = y_train,
    preProcess = c("center", "scale"),
                           method = "svmLinear3")
y_pred = predict(myclassifier, newdata = X_test)
```

Example 8.7 A simple Support Vector Machine classifier.

8.3.4 Decision Trees and Random Forests

In the models we have discussed so far, we were essentially modeling linear relationships. If the value of a feature is twice as high, its influence on the outcome will be twice as high as well. Sure, we can (and do, as in the case of the sigmoid function or the SVM kernel trick) apply some transformations, but we have not really considered yet how we can model situations in which, for instance, we care about whether the value of a feature is above (or below) a specific threshold. For instance, if we have a set of social media messages and want to model the medium from which they most likely come, then its length is very important information. If it is longer than 280 characters (or, historically, 140), then we can be *very* sure it is not from Twitter, even though the reverse is not necessarily true. But it does not matter at all whether it is 290 or 10 000 characters long.

Entering this variable into a logistic regression, thus, would not be a smart idea. We could, of course, dichotomize it, but that would only partly solve the problem, as its effect can still be overridden by other variables. In this example, we *know* how to dichotomize it based on our prior knowledge about the number of characters in a tweet, but this does not necessarily need to be the case; it might be something we need to estimate.

A step-wise decision, in which we first check one feature (the length), before checking another feature, can be modeled as a decision tree. Figure 8.4 depicts a (hypothetical) decision tree with three *leaves*.

Faced with the challenge to predict whether a social media message is a tweet or a Facebook post, we could predict 'Facebook post' if its length is greater than 280 characters. If not, we check whether it includes hashtags, and if so, we predict 'tweet', otherwise, 'Facebook post'. Of course, this simplistic model will be wrong at some times, because not all tweets have hashtags, and some Facebook posts actually do include hashtags.

While we constructed this hypothetical decision tree by hand, usually, we are more interested in learning such non-linear relationships from the data. This means that we

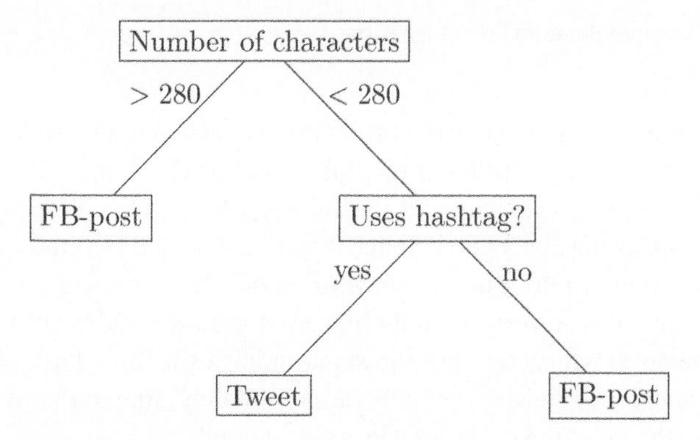

Figure 8.4 A simple decision tree.

do not have to determine the cutoff point ourselves, but also that we do not determine the order in which we check multiple variables by hand.

Decision trees have two nice properties. First, they are very easy to explain. In fact, a figure like Figure 8.4 is understandable for non-experts, which can be important in scenarios where for accountability reasons, the decision of a classifier must be as transparent as possible. Second, they allow us to approximate almost all non-linear relationships (be it not necessarily very accurately).

However, this comes at large costs. Formulating a model as a series of yes/no questions, as you can imagine, inherently loses a lot of nuance. More importantly, in such a tree, you cannot "move up" again. In other words, if you make a wrong decision early on in the tree (i.e., close to its root node), you cannot correct it later. This rigidity makes decision trees also prone to overfitting: they may fit the training data very well, but may not generalize well enough to slightly different (test) data.

Because of these drawbacks, decision trees are seldom used in real-life classification tasks. Instead, one uses an *ensemble* model: so-called random forests. Drawing random samples from the data, we estimate multiple decision trees – hence, a forest. To arrive at a final prediction, we can then let the trees "vote" on which label we should predict. This procedure is called "majority voting", but there are also other methods available. For example, *scikit-learn* in Python by default uses a method called `probabilistic prediction`, which takes into account probability values instead of simple votes.

In Example 8.8 we create a random forest classifier with 100 trees using the *caret* method `rf` in R and the *sklearn* (module ensamble) function `RandomForestClassifier` in Python.

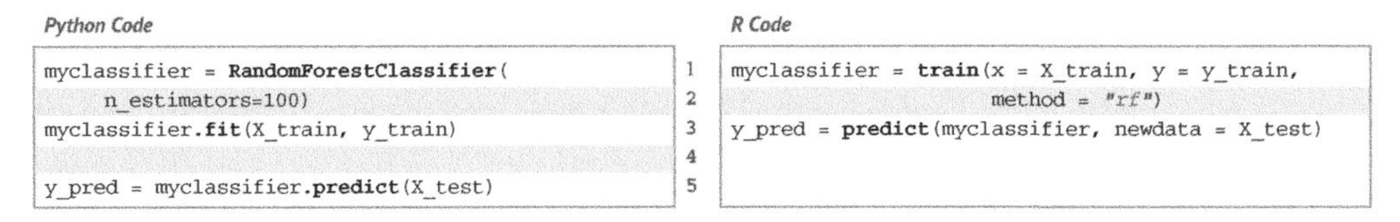

Python Code

```
myclassifier = RandomForestClassifier(
    n_estimators=100)
myclassifier.fit(X_train, y_train)

y_pred = myclassifier.predict(X_test)
```

R Code

```
myclassifier = train(x = X_train, y = y_train,
                     method = "rf")
y_pred = predict(myclassifier, newdata = X_test)
```

Example 8.8 A simple Random Forest classifier.

Because random forests alleviate the problems of decision trees, but keep the advantage of being able to model non-linear relationships, they are frequently used when we expect such relationships (or have no idea about what the relationship looks like). Also, random forests may be a good choice if you have very different types of features (some nominal, some continuous, etc.) in your model. The same holds true if you have a lot (really a lot) of features: methods like SVM would require constructing large matrices in memory, which random forests do not. But if the relationships between your features and your labels are actually (approximately) linear, then you are probably better off with one of the other models we discussed.

8.3.5 Neural Networks

Inspired by the neurons in the brains of humans (and other animals), neural networks consist of connections between neurons that are activated if the total input is above a certain threshold.

Figure 8.5 shows the simplest type of neural network, sometimes called a *perceptron*. This neural network consists only of a series of input neurons (representing the features or independent variables) which are directly connected with the output neuron or neurons (representing the output class(es)). Each of the connections between neurons has a weight, which can be positive or negative. For each output neuron, the weighted sum of inputs is calculated and a function is applied to determine the result. An example output function is the sigmoid function (Figure 8.3) which transforms the output to a value between zero and one, in which case the resulting model is essentially a form of logistic regression.

If we consider a neural network for sentiment analysis of tweets, the input neurons could be the frequencies of words such as "great" or "terrible", and we would assume that the weight of the first would be positive while the second would be negative. Such a network cannot take combinations into account, however: the result of "not great" will simply be the addition of the results of "not" and "great".

To overcome this limitation, it is possible to add a *hidden layer* of latent variables between the input and output layer, such as shown in Figure 8.6. This allows for combinations of neurons, with for example both "not" and "great" loading onto a hidden neuron, which can then override the direct effect of "great". An algorithm called

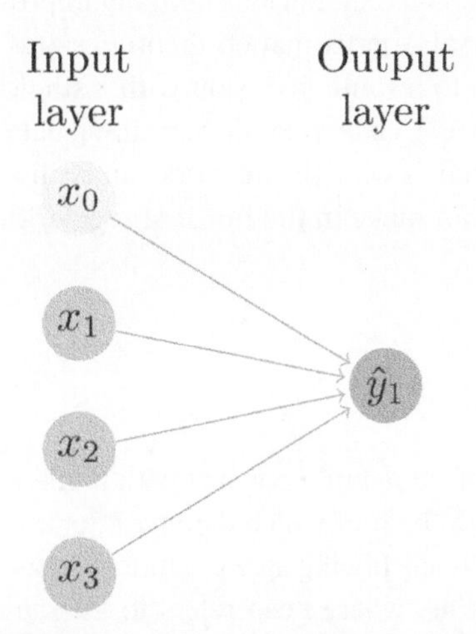

Figure 8.5 Schematic representation of a typical classical machine learning model.

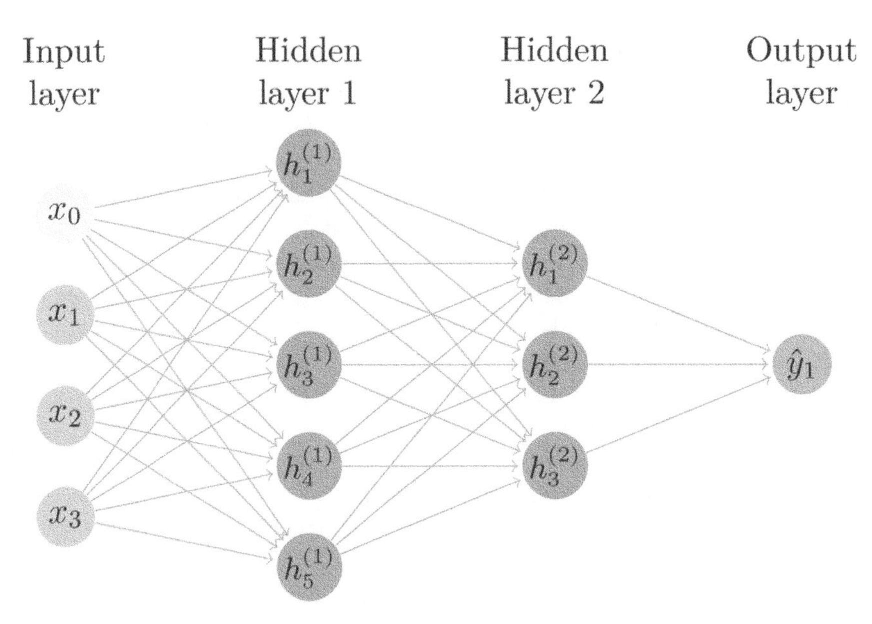

Figure 8.6 A neural network.

backpropagation can be used to iteratively approximate the optimal values for the model. This algorithm starts from a random state and optimizes the second layer while keeping the first constant, then optimizing the first layer, and repeating until it converges.

Although such hidden layers, which can easily contain thousands of neurons, are hard to interpret substantively, they can substantially improve the performance of the model. In fact, the Universal Approximation theorem states that every decision function can be approximated to infinite precision with a single (but possibly very large) hidden layer (Goldberg, 2017). Of course, since training data is always limited there is a practical limit to how deep or wide the network can be, but this shows the big difference that a hidden layer can make in the range of regularities that can be "captured" in the model.

8.4 Deep Learning

In Section 8.3.5, we introduced neural networks with hidden layers and the backpropagation algorithm to fit them, both of which date back to at least the 1970's. In the past decade, however, the Artificial Intelligence community has been transformed by the introduction of *deep learning*, where deep refers to a large amount of hidden layers between the input and output layers. Many of the recent advances in AI, from self-driving cars to automatic translation and voice assistants, are made possible by the application of deep learning techniques to the enormous amounts of digital data now becoming available.

An extensive treatment of deep learning is beyond the scope of this book (we recommend Géron (2019) instead). However, in this section we will give you a brief introduction that should help you understand deep learning at a conceptual level, and in Section 11.4.4 and Chapter 14 we will explain how these techniques can be applied to text analysis and visual analysis, respectively.

In principle, there is no clear demarcation between a "classical" neural network with hidden layers and a "deep" neural network. There are three properties, however, that distinguish deep learning and explain why it is so successful: scale, structure, and feature learning.

Scale. First, and perhaps most importantly, deep learning models are many orders of magnitude larger and more complex than the models trained in earlier decades. This has been made possible by the confluence of unprecedented amounts of digital training data and increased computer processing power. Partly, this has been enabled by the use of graphical processing units (GPUs), hardware designed for rendering the three-dimensional worlds used in games, but that can also be used very efficiently for the computations needed to train neural networks (and mine bitcoins, but that's another story).

Structure. Most classical neural networks have only "fully connected" hidden layers with forward propagation, meaning that each neuron in one layer is connected to each neuron in the next layer. In deep learning, many specific architectures (some of which will be discussed below) are used to process information in certain ways, limiting the number of parameters that need to be estimated.

Feature Learning. In all models described so far with the exception of neural networks with hidden layers, there was a direct relationship between the input features and the output class. This meant that it is important to make sure that the required information the model needs to distinguish the classes is directly encoded in the input features. In the example used earlier, if "not" and "good" are separate features, a single-layer network (or a Naïve Bayes model) cannot learn that these words together have a different meaning than the addition of their separate meanings. However, similar to regression analysis, where you can create an interaction term or squared term to model a non-linear relationship, the researcher can create input features for e.g. word pairs, for example including bigrams (word pairs) such as "not_good". In fact, engineering the right features was the main way in which a researcher could improve model performance. In deep learning, however, this feature learning step is generally included in the model itself, with subsequent layers encoding different aspects of the raw data.

The properties of scale, structure, and feature learning are intertwined in deep learning: the much larger networks enable structures with beautiful names such as "recurrent networks", "convolutional layers" or "long short-term memory", which are used to encode specific relationships and dependencies between features. In this book, we will focus on convolutional networks as our only example of deep learning, mostly because these networks are widely used in both text and image analysis. Hopefully, this will give you insight into the general idea behind deep learning, and you can learn about this and other models in more detail in the specialized resources cited above.

8.4.1 Convolutional Neural Networks

One challenge in many machine learning problems is a mismatch between the level of measurement of the output and the input. For example, we normally want to assign a single code such as sentiment or topic to a document or image. The raw input, however, is at the word or pixel level. In classical machine learning, this is generally solved by summarizing the input at the higher level of abstraction, for example by using the total frequency of each word per document as input feature. The problem is, however, that this summarization process removes a lot information that could be useful to the machine learning model, for example combinations of words ("not good") or their ordering ("John voted for Mary" versus "Mary voted for John"), unless the researcher engineers features such as word pairs to add this information.

Convolutional Neural Networks are one way in which deep learning can overcome this limitation. Essentially, the model internalizes the feature learning as a first part or "layer" of the model, using a specialized network to summarize the raw input values into document (or image) level features.

Figure 8.7 shows a highly simplified example of this for text analysis of a sentence fragment "Would not recommend". The left hand side shows how each word is encoded as a binary vector (e.g. 010 for "not", and 001 for "recommend"). In the second column, a shifting window concatenates these values for word pairs (so 010001 for "not recommend"). Next, a feature map layer detects interesting features in these concatenated values, for example a feature for a negated positive term that

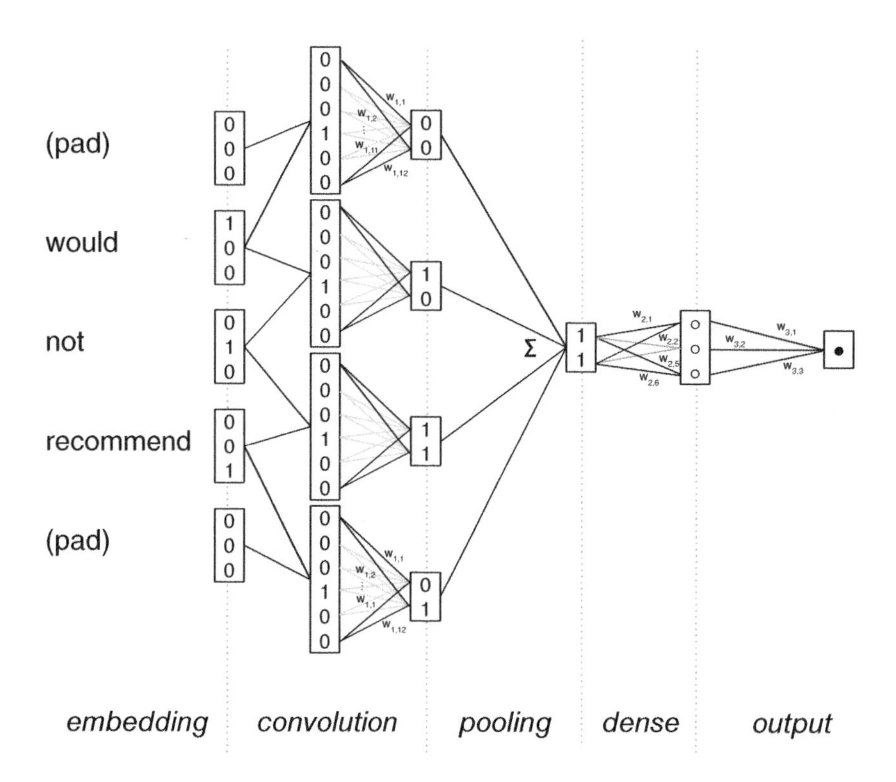

Figure 8.7 Simplified example of a Convolutional Network applied to text analysis.

has positive weights for negators in the first half and for positive words in the second. These features are then pooled together to create document-level features, for example by taking the maximum value per feature, which means that a feature is present in a document if it is present in any of the word windows in the document. Finally, these document-level features are then used in a regular (dense) neural network which is connected to the output value, e.g. the document sentiment. Since the convolutional layer is now connected with the output class, the feature maps can be automatically learned using the backpropagation algorithm explained above. This means that the model can find the features in the word windows that are most helpful in predicting the document class, bringing the feature learning into the modeling process.

Of course, this is a highly simplified example, but it shows how local dependencies can be detected automatically using the convolutional network, as long as the interesting features are found within the specified word window. Other architectures, such as the Long Short Term Memory, can also be used to find non-local dependencies, but a full discussion of different architectures is well beyond the scope of this book. Chapter 10 will give a more detailed example of deep learning for text analysis, where an embedding layer is combined with a convolutional network to build a sentiment analysis model. Similarly, Chapter 14 will show how a similar technique can be used to extract features from small areas of images which are then used in automatic image classification. This involves creating a two-dimensional window over pixels rather than a unidimensional window over words, and often multiple convolutional layers are chained to detect features in increasingly large areas of the image. The underlying technique of convolutional networks, however, is the same in both cases.

8.5 Validation and Best Practices

8.5.1 Finding a Balance Between Precision and Recall

In the previous sections, we have learned how to fit different models: Naïve Bayes, logistic regressions, support vector machines, and random forests. We have also had a first look at confusion matrices, precision, and recall.

But how do we find the best model? "Best", here, should be read as "best for our purposes" – some models may be bad, and some may be good, but which one is really the best may depend on what matters most for us: do we care more about precision or about recall? Are all classes equally important to us? And of course, other factors, such as explainability or computational costs may factor into our decision.

But in any event, we need to decide which metrics to focus on. We can then either manually inspect them and look, for instance, which model has the highest *accuracy*, or the best balance of precision and recall, or a recall higher than some threshold you are willing to accept.

If we build a classifier to distinguish spam messages from legitimate messages, we could ask the following questions:

Precision Which percentage of what our classifier predicts to be spam really is spam?
Recall What percentage of all spam messages has our classifier found?
Accuracy In which percentage of all cases was our classifier right?

We furthermore have:

F_1-score The harmonic mean of precision and recall: $F_1 = 2 \cdot \frac{\text{precision} \cdot \text{recall}}{\text{precision} + \text{recall}}$

AUC The AUC (Area under Curve) is the area under the curve that one gets when plotting the True Positive Rate (TPR) against the False Positive Rate (FPR) at various threshold settings. A perfect model will receive a value of 1.0, while random guessing between two equally probable classes will result in a value of 0.5

Micro- and macro-average Especially when we have more than two classes, we can calculate the average of measures such as precision, recall, or F_1-score. We can do so based on the separately calculated measures (macro), or based on the underlying values (TP, FP, etc.) (micro), which has different implications in the interpretation – especially if the classes have very different sizes.

So, which one to choose? If we really do not want to be annoyed by any spam in our inbox, we need a high recall (we want to find all spam messages). If, instead, we want to be sure that we do not accidentally throw away legitimate messages, we need a high precision (we want to be sure that all spam really is spam).

Maybe you say: well, I want both! You could look at the accuracy, a very straightforward to interpret measure. However, if you get many more legitimate messages than spam (or the other way round), this measure can be misleading: after all, even if your classifier finds almost none of the spam messages (it has a recall close to zero), you still get a very high accuracy, simply because there are so many legitimate messages. In other words, the accuracy is not a good measure when working with highly unbalanced classes. Often, it is therefore a better idea to look at the harmonic mean of precision and recall, the F_1-score, if you want to find a model that gives you a good compromise between precision and recall.

In fact, we can even fine-tune our models in such a way that they are geared towards either a better precision or a better recall. As an example, let us take a logistic regression model. It predicts a class label (such as "spam" versus "legitimate"), but it can also return the assigned probabilities. For a specific message, we can thus say that we estimate its probability of being spam as, say, 0.65. Unless we specify otherwise, everything above 0.5 will then be judged to be spam, everything below as legitimate. But we could specify a different cutoff point: we could, for instance, decide to classify everything above 0.7 as spam. This would give us a more conservative spam filter, with probably a higher precision at the expense of a lower recall.

We can visualize this with a so-called ROC (receiver operator characteristic), a plot in which we plot true positives against false positives at different thresholds (Figure 8.8). A good model extends until close to the upper left corner, and hence has a large area under the curve (AUC). If we choose a threshold at the left end of the curve, we get few false positives (good!), but also few true positives (bad!), if we go too far to the right, we get the other extreme. So, how can we find the best spot?

One approach is to print a table with three columns: the false positive rate, the true positive rate, and the threshold value. You then decide which FPR–TPR combination is most appealing to you, and use the corresponding threshold value. Alternatively, you can find the threshold value with the maximum distance between TPR and FPR, an approach also known as Yoden's J (Example 8.9). Plotting the ROC curve can also help interpreting which TPR/FPR combination is most promising (i.e., closest to the upper left corner).

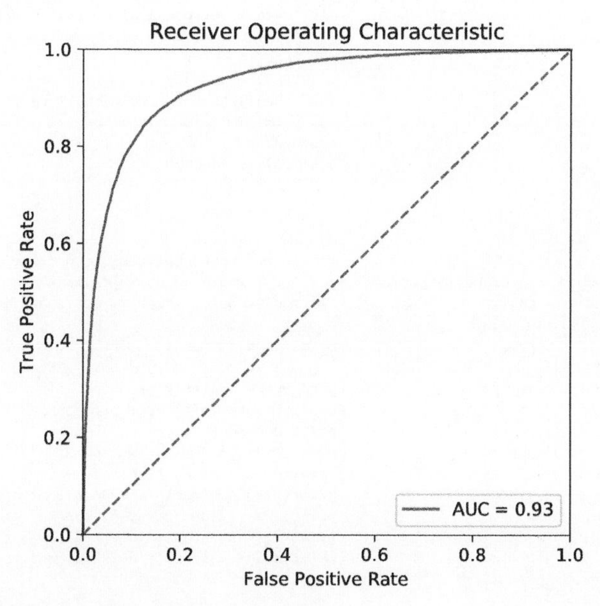

Figure 8.8 A (pretty good) ROC curve.

Python Code

```
myclassifier = LogisticRegression(solver="lbfgs")
myclassifier.fit(X_train, y_train)

print("With default cutoff point (.5):")
y_pred = myclassifier.predict(X_test)
print(classification_report(y_test, y_pred))
print(confusion_matrix(y_test, y_pred))
```

R Code

```
m = glm(usesinternet ~ age + gender + education,
        data=traindata, family="binomial")
y_pred = predict(m, newdata = testdata,
                 type = "response")
pred_default = as.factor(ifelse(y_pred>0.5,
                                "user", "non-user"))

print("Confusion matrix, default threshold (0.5)")
confmat = table(y_test, pred_default)
print(confmat)
print("Recall for predicting True internet
users and non-internet-users:")
print(diag(confmat) / rowSums(confmat))
print("Precision for predicting True internet
users and non-internet-users:")
print(diag(confmat) / colSums(confmat))
```

Python Output

```
With default cutoff point (.5):
                  precision   recall f1-score support
non-user              0.58     0.37     0.45     161
user                  0.67     0.83     0.74     252

accuracy                                0.65     413
macro avg             0.63     0.60     0.60     413
weighted avg          0.64     0.65     0.63     413

[[ 59 102]
 [ 42 210]]
```

R Output

```
[1] "Confusion matrix with default threshold
     (0.5):"
              pred_default
y_test      non-user user
non-user          52 118
user              32 210
[1] "Recall for predicting True internet users and
     non-internet-users:"
non-user      user
0.3058824 0.8677686
[1] "Precision for predicting True internet users
     and non-internet-users:"
non-user      user
0.6190476 0.6402439
```

Python Code

```
# get all predicted probabilities and ROC curve
predprobs = myclassifier.predict_log_proba(X_test)
fpr,tpr, thresholds = roc_curve(y_test,
                  predprobs[:,1], pos_label="user")

# determine the cutoff point
opt_threshold = thresholds[np.argmax(tpr-fpr)]

print("With the optimal probability threshold is"\
      f"{opt_threshold}, which is equivalent to"\
      f"a cutoff of {np.exp(optimal_threshold)},"\
      "we get:")
y_pred_alt = np.where(predprobs[:,1] >
         optimal_threshold, "user", "non-user")
print(classification_report(y_test, y_pred_alt))
print(confusion_matrix(y_test, y_pred_alt))
```

R Code

```
roc_ = roc(testdata$usesinternet ~ y_pred)
opt = roc_$thresholds[which.max(
   roc_$sensitivities + roc$_specificities)]

print(glue("Confusion matrix with optimal",
           "threshold ({opt}):"))
pred_opt = ifelse(y_pred>opt, "user", "non-user")
confmat = table(y_test, pred_opt)
print(confmat)
print("Recall for predicting True internet")
print("users and non-internet-users:")
print(diag(confmat) / rowSums(confmat))
print("Precision for predicting True internet")
print("users and non-internet-users:")
print(diag(confmat) / colSums(confmat))
```

Python Output

```
With the optimal probability threshold is
     -0.3880564601306907,
      which is equivalent to a cutoff of
     0.6783740410958241, we
       get:
                  precision   recall f1-score support
non-user              0.50     0.80     0.61     161
    user              0.79     0.49     0.61     252
accuracy                                0.61     413
macro avg             0.64     0.64     0.61     413
weighted avg          0.68     0.61     0.61     413
[[128 33]
 [128 124]]
```

R Output

```
Confusion matrix with optimal threshold
     (0.629312922296396):
            pred_optimal
y_test   non-user user
non-user     109 61
user          89 153
[1] "RecallforpredictingTrueinternetusersandnon-internet-
     users:"
non-user  user
0.6411765 0.6322314
[1] "Precision for predicting True internet users
     and non-
      internet-users:"
non-user     user
0.5505051 0.7149533
```

Example 8.9 Choosing a different cutoff point for predictions with logistic regression. In this case, we make a trade-off and maximize the difference between false positive rate and true positive rate to improve the precision for the the second category at the expense of the precision for the first category. Python and R have slightly different outcomes because the underlying implementation is different, but for this example that may be ignored.

Python Code

```python
plt.figure(figsize=(5,5))
plt.title("Receiver Operating Characteristic")
plt.plot(fpr,tpr,"b",
         label=f"AUC = {auc(fpr,tpr):0.2f}")
plt.legend(loc="lower right")
plt.plot([0,1],[0,1],"r--")
plt.xlim([0,1])
plt.ylim([0,1])
plt.ylabel("True Positive Rate")
plt.show()
```

R Code

```r
roc_ = roc(testdata$usesinternet ~ y_pred, plot=T,
   print.auc=T, print.thres="best",
   print.thres.pattern="Best threshold: %1.2f")
```

Python Output

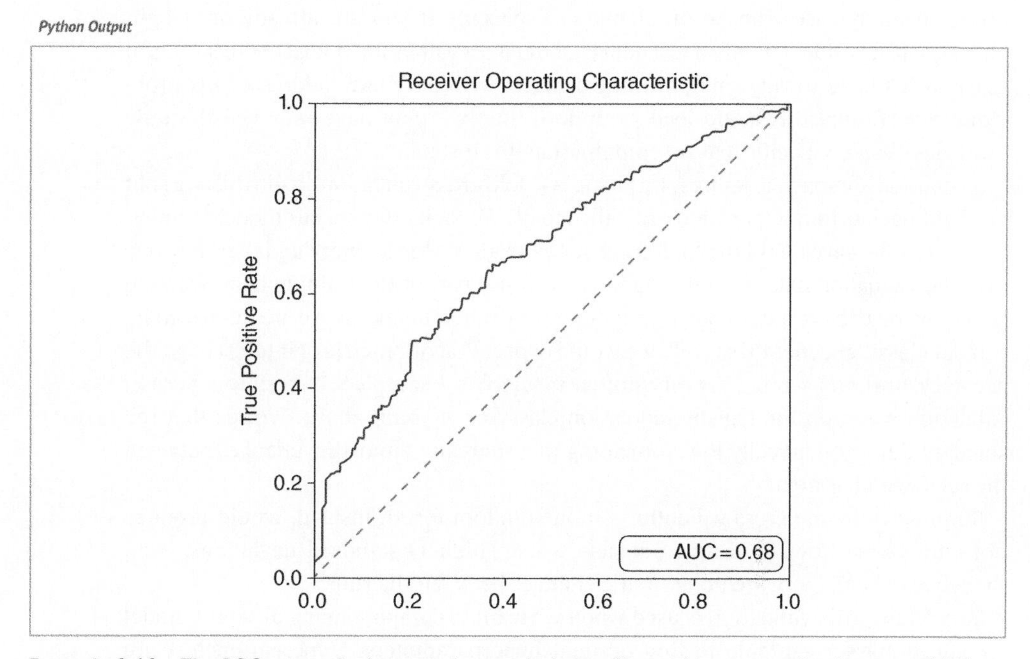

Example 8.10 The ROC curve of a (not very impressive) classifier and its area under the curve (AUC).

8.5.2 Train, Validate, Test

By now, we have established which measures we can use to decide which model to use. For all of them, we have assumed that we split our labeled dataset into two: a training dataset and a test dataset. The logic behind it was simple: if we calculate precision and recall on the training data itself, our assessment would be too optimistic – after all, our models have been trained on exactly these data, so predicting the label isn't too hard. Assessing the models on a different dataset, the test dataset, instead, gives us an assessment of what precision and recall look like if the labels haven't been seen earlier – which is exactly what we want to know.

Unfortunately, if we calculate precision and recall (or any other metric) for multiple models on the same test dataset, and use these results to determine which metric to use, we can run into a problem: we may avoid overfitting of our model on the training data, but we may now overfit it on the test data! After all, we could tweak our models until they fit our test data perfectly, even if this makes the predictions for other cases worse.

One way to avoid this is to split the original data into three datasets instead of two: a training dataset, a validation dataset, and a test dataset. We train multiple model configurations on the training dataset and calculate the metrics of interest for all of them on the validation dataset. Once we have decided on a final model, we calculate its performance (once) on the test dataset, to get an unbiased estimate of its performance.

8.5.3 Cross-validation and Grid Search

In an ideal world, we would have a huge labeled dataset and would not need to worry about the decreasing size of our training dataset as we set aside our validation and test datasets.

Unfortunately, our labeled datasets in the real world have a limited size, and setting aside too many cases can be problematic. Especially if you are already on a tight budget, setting aside not only a test dataset, but also a validation dataset of meaningful size may lead to critically small training datasets. While we have addressed the problem of overfitting, this could lead to underfitting: we may have removed the only examples of some specific feature combination, for instance.

A common approach to address this issue is k-fold cross-validation. To do this, we split our training data into k partitions, so-called folds. We then estimate our model k times, and each time leave *one* of the folds aside for validation. Hence, every fold is exactly one time the validation dataset, and exactly $k-1$ times part of the training data. We then simply average the results of our k values for the evaluation metric we are interested in.

If our classifier generalizes well, we would expect that our metric of interest (e.g., the accuracy, or the F_1-score, ...) is very similar in all folds. Example 8.11 performs a cross-validation based on the logistic regression classifier we built above. We see that the standard deviation is really low, indicating that there are almost no changes between the runs, which is great.

Running the same cross-validation on our random forest, instead, would produce not only worse (lower) means, but also worse (higher) standard deviations, even though also here, there are no dramatic changes between the runs.

Very often, cross-validation is used when we want to compare many different model specifications, for example to find optimal hyperparameters. Hyperparameters are parameters of the model that are not estimated from the data. These depend on the model, but could for example be the estimation method to use, the number of times a bootstrap should be repeated, etc. Very good examples are the hyperparameters of support vector machines (see above): it is hard to know how soft our margins should be (the C), and we may also be unsure about the right kernel (Example 8.13), or in the case of a polinomial kernel, how many degrees we want to consider.

Using the help function (e.g., `RandomForestClassifier?` in Python), you can look up which hyperparameters you can specify. For a random forest classifier, for instance, this includes the number of estimators in the model, the criterion, and whether or not to use bootstrapping. Example 8.12, 8.13, and 8.14 illustrate how you can automatically assess which values you should choose.

> Supervised machine learning is one of the areas where you really see differences between Python and R. While in Python, virtually all you need is available via *scikit-learn*, in R, we often need to combine *caret* with various libraries providing the actual models. In contrast, all components we need for machine learning in Python are developed within one package, which leads to less friction. This is what you see in the gridsearch examples in this section. In scikit-learn, *any* hyperparameter can

be part of the grid, but no hyperparameter has to be. Note that in R, in contrast, you cannot (at least, not easily) put any parameter of the model in the grid. Instead, you can look up the "tunable parameters" which *must* be present as part of the grid in the caret documentation. This means that an exact replication of the grid searches in Example 8.12 and Example 8.13 is not natively supported using *caret* and requires either manual testing or writing a so-called caret extension.

While in the end, you can find a supervised machine learning solution for all your use cases in R as well, if supervised machine learning is at the core of your project, it may save you a lot of cursing to do this in Python.

Python Code

```
myclassifier = LogisticRegression(solver="lbfgs")
acc = cross_val_score(estimator=myclassifier,
   X=X_train, y=y_train, scoring="accuracy", cv=5)
print(acc)
print(f"M={acc.mean():.2f}, SD={acc.std():.3f}")
```

R Code

```
myclassifier = train(x = X_train, y = y_train,
      method = "glm", family="binomial",
      metric="Accuracy", trControl = trainControl(
        method = "cv", number = 5,
        returnResamp ="all", savePredictions=TRUE),)
print(myclassifier$resample)
print(myclassifier$results)
```

Python Output

```
[0.64652568 0.64048338 0.62727273 0.64242424
    0.63636364]
M = 0.64, SD = 0.007
```

R Output

```
Accuracy Kappa parameter Resample
1 0.6646526 0.2376911 none Fold1
2 0.6042296 0.0767380 none Fold2
3 0.6283988 0.1552794 none Fold3
4 0.6636364 0.2225901 none Fold4
5 0.6484848 0.1889143 none Fold5
 parameter Accuracy Kappa AccuracySD KappaSD
1     none 0.6418804 0.1762426 0.02566539 0.06408044
```

Example 8.11 Crossvalidation.

Python Code

```
myclassifier = RandomForestClassifier()
f1scorer = make_scorer(f1_score, pos_label="user")

grid = {
    "n_estimators" : [10, 50, 100, 200],
    "criterion": ["gini", "entropy"],
    "bootstrap": [True, False],
    }
search = GridSearchCV(estimator=myclassifier,
                      param_grid=grid,
                      scoring=f1scorer,
                      cv=5)
search.fit(X_train, y_train)
print(f"Hyperparameters {search.best_params_} "
    "give the best performance:")
print(classification_report(y_test,
    search.predict(X_test)))
```

Python Output

```
Using these hyperparameters {'bootstrap': True, 'criterion': 'gini', 'n_estimators': 50}, we get the best
     performance:
           precision   recall f1-score support
non-user        0.42     0.37     0.39     161
    user        0.63     0.68     0.65     252
  accuracy                        0.56     413
 macro avg      0.52     0.52     0.52     413
weighted avg    0.55     0.56     0.55     413
```

Example 8.12 A simple gridsearch in Python.

Python Code

```
myclassifier = SVC(gamma="scale")

grid = {
    "C" : [100, 1e4],
    "kernel": ["linear","rbf", "poly"],
    "degree": [3,4]
}

search = GridSearchCV(estimator=myclassifier,
                      param_grid=grid,
                      scoring=f1scorer,
                      cv=5,
                      n_jobs=-1,  # use all cpus
                      verbose=10)
search.fit(X_train_scaled, y_train)
print(f"Hyperparameters {search.best_params_} "
      "give the best performance:")
print(classification_report(y_test,
      search.predict(X_test_scaled)))
```

Python Output

```
Fitting 5 folds for each of 12 candidates, totalling 60 fits
Using these hyperparameters {'C': 100, 'degree': 3, 'kernel': 'poly'}, we get the best performance:
             precision  recall f1-score support

non-user         0.58     0.04     0.08      161
    user         0.62     0.98     0.76      252
  accuracy                         0.62      413
 macro avg       0.60     0.51     0.42      413
weighted avg     0.60     0.62     0.49      413
```

Example 8.13 A simple gridsearch in Python using multiple CPUs.

R Code

```
# Create the grid of parameters
grid = expand.grid(Loss=c("L1","L2"),
                   cost=c(100,1000))

# Train the model using our previously defined
# parameters
gridsearch = train(x = X_train, y = y_train,
    preProcess = c("center", "scale"),
    method = "svmLinear3",
    trControl = trainControl(method = "cv",
               number = 5),
    tuneGrid = grid)
gridsearch
```

R Output

```
L2 Regularized Support Vector Machine (dual) with Linear Kernel

1652 samples
   3 predictor
   2 classes: 'non-user', 'user'

Pre-processing: centered (3), scaled (3)
Resampling: Cross-Validated (5 fold)
Summary of sample sizes: 1321, 1322, 1322, 1322, 1321)
Resampling results across tuning parameters:)

        Loss  cost    Accuracy     Kappa
         L1    100    0.6616204    0.20997692
         L1   1000    0.5538900    0.03503642
         L1    100    0.6525314    0.21030587
         L2   1000    0.6525314    0.21030587

Accuracy was used to select the optimal model using the largest value.
The final values used for the model were cost = 100 and Loss = L1.
```

Example 8.14 A gridsearch in R. Note that in R, not all parameters are "tunable" using standard caret. Therefore, an exact replication of the grid searches in Example 8.12 and Example 8.13 would requires either manual comparisons or writing a so-called caret extension.

9

Processing Text

Abstract

Many datasets that are relevant for social science consist of textual data, from political discussions and newspaper archives to open-ended survey questions and reviews. This chapter gives an introduction to dealing with textual data using base functions in Python and (mostly) the *stringr* package in R.

Keywords text representation, text cleaning, regular expressions

- Understand how text is represented in the computer
- Be able to clean up and alter text
- Understand and be able to use regular expressions

Packages used in this chapter

This chapter introduces the packages for handling textual data. For R, this is mainly the *stringr* package (included in *tidyverse*). In Python, most functions are built-in, but will show how to use these functions in *pandas* and also introduce the *regex* package, an alternative to built-in regular expressions. You can install these packages with the code below if needed (see Section 1.4 for more details):

Python Code

```
!pip3 install regex pandas
```

R Code

```
install.packages(c("glue", "tidyverse"))
```

After installing, you need to import (activate) the packages every session:

Python Code

```
import regex
import re
import pandas as pd
```

R Code

```
library(glue)
library(tidyverse)
```

When dealing with textual data, an important step is to normalize the data. Such preprocessing ensures that noise is removed, and reduces the amount of data to deal with. In Section 5.2.2 we explained how to read data from different formats, such as txt, csv or json that can include textual data, and we also mentioned some of the challenges when reading text (such as encoding/decoding from/to Unicode). In this section we cover typical cleaning steps such as lowercasing and removing punctuation, HTML tags and boilerplate.

As a computational communication scientist you will come across many sources of text that range from electronic versions of newspapers in HTML to parliamentary speeches in PDF. Moreover, most of the contents in their original shape will include data that will not be of interest for the analysis but, instead, will produce noise that might negatively affect the quality of the research. You have to decide which parts of the raw text should be considered for analysis and determine the shape of these contents in order to have a good input in the analytical process.

As the difference between useful information and noise is determined by your research question, there is not a fixed list of steps to take that can guide you in this preprocessing stage. It is highly likely that you will have to test different combinations of steps and assess what the best options are. For example, in some cases keeping capital letters within a chat conversation or a news comment might be valuable to detect the tone of the message, but in more formal speeches transforming the whole text to lowercases would help to normalize the content. However, it is true that there are some typical challenges to reducing the noise from the text.

This chapter and the next will show you how to clean and manipulate text to transform the raw strings of letters into useful data. This chapter focuses on dealing with the text as characters and especially shows you how to use *regular expressions* to search and replace textual content. The next chapter will focus on text as words and shows how you can represent text in a suitable format for further computational analysis.

9.1 Text as a String of Characters

Important: Unicode and Encodings

Technically speaking, text is represented as bytes (numbers) rather than characters. The Unicode standard determines how these bytes should be interpreted or "decoded". This chapter assumes that the bytes in a file are already "decoded" into characters (or Unicode code points), and we can just work with the characters. Especially if you are not working with English text, it is very important to make sure you understand Unicode and encodings and check that the texts you work with are decoded properly. Please see Section 5.2.2 for more information on how this works.

When we think about text, we might think of sentences or words, but the computer only "thinks" about letters: text is represented internally as a string of characters. This is reflected of course in the type name, with R calling it a *character* vector and Python a *string*.

As a simple example, the figure at the top of Example 9.1 shows how the text "This is text." is represented. This text is split into separate characters, with each character representing a letter (or space, punctuation, emoji, or Chinese character). These characters are indexed starting from the first one, with (as always) R counting from one, but Python counting from zero.

In Python, texts are represented as `str` (string) objects, in which we can directly address the individual characters by their position: `text [0]` is the first character of `text`, and so on. In R, however, texts (like all objects) represent columns (or vectors) rather than individual values. Thus, `text [1]` in R is the first text in a series of texts. To access individual characters in a text, you have to use a function such as `str_length` and `str_sub` that will be discussed in more detail below. This also means that in Python, if you have a column (or list) of strings that you need to apply an operation to, you either need to use one of Pandas methods shown below or use a *for loop* or *list comprehension* to iterate over all the strings (see also section 3.2).

R index	1	2	3	4	5	6	7	8	9	10	11	12	13
	T	h	i	s		i	s		t	e	x	t	.
Python Index	0	1	2	3	4	5	6	7	8	9	10	11	12
(reverse)	-13	-12	-11	-10	-9	-8	-7	-6	-5	-4	-3	-2	-1

Python Code

```python
text = "This is text."
print(f"type(text): {type(text)}")
print(f"len(text): {len(text)}")
print(f"text[0]: '{text[0]}'")
print(f"text[5:7]: '{text[5:7]}'")
print(f"text[-1]: '{text[-1]}'")
print(f"text[-4:]: '{text[-5:]}'")
```

R Code

```r
text = "This is text."
glue("class(text): {class(text)}")
glue("length(text): {length(text)}")
glue("text[1]: {text[1]}")
glue("str_length(text): {str_length(text)}")
glue("str_sub(text, 6,7): {str_sub(text, 6,7)}")
```

Python Output

```
type(text): <class 'str'>
len(text): 13
text[0]: 'T'
text[5:7]: 'is'
text[-1]: '.'
text[-4:]: 'text.'
```

R Output

```
'class(text): character'
'length(text): 1'
'text[1]: This is text.'
'str_length(text): 13'
'str_sub(text, 6,7): is'
```

Python Code

```python
words = ["These", "are", "words"]
print(f"type(words): {type(words)}")
print(f"len(words): {len(words)}")
print(f"words[0]: '{words[0]}'")
print(f"words[1:3]: '{words[1:3]}'")
```

R Code

```r
words = c("These", "are", "words")
glue("class(words): {class(words)}")
print("length(words): {length(words)}")
glue("words[1]: {words[1]}")
# Note: use collapse to convert to single value
words_2_3 = str_c(words[2:3], collapse=", ")
glue("words[2:3]: {words_2_3}")
```

Python Output

```
type(words): <class 'list'>
len(words): 3
words[0]: 'These'
words[1:3]: '['are', 'words']'
```

R Output

```
class(words): character'
[1] "length(words): {length(words)}"
'words[1]: These'
'words[2:3]: are, words'
```

Example 9.1 Internal representation and of single and multiple texts.

9.1.1 Methods for Dealing With Text

Note: Stringi, stringr, and base string operations in R

As is so often the case, R has multiple packages that partially replicate functionality for basic text handling. In this book we will mainly use the *stringr* package, which is part of *tidyverse*. This is not because that package is necessarily better or easier than the alternative *stringi* package or the built-in (*base*) methods. However, the methods are well-documented, clearly named, and consistent with other tidyverse functions, so for now it is easiest to stick to *stringr*. In particular, *stringr* is very similar to *stringi* (and in fact is partially based on it). So, to give one example, the function `str_detect` is more or less the same as `stringi::str_detect` and `base::grepl`.

The first thing to keep in mind is that once you load any text in R or Python you usually store this content as a *character* or *string* object (you may also often use *lists* or *dictionaries*, but they will have strings inside them), which means that basic operations and conditions of this data type apply, such as indexing or slicing to access individual characters or substrings (see Section 3.1). In fact, base strings operations are very powerful to clean your text and eliminate a large amount of noise. Table 9.1 summarizes some useful operations on strings in R and Python that will help you in this stage.

Let us apply some of these functions/methods to a simple Wikipedia text that contains HTML tags or boilerplate and upper/lower case letters (Example 9.2). Using the *stringr* function `str_replace_all` in R and `replace` in Python we can do a find-and-replace and replace substrings by others (in our case, replace <b> with a space, for instance). To remove unnecessary double spaces we apply the `str_squish` function provided by *stringr* and in Python, we first chunk our string into a list of words by using the `split` string method, before we use the `join` method to join them again with now a single space. In the case of converting letters from upper to lower case, we use the base R function `tolower` and the string method `lower` in Python. Finally, the base R function `trimws` and the Python string method `strip` remove the white space from the

Table 9.1 Useful strings operations in R and Python to clean noise.

String operation	R (*stringr*) (whole column)	Python (single string)	Pandas (whole column)
Count characters in s	`str_length(s)`	`len(s)`	`s.str.len()`
Extract a substring	`str_sub(s, n1, n2)`	`s[n1:n2]`	`s.str.slice(n1, n2)`
Test if s contains s2	`str_detect(s, s2)*`	`s2 in s`	`s.str.match(s2)*`
Strip spaces	`trimws(s)`	`s.strip()`	`s.str.strip()`
Convert to lowercase	`tolower(s)`	`s.lower()`	`s.str.lower()`
Convert to uppercase	`toupper(s)`	`s.upper()`	`s.str.upper()`
Find s1 and replace by s2	`str_replace(s, s1, s2)*`	`s.replace(s1, s2)`	`s.str.replace(s1, s2)*`

* The R functions `str_detect` and `str_replace` and the Pandas function `s.str.match` and `s.str.replace` use *regular expressions* to define what to find and replace. See Section 9.2 below for more information.

beginning and end of the string. Example 9.2 shows how to conduct this cleaning process.

While you can get quite far with these techniques, there are more advanced and flexible approaches possible. For instance, you probably do not want to list *all possible* HTML tags in separate `replace` methods or `str_replace_all` functions. In the next section, we therefore show how to use so-called *regular expressions* to formulate such generalizable patterns.

Python Code

```python
text = """   <b>Communication</b>
    (from Latin communicare, meaning to share) """
# remove tags:
cleaned = text.replace("<b>","").replace("</b>","")
# normalize white space
cleaned = " ".join(cleaned.split())
# lower case
cleaned = cleaned.lower()
# trim spaces from start and end
cleaned = cleaned.strip()

print(cleaned)
```

R Code

```r
text = "   <b>Communication</b>
    (from Latin communicare, meaning to share) "
cleaned = text %>%
  # remove HTML tags:
  str_replace_all("<b>", " ") %>%
  str_replace_all("</b>", " ") %>%
  # normalize white space
  str_squish() %>%
  # lower case
  tolower() %>%
  # trim spaces at start and end
  trimws()

glue(cleaned)
```

Output

```
communication (from latin communicare, meaning to share)
```

Example 9.2 Some basic text cleaning approaches.

Note: To regex or not to regex in Python

You may wonder why we introduce basic string methods like `replace` or the `split`-then-`join` trick, if everything can be done with regular expressions anyway. There are a couple of reasons for still using these methods: first, they are easy and don't have any dependencies. If you just want to replace a single thing, then you don't need to import any additional module. Second, regular expressions are considerably slower than string methods – in most cases, you won't notice, but if you do a lot of replacements (think in thousands per news article, for a million articles), then this may matter. Third, you can use the `join` trick also for other things like punctuation removal – in this case, by generating a list of all characters in a string called `text` provided they are no punctuation characters, and then joining them directly to each other: `from string import punctuation; "".join([c for c in text if c not in punctuation])`

9.2 Regular Expressions

A *regular expression* or *regex* is a powerful language to locate strings that conform to a given pattern. For instance, we can extract usernames or email-addresses from text, or normalize spelling variations and improve the cleaning methods covered in the previous section. Specifically, regular expressions are a sequence of characters that we can use to design a pattern and then use this pattern to *find* strings (identify or extract) and also *replace* those strings by new ones.

Regular expressions look complicated, and in fact they take some time to get used to initially. For example, a relatively simple (and not completely correct) expression to match an email address is `[\w\.-]+@[\w\.-]+\.\w\w+`, which doesn't look like anything at all unless you know what you are looking for. The good news is that regular expression syntax is the same in R and Python (and many other languages), so once you learn regular expressions you will have acquired a powerful and versatile tool for text processing.

In the next section, we will first review general expression syntax without reference to running them in Python or R. Subsequently, you will see how you can apply these expressions to inspect and clean texts in both languages.

9.2.1 Regular Expression Syntax

At its core, regular expressions are patters for matching sequences of characters. In the simplest case, a regular letter just matches that letter, so the pattern "cat" matches the text "cat". Next, there are various wildcards, or ways to match different letters. For example, the period (`.`) matches any character, so `c.t` matches both "cat" and "cot". You can place multiple letters between square brackets to create a *character class* that matches all the specified letters, so `c[au]t` matches "cat" and "cut", but not "cot". There are also a number of pre-defined classes, such as `\w` which matches "word characters" (letters, digits, and (curiously) underscores).

Finally, for each character or group of characters you can specify how often it should occur. For example, `a+` means one or more a's while `a?` means zero or one a, so `lo+l` matches 'lol', 'lool', etc., and `lo?l` matches 'lol' or 'll'. This raises the question, of course, of how to look for actual occurrences of a plus, question mark, or period. The solution is to *escape* these special symbols by placing a backslash (\) before them: `a\+` matches the literal text "a+", and `\\w` (with a double backslash) matches the literal text "\w".

Now we can have another look at the example emails address pattern given above. The first part, `[\w\.-]` creates a character class containing word characters, (literal) periods, and dashes. Thus, `[\w\.-]+@[\w\.-]+` means one or more letters, digits, underscores, periods, or dashes, followed by an at sign, followed by one or more letters, digits, etc. Finally, the last part `\.\w\w+` means a literal period, a word character, and one or more word characters. In other words, we are looking for a name (possibly containing dashes or periods) before the at sign, followed by a domain, followed by a top level domain (like `.com`) of at least two characters.

In essence, thinking in terms of what you want to match and how often you want to match it is all there is to regular expressions. However, it will take some practice to get comfortable with turning something sensible (such as an email address) into a correct regular expression pattern. The next subsection will explain regular expression syntax in more detail, followed by an explanation of grouping, and in the final subsection we will see how to use these regular expressions in R and Python to do text cleaning.

In Table 9.2 you will find an overview of the most important parts of regular expression syntax.[1] The first part shows a number of common specifiers for determining what to match, e.g. letters, digits, etc., followed by the quantifiers available to determine how often something should be matched. These quantifiers always follow a specifier, i.e. you first say what you're looking for, and then how many of those you need. Note that by default quantifiers are *greedy*, meaning they match as many characters as possible. For example, <.*> will match everything between angle brackets, but if you have something like '<p>a paragraph</p>' it will happily match everything from the first opening bracket to the last closing bracket. By appending a question mark (?) to the quantifier, it becomes *non-greedy*. so, <.*?> will match the individual '<p>' and '</p>' substrings.

The third section discusses other constructs. *Groups* are formed using parentheses () and are useful in at least three ways. First, by default a quantifier applies to the letter directly before it, so no+ matches "no", "nooo", etc. If you group a number of characters you can apply a quantifier to the group. So, that's (not)? good matches either "that's not good" or "that's good". Second, when using a vertical bar (|) to have multiple options, you very often want to put them into a group so you can use it as part of a larger pattern. For example, a(great| fantastic)? victory matches either "a victory", "a great victory", or "a fantastic victory". Third, as will be discussed below in Section 9.3, you can use groups to capture (extract) a specific part of a string, e.g. to get only the domain part of a web address.

The other important construct are *character classes*, formed using square brackets []. Within a character class, you can specify a number of different characters that you want to match, using a dash (-) to indicate a range. You can add as many characters as you want: [A-F0-9] matches digits and capital letters A through F. You can also invert this selection using an initial caret: [^a-z] matches everything except for lowercase Latin letters. Finally, you sometimes need to match a control character (e.g. +, ?, \). Since those characters have a special meaning within a regular expressing, they cannot be used directly. The solution is to add a backslash (\) behind them to *escape* them: . matches any character, but \. matches an actual period. \\ matches an actual backslash.

9.2.2 Example Patterns

Using the syntax explained in the previous section, we can now make patterns for common tasks in cleaning and analyzing text. Table 9.3 lists a number of regular expressions for common tasks such as finding dates or stripping HTML artifacts.

We start with a number of relatively simple patterns for Zip codes and phone numbers. Starting with the simplest example, US Zip codes are simply five consecutive numbers. Next, a US phone number can be written down as three groups of numbers separated by parentheses, where the first group is made optional for

---

[1] Note that this is not a full review of everything that is possible with regular expressions, but this includes the most used options and should be enough for the majority of cases. Moreover, if you descend into the more specialized aspects of regular expressions (with beautiful names such as "negative lookbehind assertions") you will also run into differences between Python, R, and other languages, while the features used in this chapter should function in most implementations you come across unless specifically noted.

Table 9.2 Regular expression syntax.

Function	Syntax	Example	Matches
Specifier: What to match			
All characters except for new lines	.	d.g	dig, d!g
Word characters* (letters, digits, _)	\w	d\wg	dig, dog
Digits* (0 to 9)	\d	202\d	2020, 2021
Whitespace* (space, tab, newline)	\s		
Newline	\n		
Beginning of the string	^	^go	go go go
Ending of the string	$	go$	go go go
Beginning or end of word	\b	\bword\b	a word!
Either first or second option	...|...	cat|dog	cat, dog
Quantifier: How many to match			
Zero or more	*	d.*g	dg, drag, d = g
Zero or more (non-greedy)	*?	d.*?g	dogg
One or more	+	\d+%	1%, 200%
One or more (non-greedy)	+?	\d+%	200%
Zero or one	?	colou?r	color, colour
Exactly *n* times	{n}	\d{4}	1940, 2020
At least *n* times	{n,}		
Between *n* and *m* times	{n,m}		
Other constructs			
Groups	(...)	'(bla)+'	'bla bla bla'
Selection of characters	[...]	d[iuo]g	dig, dug, dog
Range of characters in selection	[a-z]		
Everything except selection	[^...]		
Escape special character	\	3\.14	3.14
Unicode character properties†			
Letters*	\p{LETTER}		words, 巣滛
Punctuation*	\p{PUNCTUATION}		.,:
Quotation marks*	\p{QUOTATION MARK}		' ' " «
Emoji*	\p{EMOJI}		☺
Specific scripts, e.g. Hangul*	\p{HANG}		한국

* These selectors can be inverted by changing them into capital letters. Thus, \W matches everything except word characters, and \P{PUNCTUATION} matches everything except punctuation.

† See https://www.unicode.org/reports/tr44/#Property_Index for a full list of Unicode properties. Note that when using Python, these are only available if you use *regex*, which is a drop-in replacement for the more common *re*.

Table 9.3 Regular expression's syntax.

Goal	Pattern	Example
US Zip Code	\d{5}	90210
US Phone number	(\d{3}-)?\d{3}-\d{4}	202-456-1111, 456-1111
Dutch Postcode	\d{4} ?[A-Za-z]{2}	1015 GK
ISO Date	\d{4}-\d{2}-\d{2}	2020-07-20
German Date	\d{1,2}\.\d{1,2}\.\d{4}	25.6.1988
International phone number	\+(\d[-]?){7,}\d	+1 555-1234567
URL	https?://\S+	https://example.com?a=b
E-mail address	[\w\.-]+@[\w\.-]+\.\w+	me@example.com
HTML tags	</?\w[^>]*>	</html>
HTML Character escapes	&[^;]+;	

Please note that most of these patterns do not correctly distinguish all edge cases (and hence may lead to false negatives and/or false positives) and are provided for educational purposes only.

local phone numbers using parentheses to group these numbers so the question mark applies to the whole group. Next, Dutch postal codes are simply four numbers followed by two letters, and we allow an optional space in between. Similarly simple, dates in ISO format are three groups of numbers separated by dashes. German dates follow a different order, use periods as separator, and allow for single-digit day and month numbers. Note that these patterns do not check for the validity of dates. A simple addition would be to restrict months to 01-12, e.g. using `(0[1-9]|1[0-2])`. However, in general validation is better left to specialized libraries, as properly validating the day number would require taking the month (and leap years) into account.

A slightly more complicated pattern is the one given for international phone numbers. They always start with a plus sign and contain at least eight numbers, but can contain dashes and spaces depending on the country. So, after the literal + (which we need to escape since + is a control character), we look for seven or more numbers, optionally followed by a single dash or space, and end with a single number. This allows dashes and spaces at any position except the start and end, but does not allow for e.g. double dashes. It also makes sure that there are at least eight numbers regardless of how many dashes or spaces there are.

The final four examples are patterns for common notations found online. For URLs, we look for `http://` or `https://` and take everything until the next space or end of the string. For email addresses, we define a character class for letters, periods, or dashes and look for it before and after the at sign. Then, there needs to be at least one period and a top level domain containing only letters. Note that the dash within the character class does not need to be escaped because it is the final character in the class, so it cannot form a range. For HTML tags and character escapes, we anchor the start (< and &) and end (> and ;) and allow any characters except for the ending character in between using an inverted character class.

Note that these example patterns would also match if the text is enclosed in a larger text. For example, the zip code pattern would happily match the first five numbers of a 10-digit number. If you want to check that an input value is a valid zip code (or email address, etc.), you probably want to check that it only contains that code by surrounding it with start-of-text and end-of-text markers: `^\d{5}$`. If you want to extract e.g. zip codes from a longer document, it is often useful to surround them with word boundary markers: `\b\d{5}\b`.

Please note that many of those patterns are not necessarily fully complete and correct, especially the final patterns for online notations. For example, email addresses can contain plus signs in the first part, but not in the domain name, while domain names are not allowed to start with a dash – a completely correct regular expression to match email addresses is over 400 characters long! Even worse, a complete HTML tag expression is probably not even possible to describe with a regular expression because HTML tags can contain comments and nested escapes within attributes. For a better way to deal with analyzing HTML, please see Chapter 12. In the end, patterns like these are fine for a (somewhat) noisy analysis of (often also somewhat noisy) source texts as long as you understand the limitations.

9.3 Using Regular Expressions in Python and R

Now that you hopefully have a firm grasp of regular expression syntax, it is relatively easy to use these patterns in Python or R (or most other languages). Table 9.4 lists the commands for four of the most common use cases: identifying matching texts, removing and replacing all matching text, extracting matched groups, and splitting texts.

For R, we again use the functions from the *stringr* package. For Python, you can use either the *re* or *regex* package, which both support the same functions and syntax so you can just import one or the other. the *re package* is more common and substantially faster, but does not support Unicode character properties `(\p)`. We also list the corresponding commands for *pandas*, which are run on a whole column instead of a single text (but note that *pandas* does not support Unicode character properties.)

Finally, a small but important note about escaping special characters by placing a backslash `(\)` before them. The regular expression patterns are used *within* another language (in this case, Python or R), but these languages have their own special characters which are also escaped. In Python, you can create a *raw string* by putting a single `r` before the opening quotation mark: `r"\d+"` creates the regular expression pattern `\d`. From version 4.0 (released in spring 2020), R has a similar construct: `r"(\d+)"`. In R, the parentheses are part of the string delimiters, but you can use more parentheses within the string without a problem. The only thing you cannot include in a string is the closing sequence `)"`, but as you are also allowed to use square or curly brackets instead of parentheses and single instead of double quotes to delimit the raw string you can generally avoid this problem: to create the pattern `"(cat|dog)"` (i.e. cat or dog enclosed in quotation marks), you can use `r"{"(cat|dog)"}" or r'("(cat|dog)")'` (or even more legible: `r'{"(cat|dog)"}')`.

Unfortunately, in earlier versions of R (and in any case if you don't use raw strings), you need to escape special characters twice: first for the regular expression, and then for R. So, the pattern `\d` becomes `"\\d"`. To match a literal backslash you would use the pattern `\\`, which would then be represented in R as `"\\\\"`!

Table 9.4 Regular expression functions and methods.

Operation	**R (*stringr*) (whole column)**	**Python (single string)**	**Pandas (whole column)**
Does pattern p occur in text t?	`str_detect(t, p)`	`re.search(p, t)`	`t.str.contains(p)`
Does text t start with pattern p?	`str_detect(t, "^p")`	`re.match(p, t)`	`t.str.match(p)`
Count occurrences of p in t	`str_count(t, "^p")`	`re.match(p, t)`	`t.str.count(p)`
Remove all occurences of p in t	`str_remove_all(t, p)`	`re.sub(p, "", t)`	`t.str.replace(p, "")`
Replace p by r in text t	`str_replace_all(t, p, r)`	`re.sub(p, r, t)`	`t.str.replace(p, r)`
Extract the first match of p in t	`str_extract(t, p)`	`re.search(p, t).group(1)`	`t.str.extract(p)`
Extract all matches of p in t	`str_extract_all(t, p)`	`re.findall(p, t)`	`t.str.extractall(p)`
Split t on matches of p	`str_split(t, p)`	`re.split(p, t)`	`t.str.split(p)`

Note: In Python, if using Unicode character properties (`\p`), use the same functions in package *regex* instead of *re*

Example 9.3 cleans the same text as Example 9.2 above, this time using regular expressions. First, it uses `<[^>+]>` to match all HTML tags: an angular opening bracket, followed by anything except for a closing angular bracket (`[^>]`), repeated one or more times (+), finally followed by a closing bracket. Next, it replaces one or more whitespace characters (`\s+`) by a single space. Finally, it uses a vertical bar to select either space at the start of the string (`^\s+`), or at the end (`\s+$`), and removes it. As you can see, you can express a lot of patterns using regular expressions in this way, making for more generic (but sometimes less readable) clean-up code.

Finally, Example 9.4 shows how you can run the various commands on a whole column of text rather than on individual strings, using a small set of made-up tweets to showcase various operations. First, we determine whether a pattern occurs, in this case for detecting hashtags. This is very useful for e.g. subsetting a data frame to only rows that contain this pattern. Next, we count how many at-mentions are contained in the text, where we require that the character before the mention needs to be either whitespace or the start of the string (^), to exclude email addresses and other non-mentions that do contain at signs. Then, we extract the (first) url found in the text, if any, using the pattern discussed above. Finally, we extract the plain text of the tweet in two chained operations: first, we remove every word starting with an at-sign, hash, or http, removing everything up to the next whitespace character. Then, we replace everything that is not a letter by a single space.

9.3.1 Splitting and Joining Strings, and Extracting Multiple Matches

So far, the operations we used all took a single string object and returned a single value, either a cleaned version of the string or e.g. a boolean indicating whether there

Python Code

```python
text = """   <b>Communication</b>
    (from Latin communicare, meaning to share) """
# remove tags:
cleaned = re.sub("<[^>]+>", "", text)
# normalize white space
cleaned = re.sub("\s+", " ", cleaned)
# trim spaces from start and end
cleaned = re.sub("^\s+|\s+$", "", cleaned)
cleaned = cleaned.strip()

print(cleaned)
```

R Code

```r
text = "   <b>Communication</b>
    (from Latin communicare, meaning to share) "
cleaned = text %>%
  # remove HTML tags:
  str_replace_all("<[^>]+>", " ") %>%
  # normalize white space
  str_replace_all("\\p{space}+", " ") %>%
  # trim spaces at start and end
  str_remove_all("^\\s+|\\s+$")

cleaned
```

Python Output

```
Communication (from Latin communicare, meaning to share)
```

Example 9.3 Using regular expressions to clean a text.

Python Code

```python
import pandas as pd
url = "https://cssbook.net/d/example_tweets.csv"
tweets = pd.read_csv(url, index_col="id")
# identify tweets with hashtags
tweets["tag"]=tweets.text.str.contains(r"#\w+")
# How many at-mentions are there?
tweets["at"]=tweets.text.str.count(r"(^|\s)@\w+")
# Extract first url
tweets["url"]=tweets.text.str.extract(
    r"(https?://\S+)")
# Remove urls, tags, and @-mentions
expr=r"(^|\s)(@|#|https?://)\S+"
tweets["plain2"]=(tweets.text.str
                  .replace(expr, " ", regex=True)
                  .replace(r"\W+", " "))
tweets
```

R Code

```r
library(tidyverse)
url="https://cssbook.net/d/example_tweets.csv"
tweets = read_csv(url)
tweets = tweets %>% mutate(
    # identify tweets with hashtags
    has_tag=str_detect(text, "#\\w+"),
    # How many at-mentions are there?
    n_at = str_count(text, "(^|\\s)@\\w+"),
    # Extract first url
    url = str_extract(text, "(https?://\\S+)"),
    # Remove at-mentions, tags, and urls
    plain2 = str_replace_all(text,
        "(^|\\s)(@|#|https?://)\\S+", " ") %>%
          str_replace_all("\\W+", " ")
    )
tweets
```

R Output:

id	text	has_tag	n_at	url	plain2
<dbl>	<chr>	<lgl>	<int>	<chr>	<chr>
1	RT: @john_doe https://example.com/news very interesting!	FALSE	1	https://example.com/RT	news very interesting
2	tweet with just text	FALSE	0	NA	tweet with just text
3	http://example.com/pandas #breaking #mustread	TRUE	0	http://example.com/pandas	
4	@me and @myself #selfietime	TRUE	2	NA	and

Example 9.4 Using regular expressions on a data frame.

is a match. This is convenient when using data frames, as you can transform a single column into another column. There are three common operations, however, that complicate matters: you can *split* a string into multiple substrings, or *extract* multiple matches from a string, and you can *join* multiple matches together.

Example 9.5 shows the "easier" case of splitting up a single text and joining the result back together. We show three different ways to split: using a fixed pattern to

Python Code

```python
text = "apples, pears, oranges"
# Three ways to achieve the same thing:
items = text.split(", ")
items = regex.split(r"\p{PUNCTUATION}\s*", text)
items = regex.findall(r"\p{LETTER}+", text)
print(f"Split text into items: {items}")
joined = " & ".join(items)
print(joined)
```

R Code

```r
text = "apples, pears, oranges"
items=strsplit(text,", ", fixed=T)[[1]]
items=str_split(text,"\\p{PUNCTUATION}\\s*")[[1]]
items=str_extract_all(text,"\\p{LETTER}+")[[1]]
print(items)
joined = str_c(items, collapse=" & ")
print(joined)
```

Python Output

```
Split text into items: ['apples', 'pears',
'oranges']
apples & pears & oranges
```

R Output

```
[1] "apples" "pears" "oranges"
[1] "apples & pears & oranges"
```

Example 9.5 Splitting extracting and joining a single text.a

split on (in this case, a comma plus space); using a regular expression (in this case, any punctuation followed by any space); and by matching the items we are interested in (letters) rather than the separator. Finally, we join these items together again using `join` (Python) and `str_c` (R).

One thing to note in the previous example is the use of the index `[[1]]` in R to select the first element in a list. This is needed because in R, splitting a text actually splits all the given texts, returning a `list` containing all the matches for each input text. If there is only a single input text, it still returns a list, so we select the first element of the list.

In many cases, however, you are not working on a single text but rather on a series of texts loaded into a data frame, from tweets to news articles and open survey questions. In the example above, we extracted only the first url from each tweet. If we want to extract e.g. all hash tags from each tweet, we cannot simply add a "tags" column, as there can be multiple tags in each tweet. Essentially, the problem is that the urls per tweet are now nested in each row, creating a non-rectangular data structure.

Although there are multiple ways of dealing with this, if you are working with data frames our advice is to normalize the data structure to a long format. In the example, that would mean that each tweet is now represented by multiple rows, namely one for each hash tag. Example 9.6 shows how this can be achieved in both R and Pandas. One thing to note is that in pandas, `t.str.extractall` automatically returns the desired long format, but it is essential that the index of the data frame actually contains the identifier (in this case, the tweet (status) id). `t.str.split`, however, returns a data frame with a column containing lists, similar to how both R functions return a list containing character vectors. We can normalize this to a long data frame using `t.explode` (pandas) and `pivot_longer` (R). After this, we can use all regular data frame operations, for example to join and summarize the data.

A final thing to note is that while you normally use a function like `mean` to summarize the values in a group, you can also join strings together as a summarization. The only requirement for a summarization function is that it returns a single value for a group of values, which of course is exactly what joining a multiple string together does. This is shown in the final line of the example, where we split a tweet into words and then reconstruct the tweet from the individual words.

Python Code

```python
tags = tweets.text.str.extractall("(#\\w+)")
tags.merge(tweets, left_on="id", right_on="id")
```

R Code

```r
tags = tweets %>% mutate(
    tag=str_extract_all(tweets$text,"(#\\w+)"))%>%
  select(id, tag)
tags_long = tags  %>% unnest(tag)
left_join(tags_long, tweets)
```

R Output

id <dbl>	tag <chr>	text <chr>	has_tag <lgl>	n_at <int>	url <chr>	plain2 <chr>
3	#breaking	http://example.com/pandas #breaking #mustread	TRUE	0	http://example.com/ pandas	
3	#mustread	http://example.com/pandas #breaking #mustread	TRUE	0	http://example.com/ pandas	
4	#selfietime	@me and @myself #selfietime	TRUE	2	NA	and

Python Code

```python
words = tweets.text.str.split("\\W+")
words_long = words.explode()
```

R Code

```r
words = tweets %>% mutate(
    word=str_split(tweets$text, "\\W+")) %>%
  select(id, word)
words_long = words %>% unnest(word)
head(words_long)
```

R output

id <dbl>	word <chr>
1	RT
1	john_doe
1	https
1	example
1	com
1	news

Python Code

```python
words_long.groupby("id").agg("_".join)
```

R Code

```r
words_long %>%
  group_by(id) %>%
  summarize(joined=str_c(word, collapse="_"))
```

R Output

id <dbl>	joined <chr>
1	RT_john_doe_https_example_com_news_very_ interesting_
2	tweet_with_just_text
3	http_example_com_pandas_breaking_mustread
4	_me_and_myself_selfietime

Example 9.6 Applying split and extract_all on text columns.

10

Text as Data

Abstract

This chapter shows how you can analyze texts that are stored as a data frame column or variable using functions from the package *quanteda* in R and the package *sklearn* in Python. Please see Chapter 9 for more information on reading and cleaning text.

Keywords text as data, document-term matrix

- Create a document-term matrix from text
- Perform document and feature selection and weighting
- Understand and use more advanced representations such as n-grams and embeddings

This chapter introduces the packages *quanteda* (R) and *sklearn* and *nltk* (Python) for converting text into a document-term matrix. It also introduces the *udpipe* package for natural language processing. You can install these packages with the code below if needed (see Section 1.4 for more details):

Python Code

```
!pip3 install ufal.udpipe spacy nltk scikit-learn
!pip3 install gensim wordcloud
```

R Code

```
install.packages(c("glue","tidyverse","quanteda",
    "quanteda.textstats", "quanteda.textplots",
    "udpipe", "spacyr"))
```

After installing, you need to import (activate) the packages every session:

Computational Analysis of Communication: A Practical Introduction to the Analysis of Texts, Networks, and Images with Code Examples in Python and R, First Edition. Wouter van Atteveldt, Damian Trilling & Carlos Arcila Calderón.

Python Code

```python
# Standard library and basic data wrangling
import os
import sys
import urllib
import urllib.request
import re
import regex
import pandas as pd
import numpy as np

# Tokenization
import nltk
from nltk.tokenize import (TreebankWordTokenizer,
                           WhitespaceTokenizer)
from nltk.corpus import stopwords
from sklearn.feature_extraction.text import (
    CountVectorizer, TfidfVectorizer)
import nagisa

# Plotting word clouds
%matplotlib inline
from matplotlib import pyplot as plt
from wordcloud import WordCloud

# Natural language processing
import spacy
import ufal.udpipe
from gensim.models import KeyedVectors, Phrases
from gensim.models.phrases import Phraser
from ufal.udpipe import Model, Pipeline
import conllu
```

R Code

```r
library(glue)
library(tidyverse)
# Tokenization
library(quanteda)
library(quanteda.textstats)
library(quanteda.textplots)
# Natural language processing
library(udpipe)
library(spacyr)
```

10.1 The Bag of Words and the Term-Document Matrix

Before you can conduct any computational analysis of text, you need to solve a problem: computations are usually done on numerical data – but you have text. Hence, you must find a way to *represent* the text by numbers. The document-term matrix (DTM, also called the term-document matrix or TDM) is a common numerical representation of text. It represents a *corpus* (or set of documents) as a matrix or table, where each row represents a document, each column represents a term (word), and the numbers in each cell show how often that word occurs in that document.

Python Code

```python
texts = [
    "The caged bird sings with a fearful trill",
    "for the caged bird sings of freedom"]
cv = CountVectorizer()
d = cv.fit_transform(texts)
# Create a dataframe of the word counts to inspect
# - todense transforms the dtm into a dense matrix
# - get_feature_names gives a list of all words
pd.DataFrame(d.todense(),
             columns=cv.get_feature_names())
```

R Code

```r
texts = c(
    "The caged bird sings with a fearful trill",
    "for the caged bird sings of freedom")
d = tokens(texts) %>% dfm()
# Inspect by converting to a (dense) matrix
convert(d, "matrix")
```

R Output:

	the	caged	bird	sings	with	a	fearful	trill	for	of	freedom
text1	1	1	1	1	1	1	1	1	0	0	0
text2	1	1	1	1	0	0	0	0	1	1	1

Example 10.1 Example document-term matrix.

As an example, Example 10.1 shows a DTM made from two lines from the famous poem by Mary Angelou. The resulting matrix has two rows, one for each line; and 11 columns, one for each unique term (word). In the columns you see the document frequencies of each term: the word "bird" occurs once in each line, but the word "with" occurs only in the first line (text1) and not in the second (text2).

In R, you can use the `dfm` function from the *quanteda* package (Benoit et al., 2018). This function can take a vector or column of texts and transforms it directly into a DTM (which quanteda actually calls a document-*feature* matrix, hence the function name `dfm`). In Python, you achieve the same by creating an object of the `CountVectorizer` class, which has a `fit_transform` function.

10.1.1 Tokenization

In order to turn a corpus into a matrix, each text needs to be *tokenized*, meaning that it must be split into a list (vector) of words. This seems trivial, as English (and most western) text generally uses spaces to demarcate words. However, even for English there are a number of edge cases. For example, should "haven't" be seen as a single word, or two?

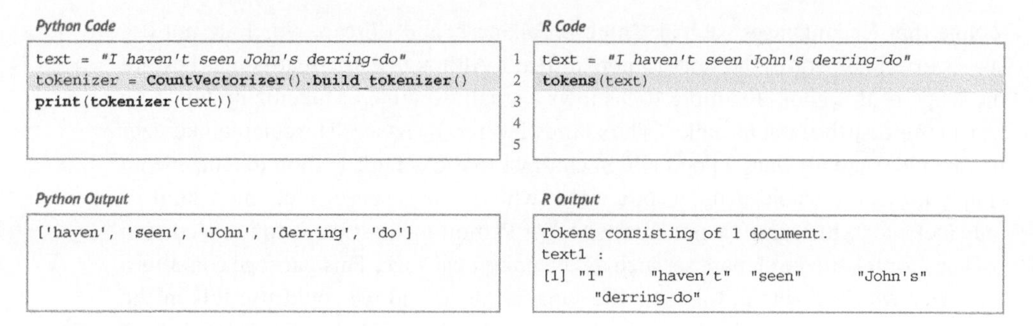

Python Code

```
text = "I haven't seen John's derring-do"
tokenizer = CountVectorizer().build_tokenizer()
print(tokenizer(text))
```

R Code

```
text = "I haven't seen John's derring-do"
tokens(text)
```

Python Output

```
['haven', 'seen', 'John', 'derring', 'do']
```

R Output

```
Tokens consisting of 1 document.
text1 :
[1] "I"        "haven't"  "seen"       "John's"
     "derring-do"
```

Example 10.2 Differences between tokenizers.

Example 10.2 shows how Python and R deal with the sentence "I haven't seen John's derring-do". For Python, we first use `CountVectorizer.build_tokenizer` to access the built-in tokenizer. As you can see in the first line of input, this tokenizes "haven't" to `haven`, which of course has a radically different meaning. Moreover, it silently drops all single-letter words, including the `'t`, `'s`, and `I`.

In the box "Tokenizing in Python" below, we therefore discuss some alternatives. For instance, the `TreebankWordTokenizer` included in the *nltk* package is a more reasonable tokenizer and splits "haven't" into `have` and `n't`, which is a reasonable outcome. Unfortunately, this tokenizer assumes that text has already been split into sentences, and it also includes punctuation as tokens by default. To circumvent this, we can introduce a custom tokenizer based on the Treebank tokenizer, which splits text into sentences (using `nltk.sent_tokenize`) – see the box for more details.

For R, we simply call the `tokens` function from the *quanteda* package. This keeps `haven't` and `John's` as a single word, which is probably less desirable than splitting the words but at least better than outputting the word `haven`.

As this simple example shows, even a relatively simple sentence is tokenized differently by the tokenizers considered here (and see the box on tokenization in Python). Depending on the research question, these differences might or might not be important. However, it is always a good idea to check the output of this (and other) preprocessing steps so you understand what information is kept or discarded.

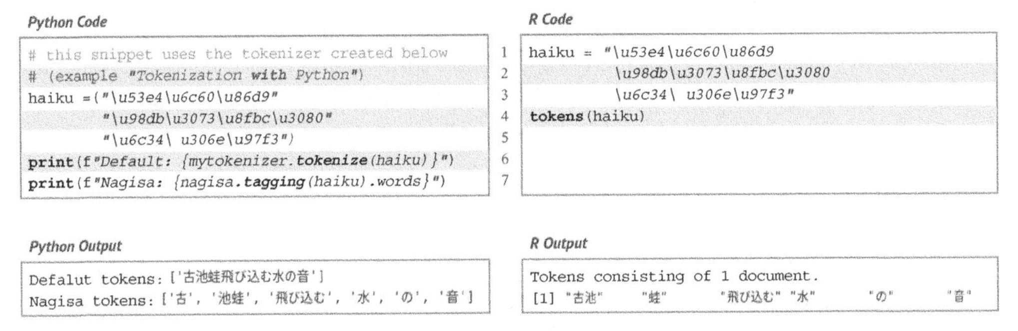

Python Code

```
# this snippet uses the tokenizer created below
# (example "Tokenization with Python")
haiku =("\u53e4\u6c60\u86d9"
        "\u98db\u3073\u8fbc\u3080"
        "\u6c34\ u306e\u97f3")
print(f"Default: {mytokenizer.tokenize(haiku)}")
print(f"Nagisa: {nagisa.tagging(haiku).words}")
```

R Code

```
haiku = "\u53e4\u6c60\u86d9
         \u98db\u3073\u8fbc\u3080
         \u6c34\ u306e\u97f3"
tokens(haiku)
```

Python Output

```
Defalut tokens: ['古池蛙飛び込む水の音']
Nagisa tokens: ['古', '池蛙', '飛び込む', '水', 'の', '音']
```

R Output

```
Tokens consisting of 1 document.
[1] "古池"  "蛙"  "飛び込む" "水"  "の"  "音"
```

Example 10.3 Tokenization of Japanese verse.

Note that for languages such as Chinese, Japanese, and Korean, which do not use spaces to delimit words, the story is more difficult. Although a full treatment is beyond the scope of this book, Example 10.3 shows a small example of tokenizing Japanese text, in this case the famous haiku "The sound of water" by Bashō. The default tokenizer in quanteda actually does a good job, in contrast to the default Python tokenizer that simply keeps the whole string as one word (which makes sense since this tokenizer only looks for whitespace or punctuation). For Python the best bet is to use a custom package for tokenizing Japanese, such as the *nagisa* package. This package contains a tokenizer which is able to tokenize the Japanese text, and we could use this in the `CountVectorizer` much like we used the `TreebankWordTokenizer` for English

Tokenization in Python

As you can see in the example, the built-in tokenizer in *scikit-learn* is not actually very good. For example, *haven't* is tokenized to *haven*, which is an entirely different word. Fortunately, there are other tokenizers in the *nltk.tokenize* package that do better.

For example, the `TreebankTokenizer` uses the tokenization rules for the Penn Treebank to tokenize, which produces better results:

The nltk.tokenize.TreebankTokenizer

```
1 text = """I haven't seen John's derring-do.
2          Second sentence!"""
3 print(TreebankWordTokenizer().tokenize(text))4
```

Another example is the `WhitespaceTokenizer`, which simply uses whitespace to tokenize, which can be useful if your input has already been tokenized, and is used in Example 10.11 below for tweets to conserve hash tags.

The nltk.tokenize.WhitespaceTokenizer

```
1 print(WhitespaceTokenizer().tokenize(text))
```

You can also write your own tokenizer if needed. For example, the `TreebankTokenizer` assumes that text has already been split into sentences (which is why the period is attached to the word *derring-do.*). The code below shows how we can make our own tokenizer class, which uses `nltk.sent_tokenize` to first split the text into sentences, and then uses the `TreebankTokenizer` to tokenize each sentence, keeping only tokens that include at least one letter character. Although a bit more complicated, this approach can give you maximum flexibility.

Writing a custom Tokenizer

```
1 nltk.download("punkt")
2 class MyTokenizer:
3     def tokenize(self, text):
4         tokenizer = TreebankWordTokenizer()
5         result = []
6         word = r"\p{letter}"
7         for sent in nltk.sent_tokenize(text):
8             tokens = tokenizer.tokenize(sent)
9             tokens = [t for t in tokens
10                 if regex.search(word, t)]
11            result += tokens
12        return result
13 mytokenizer = MyTokenizer()
14 print(mytkokenizer.tokenize(text))
```

earlier. Similarly, with heavily inflected languages such as Hungarian or Arabic, it might be better to use preprocessing tools developed specifically for these languages, but treating those is beyond the scope of this book.

10.1.2 The DTM as a Sparse Matrix

Example 10.4 shows a more realistic example. It downloads all US "State of the Union" speeches and creates a document-term matrix from them. Since the matrix is now easily too large to print, both Python and R simply list the size of the matrix. R lists 85 documents (rows) and 17 999 features (columns), and Python reports that it's size is 85 × 17 185. Note the difference in the number of columns (unique terms) due to the differences in tokenization as discussed above.

Python Code

```
# this snippet uses the tokenizer created above
# (box "Tokenization with Python")
sotu=pd.read_csv("https://cssbook.net/d/sotu.csv")
cv=CountVectorizer(tokenizer=mytokenizer.tokenize)
d=cv.fit_transform(sotu["text"])
d
```

R Code

```
1 url = "https://cssbook.net/d/sotu.csv"
2 sotu = read_csv(url) %>%
3        mutate(doc_id=paste(lubridate::year(Date),
4                            President, delivery))
5 d = corpus(sotu) %>% tokens() %>% dfm()
6 d
```

Python Output

```
<85x17185 sparse matrix of type '<class 'numpy.int64'>'
with 132900 stored elements in Compressed Sparse
Row format>
```

R Output (truncated)

```
Document-feature matrix of: 85 documents, 18,165
      features (91.1% sparse) and 6 docvars.
```

Example 10.4 Example document-term matrix.

Python Code

```
def termstats(dfm, vectorizer):
  """Helper function to calculate term and
  document frequency per term"""
  # Frequencies are the column sums of the DFM
  frequencies = dfm.sum(axis=0).tolist()[0]
  # Document frequencies are the binned count
  # of the column indices of DFM entries
  docfreqs = np.bincount(dfm.indices)
  freq_df=pd.DataFrame(
      dict(frequency=frequencies,docfreq=docfreqs),
      index=vectorizer. get_feature_names())
  return freq_df.sort_values("frequency",
                             ascending=False)

termstats(d, cv).iloc[[0, 10, 100, 1000, 10000]]
```

R Code

```
textstat_frequency(d)[c(1, 10, 100, 1000, 15000)]
```

Python Output

```
            frequency    docfreq
the             34996         85
is               5472         85
energy            707         68
scientific         73         28
escalate            2          2
```

R Output

```
      feature     frequency  rank   docfreq group
1      the         34999        1  85       all
10     our         9334        10  85       all
100    first        750       100  83       all
1000   investments 76         988  34       all
15000   tobago        1     11005  1        all
```

Python Code

```
words = ["the", "is", "energy",
         "scientific", "escalate"]
indices = [cv.vocabulary_[x] for x in words]

d[[[0], [25], [50], [75]], indices].todense()
```

R Code

```
as.matrix(d[
  c(3, 25, 50, 75),
  c("the","first","investment","defrauded")])
```

In Example 10.5 we show how you can look at the content of the DTM. First, we show the overall term and document frequencies of each word, where we showcase words at different frequencies. Unsurprisingly, the word *the* tops both charts, but further down there are minor differences. In all cases, the highly frequent words are mostly functional words like *them* or *first*. More informative words such as *investments* are by their nature used much less often. Such term statistics are very useful to check for noise in the data and get a feeling of the kind of language that is used. Second, we take a look at the frequency of these same words in four speeches from

Python Output

```
          frequency    docfreq
matrix ([[642,   78,    0,    0,    0],
         [355,   66,    1,    0,    0],
         [182,   45,    2,    0,    0],
         [326,   59,   15,    1,    0]])
```

R Output

```
                 the   first  investment defraudded
1946  Truman    2141    21       9      0
1965  Johnson    283    14       0      0
1984  Reagan     209     8       1      0
2009  Obama      269     8       4      0
```

Example 10.5 A look inside the DTM.

Truman to Obama. All use words like *the* and *first*, but none of them talk about *defrauded* – which is not surprising, since it was only used once in all the speeches in the corpus.

However, the words that ranked around 1000 in the top frequency are still used in less than half of the documents. Since there are about 17 000 even less frequent words in the corpus, you can imagine that most of the document-term matrix consists of zeros. The output also noted this *sparsity* in the first output above. In fact, R reports that the dtm is 91% sparse, meaning 91% percent of all entries are zero. Python reports a similar figure, namely that there are 132 900 non-zero entries out of a possible 85 × 17185, which boils down to a 91% sparse matrix as well.

Note that to display the matrix we turned it from a *sparse matrix* representation into a *dense matrix*. Briefly put, in a dense matrix, all entries are stored as a long list of numbers, including all the zeros. In a sparse matrix, only the non-zero entries and their location are stored. This conversion (using the function `as.matrix` and the method `todense` respectively), however, was only performed after selecting a small subset of the data. In general, it is very inefficient to store and work with the matrix in a *dense* format. For a reasonably large corpus with tens of thousands of documents and different words, this can quickly run to billions of numbers, which can cause problems even on modern computers and is, moreover, very inefficient. With sparsities of often above 99%, using a sparse matrix representation can easily reduce storage requirements by a hundred times and in the process speed up calculations by reducing the number of entries that need to be inspected. Both *quanteda* and *scikit-learn* store DTMs as sparse matrices by default, and most analysis tools are able to deal with sparse matrices very efficiently (see, however, Section 11.4.1 for problems with machine learning on sparse matrices in R).

A final note on the difference between Python and R in this example. The code in R is much simpler and produces nicer results since it also shows the words and the speech names. In Python, we wrote our own helper function to create the frequency statistics which is built into the R *quanteda* package. These differences between Python and R reflect a pattern that is true in many (but not all) cases: in Python libraries such as *numpy* and *scikit-learn* are set up to maximize performance, while in R a library such as *quanteda* or *tidyverse* is more geared towards ease of use. For that reason, the DTM in Python does not "remember" the actual words, it uses the index of each word, so it consumes less memory if you don't need to use the actual words in e.g. a machine learning setup. R, on the other hand, stores the words and also the document IDs and metadata in the DFM object. This is easier to use if you need to look up a word or document, but it consumes (slightly) more memory.

Python: Why fit_transform?

In Python, you don't have a function that directly transforms text into a DTM. Instead, you create an *transformer* called a CountVectorizer, which can then be used to "vectorize" texts (turn it into a row of numbers) by counting how often each word occurs. This uses the `fit_transform` function which is offered by all *scikit-learn* transformers. It "fits" the model on the training data, which in this case means learning the vocabulary. It can then be used to transform other data into a DTM with the exact same columns, which is often required for algorithms. Because the feature names (the words themselves) are stored in the CountVectorizer rather than the document-term matrix, you generally need to keep both objects.

10.1.3 The DTM as a "Bag of Words"

As you can see already in these simple examples, the document-term matrix discards quite a lot of information from text. Specifically, it disregards the order of words in a text: "John fired Mary" and "Mary fired John" both result in the same DTM, even though the meaning of the sentences is quite different. For this reason, a DTM is often called a *bag of words*, in the sense that all words in the document are simply put in a big bag without looking at the sentences or context of these words.

Thus, the DTM can be said to be a specific and "lossy" representation of the text, that turns out to be quite useful for certain tasks: the frequent occurrence of words like "employment", "great", or "I" might well be good indicators that a text is about the economy, is positive, or contains personal expressions respectively. As we will see in the next chapter, the DTM representation can be used for many different text analyses, from dictionaries to supervised and unsupervised machine learning.

Sometimes, however, you need information that is encoded in the order of words. For example, in analyzing conflict coverage it might be quite important to know who attacks whom, not just that an attack took place. In the Section 10.3 we will look at some ways to create a richer matrix-representation by using word pairs. Although it is beyond the scope of this book, you can also use automatic syntactic analysis to take grammatical relations into account as well. As is always the case with automatic analyses, it is important to understand what information the computer is looking at, as the computer cannot find patterns in information that it doesn't have.

10.1.4 The (Unavoidable) Word Cloud

One of the most famous text visualizations is without doubt the word cloud. Essentially, a word cloud is an image where each word is displayed in a size that is representative of its frequency. Depending on preference, word position and color can be random, depending on word frequency, or in a decorative shape.

Word clouds are often criticized since they are (sometimes) pretty but mostly not very informative. The core reason for that is that only a single aspect of the words is visualized (frequency), and simple word frequency is often not that informative: the most frequent words are generally uninformative "stop words" like "the" and "I".

For example, Example 10.6 shows the word cloud for the state of the union speeches downloaded above. In R, this is done using the *quanteda* function `textplot_wordcloud`. In Python we need to work a little harder, since it only has the counts, not the actual words. So, we sum the DTM columns to get the frequency of each word, and combine that with the feature names (words) from the `CountVectorized` object `cv`. Then we can create the word cloud and give it the frequencies to use. Finally, we plot the cloud and remove the axes.

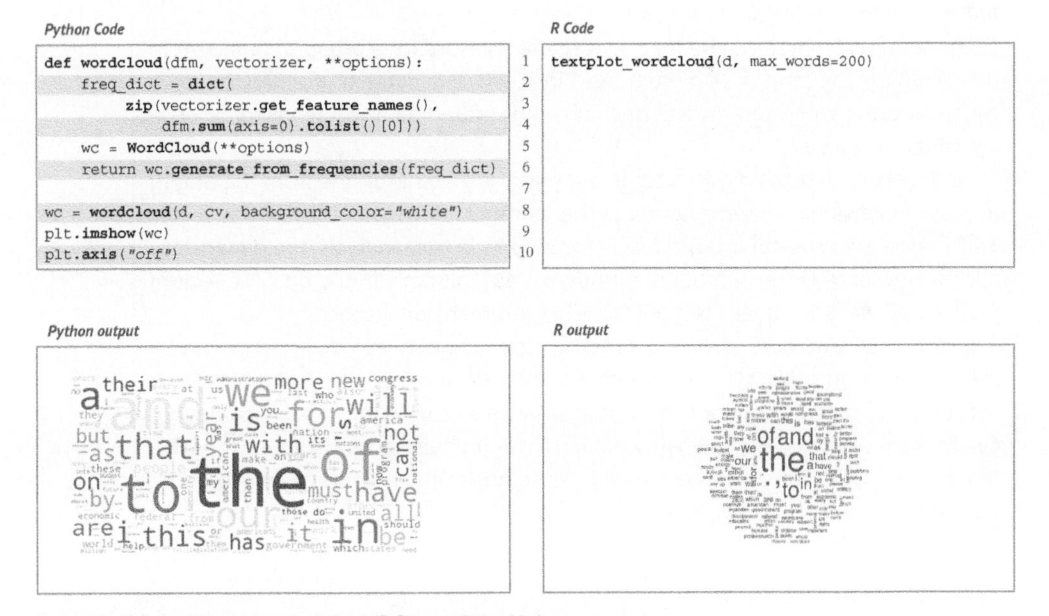

Python Code

```
def wordcloud(dfm, vectorizer, **options):
    freq_dict = dict(
        zip(vectorizer.get_feature_names(),
            dfm.sum(axis=0).tolist()[0]))
    wc = WordCloud(**options)
    return wc.generate_from_frequencies(freq_dict)

wc = wordcloud(d, cv, background_color="white")
plt.imshow(wc)
plt.axis("off")
```

R Code

```
textplot_wordcloud(d, max_words=200)
```

Python output

R output

Example 10.6 Word cloud of the US State of the Union corpus.

The results from Python and R look different at first – for one thing, R is nice and round but Python has more colors! However, if you look at the cloud you can see both are not very meaningful: the largest words are all punctuation or words like "a", "and", or "the". You have to look closely to find words like "federal" or "security" that give a hint on what the texts were actually about.

10.2 Weighting and Selecting Documents and Terms

So far, the DTMs you made in this chapter simply show the count of each word in each document. Many words, however, are not informative for many questions. This is especially apparent if you look at a *word cloud*, essentially a plot of the most frequent words in a *corpus* (set of documents).

More formally, a document-term matrix can be seen as a representation of data points about documents: each document (row) is represented as a vector containing the count per word (column). Although it is a simplification compared to the original text, an unfiltered document-term matrix contains a lot of relevant information. For example, if a president uses the word "terrorism" more often than the word "economy", that could be an indication of their policy priorities.

Note: Vectors and a geometric interpretation of document-term matrices

We said that a document is represented by a "vector" of numbers, where each number (for a document-term matrix) is the frequency of a specific word in that document. This term is also seen in the name for the tokenizer in *scikit-learn*: a *vectorizer* or function to turn texts into vectors.

The term *vector* here can be read as just a fancy word for a group of numbers. In this meaning, the term is also often used in R, where a column of a data frame is called a vector, and where functions that can be called on a whole vector at once are called *vectorized*.

More generally, however, a vector in geometry is a point (or line from the origin) in an n-dimensional space, where n is the length or dimensionality of the vector. This is also a very useful interpretation for vectors in text analysis: the dimensionality of the space is the number of unique words (columns) in the document-term matrix, and each document is a point in that n-dimensional space.

In that interpretation, various geometric distances between documents can be calculated as an indicator for how similar two documents are. Techniques that reduce the number of columns in the matrix (such as clustering or topic modeling) can then be seen as dimensionality reduction techniques since they turn the DTM into a matrix with lower dimensionality (while hopefully retaining as much of the relevant information as possible).

However, there is also a lot of *noise* crowding out this *signal*: as seen in the word cloud in the previous section the most frequent words are generally quite uninformative. The same holds for words that hardly occur in any document (but still require a column to be represented) and noisy "words" such as punctuation or technical artifacts like HTML code.

This section will discuss a number of techniques for cleaning a corpus or document-term matrix in order to minimize the amount of noise: removing stop words, cleaning punctuation and other artifacts, and trimming and weighting. As a running example in this section, we will use a collection of tweets from US president Donald Trump. Example 10.7 shows how to load these tweets into a data frame containing the ID and text of the tweets. As you can see, this dataset contains a lot of non-textual features such as hyperlinks and hash tags as well as regular punctuation and stop words. Before we can start analyzing this data, we need to decide on and perform multiple cleaning steps such as detailed below.

Please note that although tweets are perhaps overused as a source of scientific information, we use them here since they nicely exemplify issues around non-textual elements such as hyperlinks. See Chapter 12 for information on how to use the Twitter and other APIs to collect your own data.

10.2.1 Removing stop words

A first step in cleaning a DTM is often *stop word removal*. Words such as "a" and "the" are often called stop words, i.e. words that do not tell us much about the content. Both *quanteda* and *scikit-learn* include built-in lists of stop words, making it very easy to remove the most common words. Example 10.8 shows the result of specifying "English" stop words to be removed for both packages.

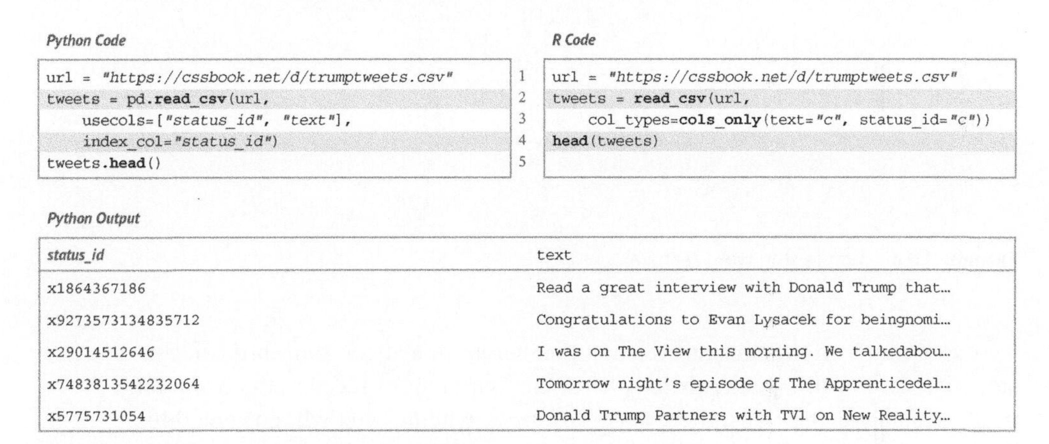

Python Code

```python
url = "https://cssbook.net/d/trumptweets.csv"
tweets = pd.read_csv(url,
    usecols=["status_id", "text"],
    index_col="status_id")
tweets.head()
```

R Code

```r
url = "https://cssbook.net/d/trumptweets.csv"
tweets = read_csv(url,
    col_types=cols_only(text="c", status_id="c"))
head(tweets)
```

Python Output

status_id	text
x1864367186	Read a great interview with Donald Trump that...
x9273573134835712	Congratulations to Evan Lysacek for beingnomi...
x29014512646	I was on The View this morning. We talkedabou...
x7483813542232064	Tomorrow night's episode of The Apprenticedel...
x5775731054	Donald Trump Partners with TV1 on New Reality...

Example 10.7 Top words used in Trump Tweets.

Note, however, that it might seem easy to list words like "a" and "and", but as it turns out there is no single well-defined list of stop words, and (as always) the best choice depends on your data and your research question.

Linguistically, stop words are generally function words or closed word classes such as determiner or pronoun, with closed classes meaning that while you can coin new nouns, you can't simply invent new determiners or prepositions. However, there are many different stop word lists around which make different choices and are compatible with different kinds of preprocessing. The Python word cloud in Example 10.8 shows a nice example of the importance of matching stop-words with the used tokenization: a central "word" in the cloud is the contraction *'s*. We are using the NLTK tokenizer, which splits *'s* from the word it was attached to, but the *scikit-learn* stop word list does not include that term. So, it is important to make sure that the words created by the tokenization match the way that words appear in the stop word list.

Python Code

```python
cv = CountVectorizer(
    stop_words=stopwords.words("english"),
    tokenizer=mytokenizer.tokenize)
d = cv.fit_transform(tweets.text)
wc = wordcloud(d, cv, background_color="white")
plt.imshow(wc)
plt.axis("off")
```

R Code

```r
d = corpus(tweets) %>%
  tokens(remove_punct=T) %>%
  dfm() %>%
  dfm_remove(stopwords("english"))
textplot_wordcloud(d, max_words=100)
```

R Output

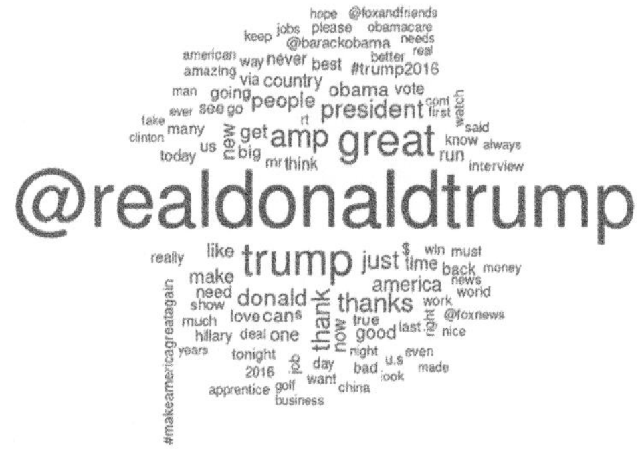

Example 10.8 Simple stop word removal.

As an example of the substantive choices inherent in using a stop word list, consider the word "will". As an auxiliary verb, this is probably indeed a stop word: for most substantive questions, there is no difference whether you will do something or simply do it. However, "will" can also be a noun (a testament) and a name (e.g. Will Smith). Simply dropping such words from the corpus can be problematic; see Section 10.3.4 for ways of telling nouns and verbs apart for more fine-grained filtering.

Moreover, some research questions might actually be interested in certain stop words. If you are interested in references to the future or specific modalities, the word might actually be a key indicator. Similarly, if you are studying self-expression on Internet forums, social identity theory, or populist rhetoric, words like "I", "us" and "them" can actually be very informative.

For this reason, it is always a good idea to understand and inspect what stop word list you are using, and use a different one or customize it as needed (see also Nothman et al., 2018). Example 10.9 shows how you can inspect and customize stop word lists. For more details on which lists are available and what choices these lists make, see the package documentation for the *stopwords* package in Python (part of NLTK) and R (part of quanteda).

Python Code

```python
mystopwords=["go","to"]+stopwords.words("english")
print(f"{len(mystopwords)} stopwords:"
      f"{', '.join(mystopwords[:5])}...")
```

R Code

```r
mystopwords = stopwords("english",
                        source="snowball")
mystopwords = c("go", "one", mystopwords)
glue("Now {length(mystopwords)} stopwords:")
mystopwords[1:5]
```

Python Output

```
181 stopwords: go, one, i, me, my...
```

R Output

```
Now 177 stopwords: go, one, i, me, my...
```

Example 10.9 Inspecting and Customizing stop word lists.

10.2.2 Removing Punctuation and Noise

Next to stop words, text often contains punctuation and other things that can be considered "noise" for most research questions. For example, it could contain emoticons or emoji, Twitter hashtags or @-mentions, or HTML tags or other annotations.

In both Python and R, we can use regular expressions to remove (parts of) words. As explained in Section 9.2, regular expressions are a powerful way to specify (sequences of) characters which are to be kept or removed. You can use this, for example, to remove things like punctuation, emoji, or HTML tags. This can be done either before or after tokenizing (splitting the text into words): in other words, we can clean the raw texts or the individual words (tokens).

In general, if you only want to keep or remove certain words, it is often easiest to do so after tokenization using a regular expression to select the words to keep or remove. If you want to remove parts of words (e.g. to remove the leading "#" in hashtags) it is easiest to do that before tokenization, that is, as a preprocessing step before the tokenization. Similarly, if you want to remove a term that would be split by the tokenization (such as hyperlinks), if can be better to remove them before the tokenization occurs.

Example 10.10 shows how we can use regular expressions to remove noise in Python and R. For clarity, it shows the result of each processing step on a single tweet that exemplifies many of the problems described above. To better understand the tokenization process, we print the tokens in that tweet separated by a vertical bar (|). As a first cleaning step, we will use a regular expression to remove hyperlinks and HTML entities like `&` from the untokenized texts. Since both hyperlinks and HTML entities are split over multiple tokens, it would be hard to remove them after tokenization.

Regular expressions are explained fully in Section 9.2, so we will keep the explanation short: the bar | splits the pattern in two parts, i.e. it will match if it finds either of the subpatterns. The first pattern looks for the literal text `http`, followed by an optional `s` and the sequence `://`. Then, it takes all non-whitespace characters it finds, i.e. the pattern ends at the next whitespace or end of the text. The second pattern looks for an ampersand (`&`) followed by one or more letters (`\\w+`), followed by a semicolon (`;`). This matches HTML escapes like `&` for an ampersand.

Python Code

```python
id = "x263687274812813312"
one_tweet=tweets.text.values[tweets.index==id][0]
print(f"Raw:\n{one_tweet}")
tweet_tokens = mytokenizer.tokenize(one_tweet)
print("\ntokenizing:")
print(" | ".join(tweet_tokens))
```

R Code

```r
id = "x263687274812813312"
single_tweet = tweets$text[tweets$status_id == id]
glue("Raw:\n{single_tweet}")
tweet_tokens = tokens(single_tweet)
glue("After tokenizing:")
paste(tweet_tokens, collapse=" | ")
```

Output

```
Raw:
Part 1 of my @jimmyfallon interview discussing my $5M offer to Obama, #TRUMP Tower atrium, my tweets &
     57th st. crane http://t.co/AvLO9Inf
After tokenizing:
Part | 1 | of | my | @jimmyfallon | interview | discussing | my | $ | 5M | offer | to | Obama | , | #TRUMP
     | Tower | atrium | , | my |tweets | & | amp | ; | 57th | st | . | crane | http://t.co/AvLO9Inf
```

Python Code

```python
one_tweet = re.sub(r"\bhttps?://\S*|&\w+;", "",
                   one_tweet)
tweet_tokens = mytokenizer.tokenize(one_tweet)
print("After pre-processing:")
print(" | ".join(tweet_tokens))
```

R Code

```r
single_tweet = single_tweet %>%
  str_remove_all("\\bhttps?://\\S*|&\\w+;")
tweet_tokens = tokens(single_tweet)
glue("After pre-processing:")
paste(tweet_tokens, collapse=" | ")
```

Output

```
After pre-processing:
Part | 1 | of | my | @jimmyfallon | interview | discussing | my | $ | 5M | offer | to | Obama | , | #TRUMP
     | Tower | atrium | , | my |tweets | 57th | st | . | crane
```

Python Code

```python
tweet_tokens = [t.lower() for t in tweet_tokens
  if not (t.lower() in stopwords.words("english")
    or regex.match(r"\P{LETTER}", t))]
print("After pruning tokens:")
print(" | ".join(tweet_tokens))
```

R Code

```r
tweet_tokens = tweet_tokens %>%
  tokens_tolower() %>%
  tokens_remove(stopwords("english")) %>%
  tokens_keep("^\\p{LETTER}", valuetype="regex")
print("After pruning tokens:")
paste(tweet_tokens, collapse=" | ")
```

Output

```
After pruning tokens:
part | interview | discussing | offer | obama | tower | atrium | tweets | st | crane
```

Example 10.10 Cleaning a single tweet at the text and token level.

In the next step, we process the tokenized text to remove every token that is either a stopword or does not start with a letter. In Python, this is done by using a list comprehension (`[process(item) for item in list]`) for tokenizing each document; and a nested list comprehension for filtering each token in each document. In R this is not needed as the `tokens_*` functions are *vectorized*, that is, they directly run over all the tokens.

Comparing R and Python, we see that the different tokenization functions mean that `#trump` is removed in R (since it is a token that does not start with a letter), but in Python the tokenization splits the # from the name and the resulting token `trump` is kept. If we would have used a different tokenizer for Python (e.g. the `WhitespaceTokenizer`) this would have been different again. This underscores the importance of inspecting and understanding the results of the specific tokenizer used, and to make sure that subsequent steps match these tokenization choices. Concretely, with the `TreebankWordtokenizer` we would have had to also remove hashtags at the text level rather than the token level.

As a final example, Example 10.11 shows how to filter tokens for the whole corpus, but rather than removing hashtags it keeps only the hashtags to produce a tag cloud. In R, this is mostly a pipeline of *quanteda* functions to create the corpus, tokenize, keep only hashtags, and create a DFM. To spice up the output we use the *RColorBrewer* package to set random colors for the tags. In Python, you can see that we now have a nested list comprehension, where the outer loop iterates over the texts and the inner loop iterates over the tokens in each text. Next, we make a `do_nothing` function for the vectorizer since the results are already tokenized. Note that we need to disable lowercasing as otherwise it will try to call `.lower()` on the token lists.

Python Code

```python
def do_nothing(x):
    return x

tokenized = [WhitespaceTokenizer().tokenize(text)
             for text in tweets.text.values]
tokens = [[t.lower() for t in tokens
           if regex.match("#", t)]
          for tokens in tokenized]

cv = CountVectorizer(tokenizer=do_nothing,
                     lowercase=False)
dtm_tags = cv.fit_transform(tokens)
wc = wordcloud(dtm_tags, cv,
               background_color="white")
plt.imshow(wc)
plt.axis("off")
```

R Code

```r
dfm_tags = tweets %>%
  corpus() %>%
  tokens() %>%
  tokens_keep("^#", valuetype="regex") %>%
  dfm()
colors = RColorBrewer::brewer.pal(8, "Dark2")
textplot_wordcloud(dfm_tags, max_words=100,
    min_size = 1, max_size=4, random_order=TRUE,
    random_color= TRUE, color=colors)
```

R Output

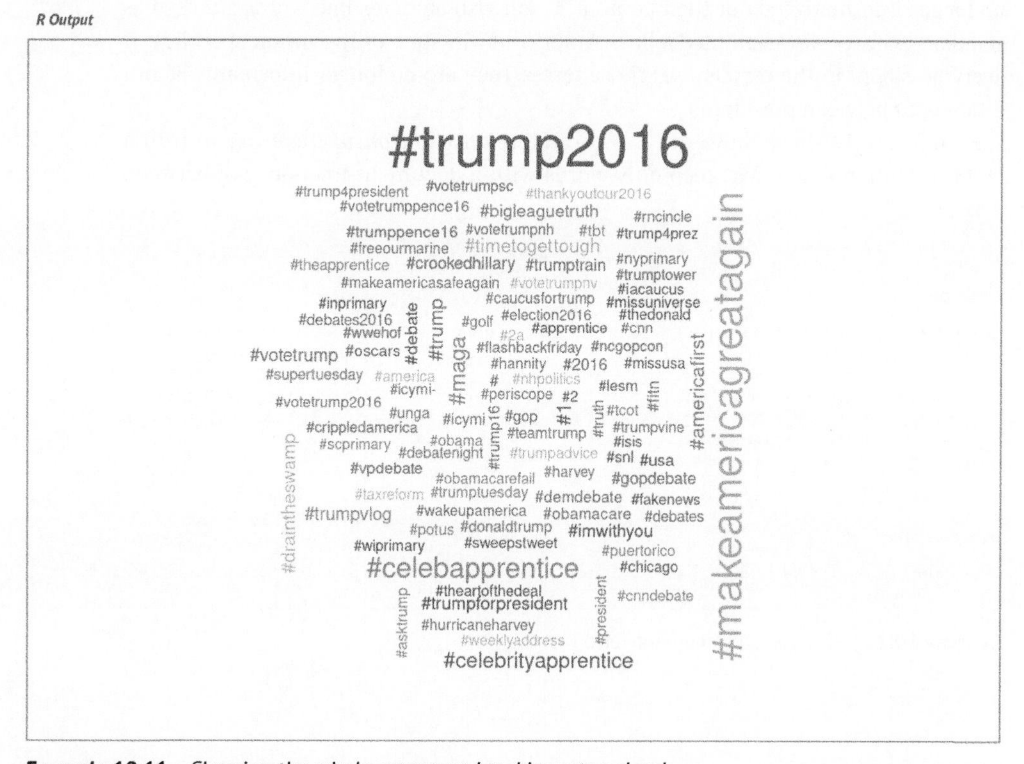

Example 10.11 Cleaning the whole corpus and making a tag cloud.

Note: `lambda` functions in Python

Sometimes, we need to define a function that is very simple and that we need only once. An example for such a throwaway function is `do_nothing` in Example 10.11. Instead of defining a reusable function with the `def` keyword and then to call it by its name when we need it later, we can therefore also directly define an unnamed function when we need it with the `lambda` keyword. The syntax is simple: `lambda argument: returnvalue`. A function that maps a value onto itself can therefore be written as `lambda x: x`. In Example 10.11, instead of defining a named function, we could therefore also simply write `cv = CountVectorizer(tokenizer=lambda x: x, lowercase=False)`. The advantages are that it saves you two lines of code here and you don't clutter your environment with functions you do not intend to re-use anyway. The disadvantage is that it may be less clear what is happening, at least for people not familiar with lambda functions.

10.2.3 Trimming a DTM

The techniques above both drop terms from the DTM based on specific choices or patterns. It can also be beneficial to trim a DTM by removing words that occur very infrequently or overly frequently. For the former, the reason is that if a word only occurs in a very small percentage of documents it is unlikely to be very informative. Overly frequent words, for example occurring in more than half or 75% of all documents, function basically like stopwords for this corpus. In many cases, this can be a result of the selection strategy. If we select all tweets containing "Trump", the word Trump itself is no longer informative about their content. It can also be that some words are used as standard phrases, for example "fellow Americans" in state of the union speeches. If every president in the corpus uses those terms, they are no longer informative about differences between presidents.

Example 10.12 shows how you can use the *relative document frequency* to trim a DTM in Python and R. We keep only words with a document frequency of between 0.5% and 75%.

Python Code

```
print(f"# of words before trimming: {d.shape[1]}")
cv_trim = CountVectorizer(
    stop_words=stopwords.words("english"),
    tokenizer=mytokenizer.tokenize,
    max_df=0.75, min_df=0.005)
d_trim = cv_trim.fit_transform(tweets.text)
print(f" after trimming: {d_trim.shape[1]}")
```

R Code

```
glue("# of words before trimming: {ncol(d)}")
d_trim = dfm_trim(d, min_docfreq = 0.005,
                  max_docfreq = 0.75,
                  docfreq_type = "prop")
glue("# of word after trimming: {ncol(d_trim)}")
```

Python Output

```
# of unique words before trimming: 45912; after
    trimming: 294
```

R Output

```
# of unique words before trimming: 44386; after
    trimming: 301
```

Example 10.12 Trimming a Document-Term Matrix.

Although these are reasonable numbers every choice depends on the corpus and the research question, so it can be a good idea to check which words are dropped. Note that dropping words that occur almost never should normally not influence the results that much, since those words do not occur anyway. However, trimming a DTM to e.g. at least 1% document frequency often radically reduces the number of words (columns) in the DTM. Since many algorithms have to assign weights or parameters to each word, this can provide a significant improvement in computing speed or memory use.

10.2.4 Weighting a DTM

The DTMs created above all use the raw frequencies as cell values. It can also be useful to weight the words so more informative words have a higher weight than less informative ones. A common technique for this is *tf·idf* weighting. This stands for *term frequency . inverse document frequency* and weights each occurrence by its raw frequency (term frequency) corrected for how often it occurs in all documents (inverse document frequency). In a formula, the most common implementation of this weight is given as follows:

$$tf \cdot idf(t, d) = tf(t, d) \cdot idf(t) = f_{t,d} \cdot - \log \frac{n_t}{N}$$

Where $f_{t,d}$ is the frequency of term t in document d, N is the total number of documents, and n_t is the number of documents in which term t occurs. In other words, the term frequency is weighted by the negative log of the fraction of documents in which that term occurs. Since log(1) is zero, terms that occur in every document are disregarded, and in general the less frequent a term is, the higher the weight will be.

Python Code

```
tfidf_vectorizer = TfidfVectorizer(
    tokenizer=mytokenizer.tokenize,
    sublinear_tf=True)
d_w = tfidf_vectorizer.fit_transform(sotu["text"])
indices = [tfidf_vectorizer.vocabulary_[x]
  for x in ["the","for","them","submit","sizes"]]
d_w[[0], [25], [50], [75]], indices].todense()
```

R Code

```
d_tf = corpus(sotu) %>%
  tokens() %>%
  dfm() %>%
  dfm_tfidf(scheme_tf="prop", smoothing=1)
as.matrix(
  d_tf[c(3, 25, 50, 75),
      c("the","first","investment","defrauded")])
```

R Output

	the	our	first	investments	defrauded
1946 Truman written	0.02084698	0.001811087	0.0002080106	3.519654e-05	0
1965 Johnson spoken	0.01746801	0.004999678	0.0008790725	0.000000e+00	0
1984 Reagan spoken	0.01132996	0.004987351	0.0004411759	0.000000e+00	0
2009 Obama spoken	0.01208793	0.005212641	0.0003657038	8.121631e-05	0

Example 10.13 Tf·Idf weighting.

tf·idf weighting is a fairly common technique and can improve the results of subsequent analyses such as supervised machine learning. As such, it is no surprise that it is easy to apply this in both Python and R, as shown in Example 10.13. This example uses the same data as Example 10.4 above, so you can compare the resulting weighted values with the results reported there. As you can see, the tf·idf weighting in both languages have roughly the same effect: very frequent terms such as *the* are made less important compared to less frequent words such as *submit*. For example, in the raw frequencies for the 1965 Johnson speech, *the* occurred 355 times compared to *submit* only once. In the weighted matrix, the weight for *submit* is four times as low as the weight for *the*.

There are two more things to note if you compare the examples from R and Python. First, to make the two cases somewhat comparable we have to use two options for R, namely to set the term frequency to proportional (`scheme_tf='prop'`), and to add smoothing to the document frequencies (`smooth=1`). Without those options, the counts for the first columns would all be zero (since they occur in all documents, and $\log\frac{85}{85}=0$), and the other counts would be greater than one since they would only be weighted, not normalized.

Even with those options the results are still different (in details if not in proportions), mainly because R normalizes the frequencies before weighting, while Python normalizes after the weighting. Moreover, Python by default uses L2 normalization, meaning that the length of the document vectors will be one, while R uses L1 normalization, that is, the row sums are one (before weighting). Both R and Python have various parameters to control these choices which are explained in their respective help pages. Although the differences in absolute values look large, the relative effect of making more frequent terms less important is the same, and the specific weighting scheme and options will probably not matter that much for the final results. However, it is always good to be aware of the specific options available and try out which work best for your specific research question.

10.3 Advanced Representation of Text

The examples above all created document-term matrices where each column actually represents a word. There is more information in a text, however, than pure word counts. The phrases: *the movie was not good, it was in fact quite bad* and *the movie was not bad, in fact it was quite good* have exactly the same word frequencies, but are quite different in meaning. Similarly, *the new residents of York* and *the residents of New York* are referring to quite different people.

Of course, in the end which aspect of the meaning of a text is important depends on your research question: if you want to know the sentiment about the movie, it is important to take a word like "not" into account; but if you are interested in the topic or genre of the review, or the extremity of the language used, this might not be relevant.

The core idea of this section is that in many cases this information can be captured in a DTM by having the columns represent different information than just words, for example word combinations or groups of related words. This is often called *feature engineering*, as we are using our domain expertise to find the right features (columns, independent variables) to capture the relevant meaning for our research question. If we are using other columns than words it is also technically more correct to use the name *document-feature matrix*, as *quanteda* does, but we will stick to the most common name here and simply continue using the name DTM.

10.3.1 *n*-grams

The first feature we will discuss are n-grams. The simplest case is a bigram (or 2-gram), where each feature is a pair of adjacent words. The example used above, *the movie was not bad*, will yield the following bigrams: *the-movie*, *movie-was*, *was-not*, and *not-bad*. Each of those bigrams is then treated as a feature, that is, a DTM would contain one column for each word pair.

As you can see in this example, we can now see the difference between *not-bad* and *not-good*. The downside of using n-grams is that there are many more unique word pairs than unique words, so the resulting DTM will have many more columns. Moreover, there is a bigger *data scarcity problem*, as each of those pairs will be less frequent, making it more difficult to find sufficient examples of each to generalize over.

Although bigrams are the most frequent use case, trigrams (3-grams) and (rarely) higher-order n-grams can also be used. As you can imagine, this will create even bigger DTMs and worse data scarcity problems, so even more attention must be paid to feature selection and/or trimming.

Example 10.14 shows how n-grams can be created and used in Python and R. In Python, you can pass the `ngram_range=(n, m)` option to the vectorizer, while R has a `tokens_ngrams(n:m)` function. Both will post-process the tokens to create all n-grams in the range of n to m. In this example, we are asking for unigrams (i.e., the words themselves), bigrams and trigrams of a simple example sentence. Both languages produce the same output, with R separating the words with an underscore while Python uses a simple space.

Python Code

```
cv = CountVectorizer(ngram_range=(1,3),
        tokenizer=mytokenizer.tokenize)
cv.fit_transform(["This is a test"])
cv.get_feature_names()
```

R Code

```
1 text = "This is a test"
2 tokens(text) %>%
3   tokens_tolower() %>%
4   tokens_ngrams(1:3)
```

Python Output

```
['a',
'a test',
'is',
'is a',
'is a test',
'test',
'this',
'this is',
'this is a']
```

R Output

```
Tokens consisting of 1 document.
text1 :
[1] "this" "is" "a" "test" "this_is" "is_a"
[7] "a_test" "this_is_a" "is_a_test"
```

Example 10.14 Generating n-grams.

Example 10.15 shows how you can generate n-grams for a whole corpus. In this case, we create a DTM of the state of the union matrix with all bigrams included. A glance at the frequency table for all words containing *government* shows that, besides the word itself and it's plural and possessive forms, the bigrams include compound words (federal and local government), phrases with the government as subject (the government can and must), and nouns for which the government is an adjective (government spending and government programs).

You can imagine that including all these words as features will add many possibilities for analysis of the DTM which would not be possible in a normal bag-of-words approach. The terms local and federal government can be quite important to understand policy positions, but for e.g. sentiment analysis a bigram like *not good* would also be insightful (but make sure "not" is not on your stop word list!).

Python Code

```
cv = CountVectorizer(ngram_range=(1,2),
    tokenizer=mytokenizer.tokenize,
    stop_words="english")
dfm = cv.fit_transform(sotu.text.values)
ts = termstats(dfm, cv)
ts.filter(like="government", axis=0).head(10)
```

R Code

```
sotu_tokens = corpus(sotu) %>%
   tokens(remove_punct=T) %>%
   tokens_remove(stopwords("english")) %>%
   tokens_tolower()
dfm_bigram = sotu_tokens %>%
   tokens_ngrams(1:2) %>%
   dfm()
textstat_frequency(dfm_bigram) %>%
   filter(str_detect(feature, "government")) %>%
   head(12)
```

R Output

	feature <chr>	frequency <dbl>	rank <int>	docfreq <dbl>	group <chr>
10	government	1424	10	84	all
198	federal_government	265	198	56	all
318	governments	188	318	50	all
652	local_governments	104	648	28	all
976	government's	71	972	25	all
1208	government_must	55	1195	28	all
1453	government_can	44	1433	26	all
1545	governmental	41	1537	19	all
1951	local_government	32	1919	16	all
2191	government_spending	28	2135	19	all
2276	self-government	26	2259	20	all
2658	government_programs	22	2589	17	all

Example 10.15 Words and bigrams containing "government".

10.2.3 Collocations

A special case of n-grams are collocations. In the strict corpus linguistic sense of the word, collocations are pairs of words that occur more frequently than expected based on their underlying occurrence. For example, the phrase *crystal clear* presumably occurs much more often than would be expected by chance given how often *crystal* and *clear* occur separately. Collocations are important for text analysis since they often have a specific meaning, for example because they refer to names such as *New York* or disambiguate a term like *sound* in *sound asleep*, a *sound proposal*, or *loud sound*.

Example 10.16 shows how to identify the most "surprising" collocations using R and Python. For Python, we use the *gensim* package which we will also use for topic modeling in Section 11.5. This package has a `Phrases` class which can identify the bigrams in a list of tokens. In R, we use the `textstat_collocations` function from *quanteda*. These packages each use a different implementation: *gensim* uses pointwise mutual information, i.e. how much information about finding the second word does seeing the first word give you? Quanteda estimates an interaction parameter in a log-linear model. Nonetheless, both methods give very similar results, with Saddam Hussein, the Iron Curtain, Al Qaida, and red tape topping the list for each.

Python Code

```
tokenized_texts = [mytokenizer.tokenize(t)
                   for t in sotu.text]
tokens = [[t.lower() for t in tokens
          if not regex.search("\P{letter}", t) ]
          for tokens in tokenized_texts]
phrases_model = Phrases(tokens, min_count=10,
             scoring="npmi", threshold=.5)
score_dict = phrases_model.export_phrases()
scores = pd.DataFrame(score_dict.items(),
                      columns=["phrase", "score"])
scores.sort_values("score", ascending=False).head()
```

R Code

```
sotu_tokens = corpus(sotu) %>%
  tokens(remove_punct=T) %>%
  tokens_tolower()

colloc = sotu_tokens %>%
  textstat_collocations(min_count=10) %>%
  as_tibble()


colloc %>% arrange(-lambda) %>% head()
```

R Output

collocation	count	count_nested	length	lambda	z
<chr>	<int>	<int>	<dbl>	<dbl>	<dbl>
saddam hussein	26	0	2	15.24283	10.344953
iron curtain	11	0	2	15.17020	9.848553
al qaida	37	0	2	14.58446	10.123585
red tape	22	0	2	13.46976	15.143997
persian gulf	31	0	2	12.90333	18.512811
line-item veto	10	0	2	12.85070	8.813419

Python Code

```
phraser = Phraser(phrases_model)
tokens_phrases = [phraser[doc] for doc in tokens]
cv = CountVectorizer(tokenizer=lambda x: x,
lowercase=False)
dtm = cv.fit_transform(tokens_phrases)
termstats(dtm, cv).filter(like="hussein", axis=0)
```

R Code

```
collocations = colloc %>%
  filter(lambda > 8) %>%
  pull(collocation) %>%
  phrase()
dfm = sotu_tokens %>%
  tokens_compound(collocations) %>%
  dfm()
textstat_frequency(dfm) %>%
  filter(str_detect(feature, "hussein"))
```

R Output

	feature	frequency	rank	docfreq	group
	<chr>	<dbl>	<int>	<dbl>	<chr>
2186	saddam_hussein	26	2120	5	all
	hussein's	3	7341	2	all

Example 10.16 Identifying and applying collocations in the US State of the Union.

The next block demonstrates how to use these collocations in further processing. In R, we filter the collocations list on *lambda* > 8 and use the `tokens_compound` function to compound bigrams from that list. As you can see in the term frequencies filtered on "Hussein", the regular terms (apart from the possessive) are removed and the compounded term now has 26 occurrences. For Python, we use the Phraser class, which uses the `phrases_model` created earlier to transform the texts. After setting a standard threshold of 0.7, we can use `fit_transform` to change the tokens. The term statistics again show how the individual terms are now replaced by their compound.

10.3.3 Word Embeddings

A recent addition to the text analysis toolbox are *word embeddings*. Although it is beyond the scope of this book to give a full explanation of the algorithms behind word embeddings, they are relatively easy to understand and use at an intuitive level.

The first core idea behind word embeddings is that the meaning of a word can be expressed using a relatively small *embedding vector*, generally consisting of around 300 numbers which can be interpreted as dimensions of meaning. The second core idea is that these embedding vectors can be derived by scanning the context of each word in millions and millions of documents.

These embedding vectors can then be used as features or DTM columns for further analysis. Using embedding vectors instead of word frequencies has the advantages of strongly reducing the dimensionality of the DTM: instead of (tens of) thousands of columns for each unique word we only need hundreds of columns for the embedding vectors. This means that further processing can be more efficient as fewer parameters need to be fit, or conversely that more complicated models can be used without blowing up the parameter space. Another advantage is that a model can also give a result for words it never saw before, as these words most likely will have an embedding vector and so can be fed into the model. Finally, since words with similar meanings should have similar vectors, a model fit on embedding vectors gets a "head start" since the vectors for words like "great" and "fantastic" will already be relatively close to each other, while all columns in a normal DTM are treated independently.

The assumption that words with similar meanings have similar vectors can also be used directly to extract synonyms. This can be very useful, for example for (semi)automatically expanding a dictionary for a concept. Example 10.17 shows how to download and use pre-trained embedding vectors to extract synonyms. First, we download a very small subset of the pre-trained Glove embedding vectors[1], wrapping the download call in a condition to only download it when needed.

---

[1] The full embedding models can be downloaded from https://nlp.stanford.edu/projects/glove/. To make the file easier to download, we took only the 10 000 most frequent words of the smallest embeddings file (the 50 dimension version of the 6B tokens model). For serious applications you probably want to download the larger files, in our experience the 300 dimension version usually gives good results. Note that the files on that site are in a slightly different format which lacks the initial header line, so if you want to use other vectors for the examples here you can convert them with the `glove2word2vec` function in the *gensim* package. For R, you can also simply omit the `skip=1` argument as apart from the header line the formats are identical.

Then, for Python, we use the excellent support from the *gensim* package to load the embeddings into a `KeyedVectors` object. Although not needed for the rest of the example, we create a pandas data frame from the internal embedding values so the internal structure becomes clear: each row is a word, and the columns (in this case 50) are the different (semantic) dimensions that characterize that word according to the embeddings model. This data frame is sorted on the first dimension, which shows that negative values on that dimension are related to various sports. Next, we switch back to the `KeyedVectors` object to get the most similar words to the word *fraud*, which is apparently related to similar words like *bribery* and *corruption* but also to words like *charges* and *alleged*. These similarities are a good way to (semi-)automatically expand a dictionary: start from a small list of words, find all words that are similar to those words, and if needed manually curate that list. Finally, we use the embeddings to solve the "analogies" that famously showcase the geometric nature of these vectors: if you take the vector for *king*, subtract the vector for *man* and add that for *woman*, the closest word to the resulting vector is *queen*. Amusingly, it turns out that soccer is a female form of football, probably showing the American cultural origin of the source material.

For R, there was less support from existing packages so we decided to use the opportunity to show both the conceptual simplicity of embeddings vectors and the power of matrix manipulation in R. Thus, we directly read in the word vector file which has a head line and then on each line a word followed by its 50 values. This is converted to a matrix with the row names showing the word, which we normalize to (Euclidean) length of one for each vector for easier processing. To determine similarity, we take the cosine distance between the vector representing a word with all other words in the matrix. As you might remember from algebra, the cosine distance is the dot product between the vectors normalized to have length one (just like Pearson's product–moment correlation is the dot product between the vectors normalized to z-scores per dimension). Thus, we can simply multiply the normalized target vector with the normalized matrix to get the similarity scores. These are then sorted, renamed, and the top values are taken using the basic functions from Chapter 6. Finally, analogies are solved by simply adding and subtracting the vectors as explained above, and then listing the closest words to the resulting vector (excluding the words in the analogy itself).

10.3.4 Linguistic Preprocessing

A final technique to be discussed here is the use of linguistic preprocessing steps to enrich and filter a DTM. So far, all techniques discussed here are language independent. However, there are also many language-specific tools for automatically enriching text developed by computational linguistics communities around the world. Two techniques will be discussed here as they are relatively widely available for many languages and easy and quick to apply: *Part-of-speech tagging* and *lemmatizing*.

In *part-of-speech tagging* or POS-tagging, each word is enriched with information on its function in the sentence: verb, noun, determiner etc. For most languages, this can be determined with very high accuracy, although sometimes text can be ambiguous: in one famous example, the word flies in fruit flies is generally a noun (fruit flies are a type of fly), but it can also be a verb (if fruit could fly). Although there are different sets of POS tags used by different tools, there is broad agreement on the core set of tags listed in Table 10.1.

Table 10.1 Overview of part-of-speech (POS) tags.

Part of speech	Example	UDPipe/Spacy Tag	Penn Treebank Tag
Noun	apple	NOUN	NN, NNS
Proper Name	Carlos	PROPN	NNP
Verb	write	VERB	VB, VBD, VBP, ..
Auxilliary verb	be, have	AUX	(same as verb)
Adjective	quick	ADJ	JJ, JJR, JJS
Adverb	quickly	ADV	RB
Pronoun	I, him	PRON	PRP
Adposition	of, in	ADP	IN
Determiner	the, a	DET	DT

POS tags are useful since they allow us for example to analyze only the *nouns* if we care about the things that are discussed, only the *verbs* if we care about actions that are described, or only the *adjectives* if we care about the characteristics given to a noun. Moreover, knowing the POS tag of a word can help disambiguate it. For example, like as a verb (I like books) is generally positive, but like as a preposition (a day like no other) has no clear sentiment attached.

Lemmatizing is a technique for reducing each word to its root or *lemma* (plural: lemmata). For example, the lemma of the verb *reads* is (to) *read* and the lemma of the noun *books* is *book*. Lemmatizing is useful since for most of our research questions we do not care about these different conjugations of the same word. By lemmatizing the texts, we do not need to include all conjugations in a dictionary, and it reduces the dimensionality of the DTM – and thus also the data scarcity.

Note that lemmatizing is related to a technique called *stemming*, which removes known suffixes (endings) from words. For example, for English it will remove the "s" from both reads and books. Stemming is much less sophisticated than lemmatizing, however, and will trip over irregular conjugations (e.g. *are* as a form of to be) and regular word endings that look like conjugations (e.g. *virus* will be stemmed to *viru*). English has relatively simple conjugations and stemming can produce adequate results. For morphologically richer languages such as German or French, however, it is strongly advised to use lemmatizing instead of stemming. Even for English we would generally advise lemmatization since it is so easy nowadays and will yield better results than stemming.

For Example 10.18, we use the *UDPipe* natural language processing toolkit (Straka and Straková, 2017), a "Pipeline" that parses text into "Universal Dependencies", a representation of the syntactic structure of the text. For R, we can immediately call the `udpipe` function from the package of the same name. This parses the given text and returns the result as a data frame with one token (word) per row, and the various features in the columns. For Python, we need to take some more steps ourselves. First, we download the English models if they aren't present. Second, we load the model and create a pipeline with all default settings, and use that to parse the same sentence.

Python Code

```python
glove_fn = "glove.6B.50d.10k.w2v.txt"
url = f"https://cssbook.net/d/{glove_fn}"
if not os.path.exists(glove_fn):
    urllib.request.urlretrieve (url, glove_fn)
```

R Code

```r
glove_fn = "glove.6B.50d.10k.w2v.txt"
url = glue("https://cssbook.net/d/{glove_fn}")
if (!file.exists(glove_fn))
    download.file(url, glove_fn)
```

Python Code

```python
wv = KeyedVectors.load_word2vec_format(glove_fn)
wvdf = pd.DataFrame(wv.vectors,
                    index=wv.index_to_key)
wvdf.sort_values(0, ascending=False).head()
```

R Code

```r
wv_tibble = read_delim(glove_fn, skip=1,
                       delim=" ", quote="",
    col_names = c("word", paste0("d", 1:50)))
wv = as.matrix(wv_tibble[-1])
rownames(wv) = wv_tibble$word
wv = wv / sqrt(rowSums(wv^2))
wv[order(wv[,1])[1:5], 1:5]
```

R Output

	d1	d2	d3	d4	d5
20003	-0.4402265	0.07209431	-0.02397687	0.18428984	0.001802660
basketball	-0.4234652	0.23817458	-0.09346347	0.17270343	-0.001520135
collegiate	-0.4232457	0.23873925	-0.28741579	0.02797958	-0.066008001
volleyball	-0.4217268	0.18378662	-0.26229465	0.31409226	-0.124286069
ncaa	-0.4131240	0.14502199	-0.06088206	0.17017979	-0.157397324

Python Code

```python
wv.most_similar("fraud")
```

R Code

```r
wvector = function(wv, word) wv[word,,drop=F]
wv_sim-ilar = function(wv, target, n=5) {
  similarities = wv %*% t(target)
  similarities %>%
    as_tibble(rownames = "word") %>%
    rename(similarity=2) %>%
    arrange(-similarity) %>%
    head(n=n)
}
wv_similar(wv, wvector(wv, "fraud"))
```

R Output

word	similarity
<chr>	<dbl>
fraud	1.0000000
charges	0.8591152
bribery	0.8559850
alleged	0.8415063
corruption	0.8299386

Python Code

```python
def analogy(a, b, c):
    result = wv.most_similar(positive=[b, c],
                             negative=[a])
    return result[0][0]

words = ["king","boy","father","pete","football"]:
for x in words:
    y = analogy("man", x, "woman")
    print(f"Man is to {x} as woman is to {y}")
```

R Code

```r
wv_analogy = function(wv, a, b, c) {
  result = (wvector(wv, b)
            + wvector(wv, c)
            - wvector(wv, a))
  matches = wv_similar(wv, result) %>%
    filter(!word %in% c(a,b,c))
  matches$word[1]
}
words=c("king","boy","father","pete","football")
for (x in words) {
  y = wv_analogy(wv, "man", x, "woman")
  print(glue("Man is to {x} as woman is to: {y}"))
}
```

R Output

```
Man is to king as woman is to: queen
Man is to boy as woman is to: girl
Man is to father as woman is to: mother
Man is to pete as woman is to: barbara
Man is to football as woman is to: soccer
```

Example 10.17 Using word embeddings for finding similar and analogous words.

Python Code

```python
m = Model.load(udpipe_model)
pipeline = Pipeline(m, "tokenize",
    Pipeline.DEFAULT, Pipeline.DEFAULT, "conllu")
text = "John bought new knives"
tokenlist = conllu.parse(pipeline.process(text))
pd.DataFrame(tokenlist[0])
```

R Code

```r
udpipe("John bought new knives", "english") %>%
  select(token_id:upos, head_token_id:dep_rel)
```

R Output

token_id	token	lemma	upos	head_token_id	dep_rel
<chr>	<chr>	<chr>	<chr>	<chr>	<chr>
1	John	John	PROPN	2	nsubj
2	bought	buy	VERB	0	root
3	new	new	ADJ	4	amod
4	knives	knife	NOUN	2	obj

Example 10.18 Using UDPipe to analyze a sentence.

Finally, we use the *conllu* package to read the results into a form that can be turned into a data frame.

In both cases, the resulting tokens clearly show some of the potential advantages of linguistic processing: the lemma column shows that it correctly deals with irregular verbs and plural forms. Looking at the upos (universal part-of-speech) column, "'John" is recognized as a proper name (PROPN), "bought" as a verb, and "knives" as a noun. Finally, the `head_token_id` and `dep_rel` columns represent the syntactic information in the sentence: "bought" (token 2) is the root of the sentence, and "John" is the subject (nsubj) while "knives" is the object of the buying.

The syntactic relations can be useful if you need to differentiate between who is doing something and whom it was done to. For example, one of the authors of this book used syntactic relations to analyze conflict coverage, where there is an important difference between attacking and getting attacked (Van Atteveldt et al., 2017). However, in most cases you probably don't need this information and analyzing dependency graphs is relatively complex. We would advise you to almost always consider lemmatizing and tagging your texts, as lemmatizing is simply so much better than stemming (especially for languages other than English), and the part-of-speech can be very useful for analyzing different aspects of a text.

If you only need the lemmatizer and tagger, you can speed up processing by setting `udpipe(.., parser='none')` (R) or setting the third argument to Pipeline (the parser) to `Pipeline.NONE` (Python). Example 10.19 shows how this can be used to

Python Code

```python
def get_nouns(text):
    result = conllu.parse(pipeline.process(text))
    for sentence in result:
        for token in sentence:
            if token["upos"] == "NOUN":
                yield token["lemma"]

parser = Pipeline.NONE
pipeline = Pipeline(m, "tokenize",
    Pipeline.DEFAULT, Pipeline.NONE, "conllu")

tokens = [list(get_nouns(text))
          for text in sotu.text[-5:]]
cv = CountVectorizer(tokenizer=lambda x: x,
                     lowercase=False, max_df=.7)
dtm_verbs = cv.fit_transform(tokens)
wc = wordcloud(dtm_verbs, cv,
               background_color="white")
plt.imshow(wc)
plt.axis("off")
```

R Code

```r
tokens = sotu %>%
  top_n(5, Date) %>%
  udpipe("english", parser="none")
nouns = tokens %>%
  filter(upos == "NOUN") %>%
  group_by(doc_id) %>%
  summarize(text=paste(lemma, collapse=" "))
nouns %>%
  corpus() %>%
  tokens() %>%
  dfm() %>%
  dfm_trim(max_docfreq=0.7,docfreq_type="prop")%>%
  textplot_wordcloud(max_words=50)
```

Python Output

Example 10.19 Nouns used in the most recent State of the Union addresses.

extract only the nouns from the most recent state of the union speeches, create a DTM with these nouns, and then visualize them as a word cloud. As you can see, these words (such as student, hero, childcare, healthcare, and terrorism), are much more indicative of the topic of a text than the general words used earlier. In the next chapter we will show how you can further analyze these data, for example by analyzing usage patterns per person or over time, or using an unsupervised topic model to cluster words into topics.

As an alternative to UDPipe, you can also use Spacy, which is another free and popular natural language toolkit. It is written in Python, but the *spacyr* package offers an easy way to use it from R. For R users, installation of *spacyr* on MacOS and Linux is easy, but note that on Windows there are some additional steps, see https://cran.r-project.org/web/packages/spacyr/readme/README.html for more details.

Example 10.20 shows how you can use Spacy to analyze the proverb "all roads lead to Rome" in Spanish. In the first block, the Spanish language model is downloaded (this is only needed once). The second block loads the language model and parses the sentence. You can see that the output is quite similar to UDPipe, but one additional feature is the inclusion of *Named Entity Recognition*: Spacy

Python Code

```python
model = "es_core_news_sm"
!{sys.executable} -m spacy download {model}
# Note: restart the kernel and re-import spacy
#       for the model to be found by python
```

R Code

```r
# Only needed once
spacy_install()
# Only needed for languages other than English:
spacy_download_langmodel("es_core_news_sm")
```

Python Code

```python
nlp = spacy.load("es_core_news_sm")
tokens = nlp("Todos los caminos llevan a Roma")
pd.DataFrame([dict(i=t.i, word=t.text,
                   lemma=t.lemma_, head=t.head,
                   dep=t.dep_, ner=t.ent_type_)
              for t in tokens])
```

R Code

```r
spacy_initialize("es_core_news_sm")
spacy_parse("Todos los caminos llevan a Roma")
# To close spacy (or switch languages), use:
spacy_finalize()
```

R Output

doc_id	sentence_id	token_id	token	lemma	pos	entity
<chr>	<int>	<int>	<chr>	<chr>	<chr>	<chr>
text1	1	1	Todos	Todos	DET	
text1	1	2	los	lo	DET	
text1	1	3	caminos	camino	NOUN	
text1	1	4	llevan	llevar	AUX	
text1	1	5	a	a	ADP	
text1	1	6	Roma	Roma	PROPN	LOC_B

Example 10.20 Using Spacy to analyze a Spanish sentence.

can automatically identify persons, locations, organizations and other entities. In this example, it identifies "Rome" as a location. This can be very useful to extract e.g. all persons from a newspaper corpus automatically. Note that in R, you can use the *quanteda* function `as.tokens` to directly use the Spacy output in quanteda.

As you can see, nowadays there are a number of good and relatively easy to use linguistic toolkits that can be used. Especially *Stanza* (Qi et al., 2020) is also a very good and flexible toolkit with support for multiple (human) languages and good integration especially with Python. If you want to learn more about natural language processing, the book *Speech and Language Processing* by Jurafsky and Martin is a very good starting point (Jurafsky and Martin, 2009)[2].

10.4 Which Preprocessing to Use?

This chapter has shown how to create a DTM and especially introduced a number of different steps that can be used to clean and preprocess the DTM before analysis. All of these steps are used by text analysis practitioners and in the relevant literature.

2 See https://web.stanford.edu/~jurafsky/slp3/ for their draft of a new edition, which is (at the time of writing) free to download.

However, no study ever uses all of these steps on top of each other. This of courses raises the question of how to know which preprocessing steps to use for your research question.

First, there are a number of things that you should (almost) always do. If your data contains noise such as boilerplate language, HTML artifacts, etc., you should generally strip these out before proceeding. Second, text almost always has an abundance of uninformative (stop) words and a very long tail of very rare words. Thus, it is almost always a good idea to use a combination of stop word removal, trimming based on document frequency, and/or tf·idf weighting. Note that when using a stop word list, you should always manually inspect and/or fine-tune the word list to make sure it matches your domain and research question.

The other steps such as n-grams, collocations, and tagging and lemmatization are more optional but can be quite important depending on the specific research. For this (and for choosing a specific combination of trimming and weighting), it is always good to know your domain well, look at the results, and think whether you think they make sense. Using the example given above, bigrams can make more sense for sentiment analysis (since *not good* is quite different from *good*), but for analyzing the topic of texts it may be less important.

Ultimately, however, many of these questions have no good theoretical answer, and the only way to find a good preprocessing pipeline for your research question is to try many different options and see which works best. This might feel like "cheating" from a social science perspective, since it is generally frowned upon to just test many different statistical models and report on what works best. There is a difference, however, between substantive statistical modeling where you actually want to understand the mechanisms, and technical processing steps where you just want the best possible measurement of an underlying variable (presumably to be used in a subsequent substantive model). Lin (2015) uses the analogy of the mouse trap and the human condition: in engineering you want to make the best possible mouse trap, while in social science we want to understand the human condition. For the mouse trap, it is OK if it is a black box for which we have no understanding of how it works, as long as we are sure that it does work. For the social science model, this is not the case as it is exactly the inner workings we are interested in.

Technical (pre)processing steps such as those reviewed in this chapter are primarily engineering devices: we don't really care how something like tf·idf works, as long as it produces the best possible measurement of the variables we need for our analysis. In other words, it is an engineering challenge, not a social science research question. As a consequence, the key criterion by which to judge these steps is validity, not explainability. Thus, it is fine to try out different options, as long as you validate the results properly. If you have many different choices to evaluate against some metric such as performance on a subsequent prediction task, using the split-half or cross-validation techniques discussed in Chapter 8 are also relevant here to avoid biasing the evaluation.

11

Automatic Analysis of Text

Abstract

In this chapter, we discuss different approaches to the automatic analysis of text; or automated content analysis. We combine techniques from earlier chapters, such as transforming texts into a matrix of term frequencies and machine learning. In particular, we describe three different approaches (dictionary-based analyses, supervised machine learning, unsupervised machine learning). The chapter provides guidance on how to conduct such analyses, and also on how to decide which of the approaches is most suitable for which types of question.

Keywords dictionary approaches, supervised machine learning, unsupervised machine learning, topic models, automated content analysis, sentiment analysis

---

- Understand different approaches to automatic analysis of text
- Be able to decide on whether to use a dictionary approach, supervised machine learning, or unsupervised machine learning
- Be able to use these techniques

Packages used in this chapter

This chapter uses the basic text and data handling that were described in Chapter 10 (*tidyverse*, *readtext*, and *quanteda* for R, *pandas* and *nltk* for Python). For supervised text analysis, we use *quanteda.textmodels* (R) and *sklearn* and keras (Python). For topic models we use *topicmodels* (R) and *gensim* (Python). You can install these packages with the code below if needed (see Section 1.4 for more details):

Python Code

```
!pip3 install nltk scikit-learn pandas
!pip3 install gensim eli5 keras tensorflow
```

R Code

```
install.packages(c("tidyverse", "readtext",
    "quanteda", "quanteda.textmodels",
    "topicmodels", "keras", "topicdoc"))
```

After installing, you need to import (activate) the packages every session:

Computational Analysis of Communication: A Practical Introduction to the Analysis of Texts, Networks, and Images with Code Examples in Python and R, First Edition. Wouter van Atteveldt, Damian Trilling & Carlos Arcila Calderón.

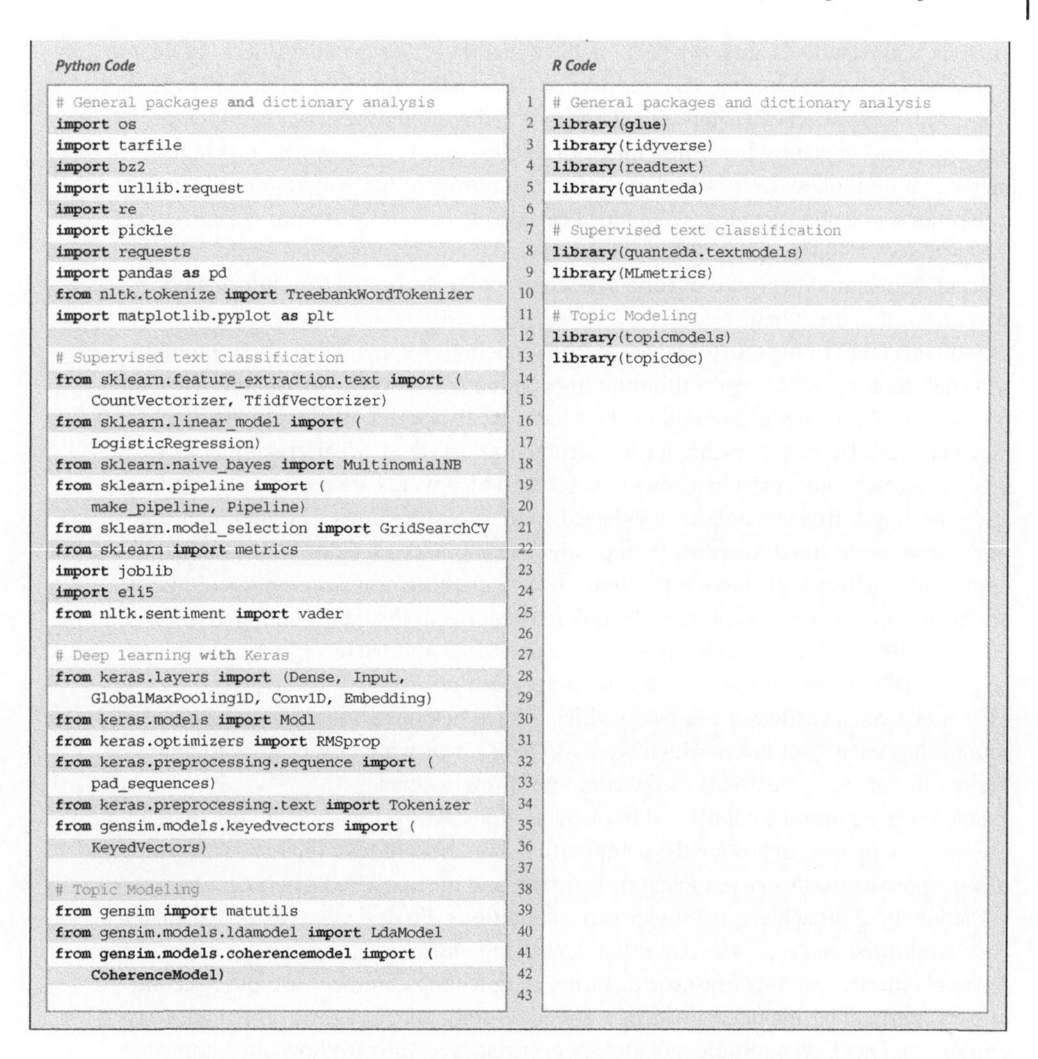

Python Code

```python
# General packages and dictionary analysis
import os
import tarfile
import bz2
import urllib.request
import re
import pickle
import requests
import pandas as pd
from nltk.tokenize import TreebankWordTokenizer
import matplotlib.pyplot as plt

# Supervised text classification
from sklearn.feature_extraction.text import (
    CountVectorizer, TfidfVectorizer)
from sklearn.linear_model import (
    LogisticRegression)
from sklearn.naive_bayes import MultinomialNB
from sklearn.pipeline import (
    make_pipeline, Pipeline)
from sklearn.model_selection import GridSearchCV
from sklearn import metrics
import joblib
import eli5
from nltk.sentiment import vader

# Deep learning with Keras
from keras.layers import (Dense, Input,
    GlobalMaxPooling1D, Conv1D, Embedding)
from keras.models import Modl
from keras.optimizers import RMSprop
from keras.preprocessing.sequence import (
    pad_sequences)
from keras.preprocessing.text import Tokenizer
from gensim.models.keyedvectors import (
    KeyedVectors)

# Topic Modeling
from gensim import matutils
from gensim.models.ldamodel import LdaModel
from gensim.models.coherencemodel import (
    CoherenceModel)
```

R Code

```r
# General packages and dictionary analysis
library(glue)
library(tidyverse)
library(readtext)
library(quanteda)

# Supervised text classification
library(quanteda.textmodels)
library(MLmetrics)

# Topic Modeling
library(topicmodels)
library(topicdoc)
```

In earlier chapters, you learned about both supervised and unsupervised machine learning as well about dealing with texts. This chapter brings together these elements and discusses how to combine them to automatically analyze large corpora of texts. After presenting guidelines for choosing an appropriate approach in Section 11.1 and downloading an example dataset in Section 11.2, we discuss multiple techniques in detail. We begin with a very simple top-down approach in Section 11.3, in which we count occurrences of words from an *a priori* defined list of words. In Section 11.4, we still use pre-defined categories that we want to code, but let the machine "learn" the rules of the coding itself. Finally, in Section 11.5, we employ a bottom-up approach in which we do not use any *a priori* defined lists or coding schemes, but inductively extract topics from our data.

11.1 Deciding on the Right Method

When thinking about the computational analysis of texts, it is important to realize that there is no method that is *the one* to do so. While there are good choices and bad choices, we also cannot say that one method is necessarily and always superior to another. Some methods are more fashionable than others. For instance, there has been a growing interest in topic models (see Section 11.5) in the last years. There are indeed very good applications for such models, they are also sometimes applied to research questions and/or data where they make much less sense. As always, the choice of method should follow the research question and not the other way round. We therefore caution you about reading Chapter 11 selectively because you want, for instance, to learn about supervised machine learning or about unsupervised topic models. Instead, you should be aware of very different approaches to make an informed decision on what to use when.

Boumans and Trilling (2016) provide useful guidelines for this. They place automatic text analysis approaches on a continuum from deductive (or top-down) to inductive (or bottom-up). At the deductive end of the spectrum, they place dictionary approaches (Section 11.3). Here, the researcher has strong *a priori* (theoretical) assumptions (for instance, which topics exist in a news data set; or which words are positive or negative) and can compile lists of words or rules based on these assumptions. The computer then only needs to execute these rules. At the inductive end of the spectrum, in contrast, lie approaches such as topic models (Section 11.5) where little or no *a priori* assumptions are made, and where we exploratively look for patterns in the data. Here, we typically do not know which topics exist in advance. Supervised approaches (Section 11.4) can be placed in between: here, we do define categories *a priori* (we do know which topics exist, and given an article, we know to which topic it belongs), but we do not have any set of rules: we do not know which words to look for or which exact rules to follow. These rules are to be "learned" by the computer from the data.

Before we get into the details and implementations, let us discuss some use cases of the three main approaches for the computational analysis of text: dictionary (or rule-based) approaches, supervised machine learning, and unsupervised machine learning.

Dictionary approaches excel under three conditions. First, the variable we want to code is *manifest and concrete* rather than *latent and abstract*: names of actors, specific physical objects, specific phrases, etc., rather than feelings, frames, or topics. Second, all synonyms to be included must be known beforehand. And third, the dictionary entries must not have multiple meanings. For instance, coding for how often gun control is mentioned in political speeches fits these criteria. There are only so many ways to talk about it, and it is rather unlikely that speeches about other topics contain a phrase like "gun control". Similarly, if we want to find references to Angela Merkel, Donald Trump, or any other well-known politician, we can just directly search for their names – even though problems arise when people have very common surnames and are referred to by their surnames only.

Sadly, most interesting concepts are more complex to code. Take a seemingly straightforward problem: distinguishing whether a news article is about the economy or not. This is really easy to do for humans: there may be some edge cases, but in general, people rarely need longer than a few seconds to grasp whether an article is about the economy rather than about sports, culture, etc. Yet, many of these articles won't directly state that they are about the economy by explicitly using the word "economy".

We may think of extending our dictionary not only with `econom.+` (a regular expression that includes economists, economic, and so on), but also come up with other words like "stock exchange", "market", "company." Unfortunately, we will quickly run into a problem that we also faced when we discussed the precision-recall trade-off in Section 8.5: the more terms we add to our dictionary, the more false positives we will get: articles about the geographical space called "market", about some celebrity being seen in "company" of someone else, and so on.

From this example, we can conclude that often (1) it is easy for humans to decide to which class a text belongs, but (2) it is very hard for humans to come up with a list of words (or rules) on which their judgment is based. Such a situation is perfect for applying supervised machine learning: after all, it won't take us much time to annotate, say, 1000 articles based on whether they are about the economy or not (probably this takes less time than thoroughly fine tuning a list of words to include or exclude); and the difficult part, deciding on the exact rules underlying the decision to classify an article as economic is done by the computer in seconds. Supervised machine learning, therefore, has replaced dictionary approaches in many areas.

Both dictionary (or rule-based) approaches and supervised machine learning assume that you know in advance which categories (positive versus negative; sports versus economy versus politics; ...) exist. The big strength of unsupervised approaches such as topic models is that you can also apply them without this knowledge. They therefore allow you to find patterns in data that you did not expect and can generate new insights. This makes them particularly suitable for explorative research questions. Using them for confirmatory tests, in contrast, is less defensible: after all, if we are interested in knowing whether, say, news site A published more about the economy than news site B, then it would be a bit weird to pretend not to know that the topic "economy" exists. Also practically, mapping the resulting topics that the topic model produces onto such *a priori* existing categories can be challenging.

Despite all differences, all approaches share one requirement: you need to "Validate. Validate. Validate" (Grimmer and Stewart, 2013). Though it has been done in the past, simply applying a dictionary without comparing the performance to manual coding of the same concepts is not acceptable; neither is using a supervised machine learning classifier without doing the same; or blindly trusting a topic model without at least manually checking whether the scores the model assigns to documents really capture what the documents are about.

11.2 Obtaining a Review Dataset

For the sections on dictionary and supervised approaches we will use a dataset of movie reviews from the IMDB database (Maas et al., 2011). This dataset is published as a compressed set of folders, with separate folders for the train and test datasets and subfolders for positive and negative reviews. Lots of other review datasets are available online, for example for Amazon review data (https://jmcauley.ucsd.edu/data/amazon/).

The IMDB dataset we will use is a relatively large file and it requires bit of processing, so it is smart to *cache* the data rather than downloading and processing it every time you need it. This is done in Example 11.1, which also serves as a nice example of how to download and process files. Both R and Python follow the same basic pattern. First, we check whether the cached file exists, and if it does we read the data from that file. For R, we use the standard *RDS* format, while for Python we use a compressed *pickle* file. The format of the data is also slightly different, following the convention for each language: In R we use the data frame returned by `readtext`, which can read files from a folder or zip archive and return a data frame containing

Python Code

```python
filename = "reviewdata.pickle.bz2"
if os.path.exists(filename):
  print(f"Using cached file {filename}")
  with bz2.BZ2File(filename, "r") as zipfile:
    data = pickle.load(zipfile)
    text_train, text_test, y_train, y_test = data
else:
  url = "https://cssbook.net/d/aclImdb_v1.tar.gz"
  print(f"Downloading from {url}")
  fn, _headers = urllib.request.urlretrieve(url,
                                             filename=None)
  t = tarfile.open(fn, mode="r:gz")
  text_train,text_test = [], []
  y_train, y_test = [], []
  for f in t.getmembers():
    m=re.match("aclImdb/(\w+)/(pos|neg)/", f.name)
    if not m:
      # skip folder names, other categories
      continue
    dataset, label = m.groups()
    text = t.extractfile(f).read().decode("utf-8")
    if dataset == "train":
      text_train.append(text)
      y_train.append(label)
    elif dataset == "test":
      text_test.append(text)
      y_test.append(label)
print(f"Saving to {filename}")
with bz2.BZ2File(filename, "w") as zipfile:
    data = text_train, text_test, y_train, y_test
    pickle.dump(data, zipfile)
```

R Code

```r
filename = "reviewdata.rds"
if (file.exists(filename)) {
    print("Using cached data")
    reviewdata= readRDS(filename)
} else {
    print("Downloading data")
    fn = "aclImdb_v1.tar.gz"
    url = glue("https://cssbook.net/d/{fn}")
    download.file(url, fn)
    untar(fn)
    reviewdata = readtext(
      file.path("aclImdb", "*", "*", "*.txt"),
      docvarsfrom = "filepaths", dvsep="[/\\]",
      docvarnames=c("i","dataset","label","fn"))
    unlink(c("aclImdb", fn), recursive=TRUE)
    reviewdata = reviewdata %>%
      filter(label %in% c("pos", "neg")) %>%
      select(-i) %>%
      corpus()
    saveRDS(reviewdata, filename)
}
head(docvars(reviewdata))
```

R Output

	dataset <chr>	label <chr>	filename <chr>
1	test	neg	0_2.txt
2	test	neg	1_3.txt
3	test	neg	10_3.txt
4	test	neg	100_4.txt
5	test	neg	1000_3.txt
6	test	neg	10000_4.txt

Example 11.1 Downloading and caching IMDB review data.

one text per row. In Python, we have separate lists for the train and test datasets and for the full texts and labels: `text_train` are the training texts and `y_train` are the corresponding labels.

If the cached data file does not exist yet, the file is downloaded from the Internet. In R, we then extract the file and call `readtext` on the resulting folder. This automatically creates columns for the subfolders, so in this case for the dataset and label. After this, we remove the download file and the extracted folder, clean up the `reviewdata`, and save it to the `reviewdata.rds` file. In Python, we can extract files from the downloaded file directly, so we do not need to explicitly extract it. We loop over all files in the archive, and use a regular expression to select only text files and extract the label and dataset name (see Section 9.2 for more information about regular expressions). Then, we extract the text from the archive, and add the text and the label to the appropriate list. Finally, the data is saved as a compressed pickle file, so the next time we run this cell it does not need to download the file again.

11.3 Dictionary Approaches to Text Analysis

A straightforward way to automatically analyze text is to compile a list of terms you are interested in and simply count how often they occur in each document. For example, if you are interested in finding out whether mentions of political parties in news articles change over the years, you only need to compile a list of all party names and write a small script to count them.

Historically, this is how sentiment analysis was done. Example 11.2 shows how to do a simple sentiment analysis based on a list of positive and negative words. The logic is straightforward: you count how often each positive word occurs in a text, you do the same for the negative words, and then determine which occur more often.

Python Code

```python
poswords = "https://cssbook.net/d/positive.txt"
negwords = "https://cssbook.net/d/negative.txt"
pos = set(requests.get(poswords).text.split("\n"))
neg = set(requests.get(negwords).text.split("\n"))
sentimentdict = {word:+1 for word in pos}
sentimentdict.update({word:-1 for word in neg})

scores = []
mytokenizer = TreebankWordTokenizer()
# For speed, we only take the first 100 reviews
for review in text_train[:100]:
    words = mytokenizer.tokenize(review)
    # we look up each word in the sentiment dict
    # and assign its value (with default 0)
    scores.append(sum(sentimentdict.get(word,0)
                      for word in words))
scores
```

R Code

```r
poswords = "https://cssbook.net/d/positive.txt"
negwords = "https://cssbook.net/d/negative.txt"
pos = scan(poswords, what="list")
neg = scan(negwords, what="list")
sentimentdict = dictionary(list(pos=pos, neg=neg))


# For speed, we only take the first 100 reviews
scores = corpus_sample(reviewdata, 100) %>%
    tokens() %>%
    dfm() %>%
    dfm_lookup(sentimentdict) %>%
    convert(to="data.frame") %>%
    mutate(sent = pos - neg)
head(scores)
```

Example 11.2 Different approaches to a simple dictionary-based sentiment analysis: counting and summing all words using a for-loop over all reviews (Python) versus constructing a term-document matrix and looking up the words in there (R). Note that both approaches would be possible in either language.

As you may already realize, there are a lot of downsides to this approach. Most notably, our bag-of-words approach does not allow us to account for negation: "not good" will be counted as positive. Relatedly, we cannot handle modifiers such as "very good". Also, all words are either positive or negative, while "great" should be more positive than "good". More advanced dictionary-based sentiment analysis packages like Vader (Hutto and Gilbert, 2014) or SentiStrength (Thelwall et al., 2012) include such functionalities. Yet, as we will discuss in Section 11.4, also these off-the-shelf packages perform very poorly in many sentiment analysis tasks, especially outside of the domains they were developed for. Dictionary-based sentiment analysis has been shown to be problematic when analyzing news content (e.g. Gonzalez-Bailon and Paltoglou, 2015, Boukes et al., 2019). They are problematic when accuracy at the sentence level is important, but may be satisfactory with longer texts for comparatively easy tasks such as movie review classification (Reagan et al., 2017), where there is clear ground truth data and the genre convention implies that the whole text is evaluative and evaluates one object (the film).

Still, there are many use cases where dictionary approaches work very well. Because your list of words can contain anything, not just positive or negative words, dictionary approaches have been used, for instance, to measure the use of racist words or swearwords in online fora (e.g., Tulkens et al., 2016). Dictionary approaches are simple to understand and straightforward, which can be a good argument for using them when it is important that the method is no black-box but fully transparent even without technical knowledge. Especially when the dictionary already exists or is easy to create, it is also a very cheap method. However, this is at the expense of their limitation to only performing well when measuring easy to operationalize concepts. To put it bluntly: it's great for measuring the visibility of parties or organizations in the news, but it's not good for measuring concepts such as emotions or frames.

What gave dictionary approaches a bit of a bad name is that many researchers applied them without validating them. This is especially problematic when a dictionary is applied in a slightly different domain than that for which it was originally made.

If you want to use a dictionary-based approach, we advise the following procedure:

1. Construct a dictionary based on theoretical considerations and by closely reading a sample of example texts.
2. Code some articles manually and compare with the automated coding.
3. Improve your dictionary and check again.
4. Manually code a validation dataset of sufficient size. The required size depends a bit on how balanced your data is – if one code occurs very infrequently, you will need more data.
5. Calculate the agreement. You could use standard intercoder reliability measures used in manual content analysis, but we would also advise you to calculate precision and recall (see Section 8.5).

Very extensive dictionaries will have a high recall (it becomes increasingly unlikely that you "miss" a relevant document), but often suffer from low precision (more documents will contain one of the words even though they are irrelevant). Vice versa, a very short dictionary will often be very precise, but miss a lot of documents. It depends on your research question where the right balance lies, but to substantially interpret your results, you need to be able to quantify the performance of your dictionary-based approach.

How many documents do you need to calculate agreement with human annotators?

To determine the number of documents one needs to determine the agreement between a human and a machine, one can follow the same standards that are recommended for traditional manual content analysis.

For instance, Krippendorff (2004) provides a convenience table to look up the required sample sizes for determining the agreement between two human coders (p. 240). Riffe et al. (2019) provide similar suggestions (p. 114). In short, the sample size depends on the level of statistical significance the researcher deems acceptable as well as on the distribution of the data. In an extreme case, if only 5 out of 100 items are to be coded as *x*, then in a sample of 20 items, such an item may not even occur. In order to determine agreement between the automated method and a human, we suggest that sample sizes that one would also use for the calculation of agreement between human coders are used. For specific calculations, we refer to content analysis books such as the two referenced here. To give a very rough ballpark figure (that shouldn't replace a careful calculation!), roughly 100 to 200 items will cover many scenarios (assuming a small amount of reasonably balanced classes).

11.4 Supervised Text Analysis: Automatic Classification and Sentiment Analysis

For many applications, there are good reasons to use the dictionary approach presented in the previous section. First, it is intuitively understandable and results can – in principle – even be verified by hand, which can be an advantage when transparency or communicability is of high importance. Second, it is very easy to use. But as we have discussed in Section 11.1, dictionary approaches in general perform less well the more abstract, non-manifest, or complex a concept becomes. In the next section, we will make the case that topics, but also sentiment, in fact, are quite a complex concepts that are often hard to capture with dictionaries (or at least, crafting a custom dictionary would be difficult). For instance, while "positive" and "negative" seem straightforward categories at first sight, the more we think about it, the more apparent it becomes how context-dependent it actually is: in a dataset about the economy and stock market returns, "increasing" may indicate something positive, in a dataset about unemployment rates the same word would be something negative. Thus, machine learning can be a more appropriate technique for such tasks.

As explained in the box on the next page, at the time of writing R has more limited support for supervised text classification. For this reason, we give one example in both Python and R, but the rest of this section uses Python examples only. Hopefully R support for text classification will be improved in the near future with the introduction of *tidymodels* or other packages.

11.4.1 Putting Together a Workflow

With the knowledge we gained in previous chapters, it is not difficult to set up a supervised machine learning classifier to automatically determine, for instance, the topic of a news article.

Let us recap the building blocks that we need. In Chapter 8, you learned how to use different classifiers, how to evaluate them, and how to choose the best settings. However, in these examples, we used numerical data as features; now, we have text. In Chapter 10, you learned how to turn text into numerical features. And that's all we need to get started!

Typical examples for supervised machine learning in the analysis of communication include the classification of topics (e.g., Scharkow, 2011), frames (e.g., Burscher et al., 2014), user characteristics such as gender or ideology, or sentiment.

Let us consider the case of sentiment analysis in more detail. Classical sentiment analysis is done with a dictionary approach: you take a list of positive words, a list of negative words, and count which occur more frequently. Additionally, one may attach a weight to each word, such that "perfect" gets a higher weight than "good", for instance. An obvious drawback is that these pure bag-of-words approaches cannot cope with negation ("not good") and intensifiers ("very good"), which is why extensions have been developed that take these (and other features, such as punctuation) into account (Thelwall et al., 2012, Hutto and Gilbert, 2014, De Smedt et al., 2012).

But while available off-the-shelf packages that implement these extended dictionary-based methods are very easy to use (in fact, they spit out a sentiment score with one single line of code), it is questionable how well they work in practice. After all, "sentiment" is not exactly a clear, manifest concept for which we can enumerate a list of words. It has been shown that results obtained with multiple of these packages correlate very poorly with each other and with human annotations (Boukes et al., 2019, Chan et al., 2021).

Consequently, it has been suggested that it is better to use supervised machine learning to automatically code the sentiment of texts (Gonzalez-Bailon and Paltoglou, 2015, Vermeer et al., 2019). However, you may need to annotate documents from your own dataset: training a classifier on, for instance, movie reviews and then using it to predict sentiment in political texts violates the assumption that training set, test set, and the unlabeled data that are to be classified are (at least in principle and approximately) drawn from the same population.

To illustrate the workflow, we will use the ACL IMDB dataset, a large dataset that consists of a training dataset of 25 000 movie reviews (of which 12 500 are positive and 12 500 are negative) and an equally sized test dataset (Maas et al., 2011). It can be downloaded at https://ai.stanford.edu/~amaas/data/sentiment/aclImdb_v1.tar.gz

These data do not come in one file, but rather in a set of textfiles that are sorted in different folders named after the dataset to which they belong (`test` or `train`) and

Sparse versus dense matrices and why it matters a lot for choosing between R and Python for machine learning

In a document-term matrix, you would typically find a lot of zeros: most words do *not* appear in any given document. For instance, the reviews in the IMDB dataset contain more than 100 000 unique words. Hence, the matrix has more than 100 000 columns. Yet, most reviews only consist of a couple of hundred words. As a consequence, more than 99% of the cells in the table contain a zero. In a sparse matrix, we do not store all these zeros, but only store the values for cells that actually contain a value. This drastically reduces the memory needed. But even if you have a huge amount of memory, this does not solve the issue: in R, the number of cells in a matrix is limited to 2 147 483 647. It is therefore impossible to store a matrix with 100000 features and 25000 documents as a dense matrix. Unfortunately, many models that you can run via *caret* in R will convert your sparse document-term matrix to a dense matrix, and hence are effectively only usable for very small datasets. An alternative is using the *quanteda* package, which does use sparse matrices throughout. However, at the time of writing this book, quanteda only provides a very limited number of models. As all of these problems do not arise in *scikit-learn*, you may want to consider using Python for many text classification tasks.

their label (pos and neg). This means that we cannot simply use a pre-defined function to read them, but we need to think of a way of reading the content into a data structure that we can use. This data was loaded in Example 11.1 above.

Let us now train our first classifier. We choose a Naïve Bayes classifier with a simple count vectorizer (Example 11.3). In the Python example, pay attention to the fitting of the vectorizer: we fit on the training data *and* transform the training data with it, but we only transform the test data *without re-fitting the vectorizer*. Fitting, here, includes the decision about which words to include (by definition, words that are not present in the training data are not included; but we could also choose additional constraints, such as excluding very rare or very common words), but also assigning an (internally used) identifier (variable name) to each word. If we fit the classifier again, these would not be compatible any more. In R, the same is achieved in a slightly different way: two term-document

Python Code

```
vectorizer = CountVectorizer(stop_words="english")
X_train = vectorizer.fit_transform(text_train)
X_test = vectorizer.transform(text_test)

nb = MultinomialNB()
nb.fit(X_train, y_train)

y_pred = nb.predict(X_test)

rep=metrics.classification_report(y_test, y_pred)
print(rep)
```

R Code

```
1  dfm_train = reviewdata %>%
2    corpus_subset(dataset == "train") %>%
3    tokens() %>%
4    dfm() %>%
5    dfm_trim(min_docfreq=0.01, docfreq_type="prop")
6
7  dfm_test = reviewdata %>%
8    corpus_subset(dataset == "test") %>%
9    tokens() %>%
10   dfm() %>%
11   dfm_match(featnames(dfm_train))
12
13 myclassifier = textmodel_nb(dfm_train,
14                             docvars(dfm_train, "label"))
15
16 predicted = predict(myclassifier,newdata=dfm_test)
17 actual = docvars(dfm_test, "label")
18
19 results = list()
20 for (label in c("pos", "neg")){
21   results[[label]] = tibble(
22     Precision=Precision(actual, predicted, label),
23     Recall=Recall(actual, predicted, label),
24     F1=F1_Score(actual, predicted, label))
25 }
26 bind_rows(results, .id="label")
```

Python Output

```
              precision    recall  f1-score   support
         neg       0.79      0.88      0.83     12500
         pos       0.86      0.76      0.81     12500

    accuracy                           0.82     25000
   macro avg       0.82      0.82      0.82     25000
weighted avg       0.82      0.82      0.82     25000
```

Example 11.3 Training a Naïve Bayes classifier with simple word counts as features.

> **Note:** A word that is not present in the training data, but is present in the test data, is thus ignored. If you want to use the information such out-of-vocabulary words can entail (e.g., they may be synonyms), you could consider using a word embedding approach (see Section 10.3.3).

matrices are created independently, before they are matched in such a way that only the features that are present in the training matrix are retained in the test matrix.

We do not necessarily expect this first model to be the best classifier we can come up with, but it provides us with a reasonable baseline. In fact, even without any further adjustments, it works reasonably well: precision is higher for positive reviews and recall is higher for negative reviews (classifying a positive review as negative happens twice as much as the reverse), but none of the values is concerningly low.

11.4.2 Finding the Best Classifier

Let us start by comparing two simple classifiers we know (Naïve Bayes and Logistic Regression, see Section 8.3) and the two vectorizers that transform our texts into two numerical representations that we know: word counts and *tf · idf* scores (see Chapter 10).

We can also tune some things in the vectorizer, such as filtering out stopwords, or specifying a minimum number (or proportion) of documents in which a word needs to occur in order to be included, or the maximum number (or proportion) of documents in which it is allowed to occur. For instance, it could make sense to say that a word that occurs in less than $n=5$ documents is probably a spelling mistake or so unusual that it just unnecessarily bloats our feature matrix; and on the other hand, a word that is so common that it occurs in more than 50% of all documents is so common that it does not help us to distinguish between different classes.

We can try all of these things out by hand by just re-running the code from Example 11.3 and only changing the line in which the vectorizer is specified and the line in which the classifier is specified. However, copy-pasting essentially the same code is generally not a good idea, as it makes your code unnecessary long and increases the likelihood of errors creeping in when you, for instance, need to apply the same changes to multiple copies of the code. A more elegant approach is outlined in Example 11.4: We define a function that gives us a short summary of only the output we are interested in, and then use a for-loop

Python Code

```
1 def short_classification_report (y_test, y_pred):
2   print("     \t precision\tRecall")
3   for label in set(y_pred):
4     pr = metrics.precision_score(y_test, y_pred,
5                                  pos_label=label)
6     re = metrics.recall_score(y_test,y_pred,
7                                  pos_label=label)
8     print(f"{label}:\t{pr:0.2f}\t\t{re:0.2f}")
9
10 configs = [
11     ("NB with Count", CountVectorizer(min_df=5, max_df=.5),
12      MultinomialNB()),
13     ("NB with TfIdf", TfidfVectorizer(min_df=5, max_df=.5),
14      MultinomialNB()),
15     ("LogReg with Count", CountVectorizer(min_df=5, max_df=.5),
16      LogisticRegression(solver="liblinear")),
17     ("LogReg with TfIdf", TfidfVectorizer(min_df=5, max_df=.5),
18      LogisticRegression(solver="liblinear"))]
19
20 for name, vectorizer, classifier in configs:
21     print(name)
22     X_train = vectorizer.fit_transform(text_train)
23     X_test = vectorizer.transform(text_test)
24     classifier.fit(X_train, y_train)
25     y_pred = classifier.predict(X_test)
26     short_classification_report(y_test, y_pred)
27     print("\n")
```

Output

```
NB with Count
                      precision           recall
positive reviews:     0.87                0.77
negative reviews:     0.79                0.88

NB with TfIdf
                      precision           recall
positive reviews:     0.87                0.78
negative reviews:     0.80                0.88

LogReg with Count
                      precision           recall
positive reviews:     0.87                0.85
negative reviews:     0.85                0.87

LogReg with TfIdf
                      precision           recall
positive reviews:     0.89                0.88
negative reviews:     0.88                0.89
```

Example 11.4 An example of a custom function to give a brief overview of the performance of four simple vectorizer-classifier combinations.

to iterate over all configurations we want to evaluate, fit them and call the function we defined before. With just a few lines more than we need to evaluate one single model (as done in Example 11.3), we can compare four different ones.

The output of this little example already gives us quite a bit of insight into how to tackle our specific classification tasks: first, we see that a $tf \cdot idf$ classifier seems to be slightly but consistently superior to a count classifier (this is often, but not always the case). Second, we see that the logistic regression performs better than the Naïve Bayes classifier (again, this is often, but not always, the case). In particular, in our case, the logistic regression improved on the excessive misclassification of positive reviews as negative, and achieves a very balanced performance.

There may be instances where one nevertheless may want to use a Count Vectorizer with a Naïve Bayes classifier instead (especially if it is too computationally expensive to estimate the other model), but for now, we may settle on the best performing combination, logistic regression with a $tf \cdot idf$ vectorizer. You could also try fitting a Support Vector Machine instead, but we have little reason to believe that our data isn't linearly separable, which means that there is little reason to believe that a SVM with a specific kernel will perform better. Given the good performance we already achieved, we decide to stick to the logistic regression for now.

We can now go as far as we like, include more models, use crossvalidation and gridsearch (see Section 8.5.3), etc. However, our workflow now consists of *two* steps: fitting/transforming our input data using a vectorizer, and fitting a classifier. To make things easier, in *scikit-learn*, both steps can be combined into a so-called pipe. Example 11.5 shows how the loop in Example 11.4 can be re-written using pipes (the result stays the same).

Such a pipeline lends itself very well to performing a gridsearch. Example 11.6 gives you an example. With `LogisticRegression?` and `TfIdfVectorizer?`, we can get a list of all possible hyperparameters that we may want to tune. For instance, these could be the minimum and maximum frequency for words to be included or whether

Python Code

```
1 for name, vectorizer, classifier in configs:
2     print(name)
3     pipe = make_pipeline(vectorizer, classifier)
4     pipe.fit(text_train, y_train)
5     y_pred = pipe.predict(text_test)
6     short_classification_report(y_test, y_pred)
7     print("\n")
```

Example 11.5 Instead of fitting vectorizer and classifier separately, they can be combined in a pipeline.

Python Code

```
1 pipeline = Pipeline(steps = [
2     ("vectorizer", TfidfVectorizer()),
3     ("classifier", LogisticRegression(solver="liblinear"))])
4 grid = {"vectorizer__ngram_range" : [(1,1), (1,2)],
5         "vectorizer__max_df": [0.5, 1.0],
6         "vectorizer__min_df": [0, 5],
7         "classifier__C": [0.01, 1, 100]
8        }
9 search = GridSearchCV(estimator=pipeline, n_jobs=-1,
10    param_grid=grid,scoring="accuracy", cv=5)
11 search.fit(text_train, y_train)
12 print(f"Best parameters: {search.best_params_}")
13 pred = search.predict(text_test)
14 print(short_classification_report(y_test, pred))
```

Output

```
Fitting 5 folds for each of 24 candidates, totalling 120 fits
Using these hyperparameters {'classifier__C': 100, 'vectorizer__max_df': 0.5, 'vectorizer__min_df': 0,
   'vectorizer__ngram_range': (1, 2)
   }, we get the best performance:
                        precision     recall
positive reviews:       0.90          0.90
negative reviews:       0.90          0.90
None
```

Example 11.6 A gridsearch to find the best hyperparameters for a pipeline consisting of a vectorizer and a classifier. Note that we can tune any parameter that either the vectorizer or the classifier accepts as an input, not only the four hyperparameters we chose in this example.

we want to use only unigrams (single words) or also bigrams (combinations of two words, see Section 10.3). For the Logistic Regression, it may be the regularization hyperparameter C, which applies a penalty for too complex models. We can put all values for these parameters that we want to consider in a dictionary, with a descriptive key (i.e., a string with the step of the pipeline followed by two underscores and the name of the hyperparameter) and a list of all values we want to consider as the corresponding value.

The gridsearch procedure will then estimate all combinations of all values, using cross-validation (see Section 8.5). In our example, we have $2 \times 2 \times 2 \times 3 = 24$ different models, and 5 folds $\times$ 24 models $= 120$ models to estimate. Hence, it may take you some time to run the code.

We see that we could further improve our model to precision and recall values of 0.90, by excluding extremely infrequent and extremely frequent words, including both unigrams and bigrams (which, we may speculate, help us to account for the "not good" versus "not", "good" problem), and changing the default penalty of $C=1$ to $C=100$.

Let us, just for the sake of it, compare the performance of our model with an off-the-shelf sentiment analysis package, in this case Vader (Hutto and Gilbert, 2014). For any text, it will directly estimate sentiment scores (more specifically, a positivity score, a negativity score, a neutrality score, and a compound measure that combines them), without any need to have training data. However, as Example 11.7 shows, such a method is clearly inferior to a supervised machine learning approach. While in almost all cases (except for $n=11$ cases), Vader was able to make a choice (getting scores of 0 is a notorious problem in very short texts), precision and recall are clearly worse than even the simple baseline model we started with, and much worse than those of the final model we finished with. In fact, we miss half (!) of the negative reviews. There are probably very few applications in the analysis of communication in which we would find this acceptable. It is important to highlight that this is not because the off-the-shelf package we chose is a particularly bad one (on the contrary, it is actually comparatively good), but because of the inherent limitations of dictionary-based sentiment analysis.

We need to keep in mind, though, that with this dataset, we chose one of the easiest sentiment analysis tasks: a set of long, rather formal texts (compared to informal short social media messages), that evaluate exactly one entity (one film), and that are not ambiguous at all. Many applications that communication scientists are interested in are much less straightforward. Therefore, however tempting it may be to use an off-the-shelf package, doing so requires a thorough test based on at least some human-annotated data.

Python Code

```
1 nltk.download("vader_lexicon")
2 analyzer = SentimentIntensityAnalyzer()
3 pred = []
4 for review in text_test:
5     sentiment = analyzer.polarity_scores(review)
6     if sentiment["compound"]>0:
7         pred.append("pos")
8     elif sentiment["compound"]<0:
9         pred.append("neg")
10    else:
11        pred.append("dont know")
12
13 print(metrics.confusion_matrix(y_test, pred))
14 print(metrics.classification_report(y_test, pred))
```

Output

```
[[   0     0     0]
 [   6  6688  5806]
 [   5  1745 10750]]
              precision recall f1-score support

   dont know      0.00    0.00    0.00        0
         neg      0.79    0.54    0.64    12500
         pos      0.65    0.86    0.74    12500

    accuracy                      0.70    25000
   macro avg      0.48    0.47    0.46    25000
weighted avg      0.72    0.70    0.69    25000
```

Example 11.7 For the sake of comparison, we calculate how an off-the-shelf sentiment analysis package would have performed in this task.

11.4.3 Using the Model

So far, we have focused on training and evaluating models, almost forgetting why we were doing this in the first place: to use them to predict the label for new data that we did not annotate.

Of course, we could always re-train the model when we need to use it – but that has two downsides: first, as you may have seen, it may actually take considerable time to train it, and second, you need to have the training data available, which may be a problem both in terms of storage space and of copyright and/or privacy if you want to share your classifier with others.

Therefore, it makes sense to save both our classifier and our vectorizer to a file, so that we can reload them later (Example 11.8). Keep in mind that you have to re-use *both* – after all, the columns of your feature matrix will be different (and hence, completely useless for the classifier) when fitting a new vectorizer. Therefore, as you see, you do not do any fitting any longer, and only use the `.transform()` method of the (already fitted) vectorizer and the `.predict()` method of the (already fitted) classifier.

In R, you have no vectorizer you could save – but because in contrast to Python, both your DTM and your classifier include the feature names, it suffices to save the classifier only (using `saveRDS(myclassifier, "myclassifier.rds")`) and using on a new DTM later on. You do need to remember, though, how you constructed the DTM (e.g., which preprocessing steps you took), to make sure that the features are comparable.

Python Code

```
1  # Make a vectorizer and train a classifier
2  vectorizer=TfidfVectorizer(min_df=5, max_df=.5)
3  classifier=LogisticRegression(solver="liblinear")
4  X_train=vectorizer.fit_transform(text_train)
5  classifier.fit(X_train, y_train)
6
7  # Save them to disk
8  with open("myvectorizer.pkl",mode="wb") as f:
9      pickle.dump(vectorizer, f)
10 with open("myclassifier.pkl",mode="wb") as f:
11     joblib.dump(classifier, f)
12
13 # Later on, re-load this classifier and apply:
14 new_texts = ["This is a great movie",
15              "I hated this one.",
16              "What an awful fail"]
17
18 with open("myvectorizer.pkl",mode="rb") as f:
19     myvectorizer = pickle.load(f)
20 with open("myclassifier.pkl",mode="rb") as f:
21     myclassifier = joblib.load(f)
22
23 new_features = myvectorizer.transform(new_texts)
24 pred = myclassifier.predict(new_features)
25
26 for review, label in zip(new_texts, pred):
27     print(f"'{review}' is probably '{label}'.")
```

Output

```
'This is a great movie' is probably 'pos'.
'I hated this one.' is probably 'neg'.
'What an awful fail' is probably 'neg'.
```

Example 11.8 Saving and loading a vectorizer and a classifier.

Another thing that we might want to do is to get a better idea of the features that the model uses to arrive at its prediction; in our example, what actually characterizes the best and the worst reviews. Example 11.9 shows how this can be done in one line of code using *eli5* – a package that aims to "*e*xplain [the model] *l*ike *I*'m 5 years old". Here, we re-use the `pipe` we constructed earlier to provide both the vectorizer and the classifier to *eli5* – if we had only provided the classifier, then the feature names would have been internal identifiers (which are meaningless to us) rather than human-readable words.

We can also use eli5 to explain how the classifier arrived at a prediction for a specific document, by using different shades of green and red to explain how much different features contributed to the classification, and in which direction (Example 11.10).

11.4.4 Deep Learning

Deep learning models were introduced in Section 8.4 as a (relatively) new class of models in supervised machine learning. Using the Python *keras* package you can define various model architectures such as Convolutional or Recurrent Neural Networks. Although it is beyond the scope of this chapter to give a detailed treatment of building and training deep learning models, in this section we do give an example of using a Convolutional Neural Network using pre-trained word embeddings. We would urge anyone who is interested in machine learning for text analysis to continue studying deep learning, probably starting with the excellent book by Goldberg (2017).

Python Code

```
1 pipe = make_pipeline(
2     TfidfVectorizer(min_df=5, max_df=.5),
3     LogisticRegression(solver="liblinear"))
4 pipe.fit(text_train, y_train)
5 eli5.show_weights(pipe, top = 10)
```

Output:

y=pos top features

Weight?	Feature
+7.173	great
+6.101	excellent
+5.055	best
+4.791	perfect
... 13663 more positive ...	
... 13574 more negative ...	
-5.337	poor
-5.733	boring
-6.315	waste
-6.349	awful
-7.347	bad
-9.059	worst

Example 11.9 Using eli5 to understand text classification results.

Python Code

```
eli5.show_prediction(classifier, txt_test[0],
                     vec=vectorizer, targets=["pos"])
```

Output:

y=pos (probability **0.140**, score **-1.817**) top features

Contribution?	Feature
+0.013	<BIAS>
-1.830	Highlighted in text (sum)

i love sci-fi and am willing to put up with a lot. sci-fi movies/tv are usually underfunded, under-appreciated and misunderstood. i tried to like this, i really did, but it is to good tv sci-fi as babylon 5 is to star trek (the original). silly prosthetics, cheap cardboard sets, stilted dialogues, cg that doesn't match the background, and painfully one-dimensional characters cannot be overcome with a 'sci-fi' setting. (i'm sure there are those of you out there who think babylon 5 is good sci-fi tv. it's not. it's clichéd and uninspiring.) while us viewers might like emotion and character development, sci-fi is a genre that does not take itself seriously (cf. star trek). it may treat important issues, yet not as a serious philosophy. it's really difficult to care about the characters here as they are not simply foolish, just missing a spark of life. their actions and reactions are wooden and predictable, often painful to watch. the makers of earth know it's rubbish as they have to always say "gene roddenberry's earth..." otherwise people would not continue watching. roddenberry's ashes must be turning in their orbit as this dull, cheap, poorly edited (watching it without advert breaks really brings this home) trudging trabant of a show lumbers into space. spoiler. so, kill off a main character. and then bring him back as another actor. jeeez! dallas all over again.

Example 11.10 Using eli5 to explain a prediction.

Impressively, in R you can now also use the *keras* package to train deep learning models, as shown in the example. Similar to how *spacyr* works (Section 10.3.4), the R package actually installs and calls Python behind the screens using the *reticulate* package. Although the resulting models are relatively similar, it is less easy to build and debug the models in R because most of the documentation and community examples are written in Python. Thus in the end, we probably recommend people who want to dive into deep learning should choose Python rather than R.

First, Example 11.11 loads a dataset described by Van Atteveldt et al. (2020b), which consists of Dutch economic news headlines with a sentiment value. Next, in Example 11.12 a model is defined consisting of several layers, corresponding roughly to Figure 8.7. First, an *embedding* layer transforms the textual input into a semantic vector for each word. Next, the *convolutional* layer defines features (filters) over windows of vectors, which are then pooled in the *max-pooling* layer. This results in a vector of detected features for each document, which are then used in a regular (hidden) *dense* layer followed by an output layer.

Python Code

```
url="https://cssbook.net/d/dutch_sentiment.csv"
h = pd.read_csv(url)
h.head()
```

R Code

```
url="https://cssbook.net/d/dutch_sentiment.csv"
d = read_csv(url)
head(d)
```

R Output:

id <dbl>	value <dbl>	lemmata <chr>
10007	0	Rabobank voorspellen flink stijging hypotheekrente
10027	0	D66 willen reserve provincie aanspreken voor groei
10037	1	UWV dit jaar veel baan
10059	1	proost op geslaagd beursgang bols
10099	0	helft werknemer gaan na 65ste met pensioen
10101	1	Europa groeien voorzichtig dankzij laag energieprijs

Example 11.11 Dutch sentiment data (Modified from Van Atteveldt et al., 2020).

Python Code

```python
# Tokenize texts
tokenizer=Tokenizer(num_words=9999)
tokenizer.fit_on_texts(h.lemmata)
word_index=tokenizer.word_index
sequences=tokenizer.texts_to_sequences(h.lemmata)
tokens=pad_sequences(sequences, maxlen=1000)

# Prepare embeddings layer
fn = "w2v_320d_trimmed"
if not os.path.exists(fn):
    url = f"https://cssbook.net/d/{fn}"
    print(f"Downloading embeddings from {url}")
    urllib.request.urlretrieve(url, fn)
embeddings = KeyedVectors.load_word2vec_format(fn)
emb_matrix = np.zeros(
    (len(tokenizer.word_index) + 1,
     embeddings.vector_size))
for word, i in tokenizer.word_index.items():
    if word in embeddings:
        emb_matrix[i] = embeddings[word]
embedding_layer = Embedding(
   emb_matrix.shape[0], emb_matrix.shape[1],
   input_length=tokens.shape[1], trainable=True,
   weights=[emb_matrix])

print("Building CNN model")
sequence_input = Input(shape=(tokens.shape[1],),
                       dtype="int32")
seq = embedding_layer(sequence_input)
m = Conv1D(filters=128, kernel_size=3,
           activation="relu")(seq)
m = GlobalMaxPooling1D()(m)
m = Dense(64, activation="relu")(m)
preds = Dense(1, activation="tanh")(m)
m = Model(sequence_input, preds)
m.summary()
```

R Code

```r
text_vectorization = layer_text_vectorization(
     max_tokens=10000, output_sequence_length=50)
adapt(text_vectorization, d$lemmata)

input = layer_input(shape=1, dtype = "string")
output = input %>%
  text_vectorization() %>%
  layer_embedding(input_dim = 10000 + 1,
                  output_dim = 16) %>%
  layer_conv_1d(filters=128, kernel_size=3,
                activation="relu") %>%
  layer_global_max_pooling_1d() %>%
  layer_dense(units = 64, activation = "relu") %>%
  layer_dense(units = 1, activation = "tanh")

model = keras_model(input, output)
model
```

Output

```
Loading embeddings (this might take a while…)
Build CNN model
Model: "functional_1"
_________________________________________________________________
Layer (type) Output Shape Param #
=================================================================
input_1 (InputLayer) [(None, 1000)] 0
_________________________________________________________________
embedding (Embedding) (None, 1000, 320) 2176640
_________________________________________________________________
conv1d (Conv1D) (None, 998, 128) 123008
_________________________________________________________________
global_max_pooling1d (Global (None, 128) 0
_________________________________________________________________
dense (Dense) (None, 64) 8256
_________________________________________________________________
dense_1 (Dense) (None, 1) 65
=================================================================
Total params: 2,307,969
Trainable params: 2,307,969
Non-trainable params: 0
_________________________________________________________________
```

Example 11.12 Deep Learning: Defining and training a Convolutional Neural Network.

Finally, in Example 11.13 we train the model on 4000 documents and test it against the remaining documents. The Python model, which uses pre-trained word embeddings (the `w2v_320d` file downloaded at the top), achieves a mediocre accuracy of about 56% (probably due to the low number of training sentences). The R model, which trains the embedding layer as part of the model, performs more poorly at 44% as this model is even more dependent on large training data to properly estimate the embedding layer.

Python Code

```python
# Split data into train and test
train_data = tokens[:4000]
test_data = tokens[4000:]
train_labels = h.value[:4000]
test_labels = h.value[4000:]

# Train model
m.compile(loss="mean_absolute_error",
          optimizer=RMSprop(lr=.004))
labels = np.asarray([[x] for x in train_labels])
m.fit(train_data, labels, epochs=5, batch_size=128)

# Validate against test data
output = m.predict(test_data)
# Bin output into -1, 0, 1
pred=[1 if x[0]>.3 else (0 if x[0]>-.3 else -1)
      for x in output]
correct=[x==y for (x,y) in zip(pred, test_labels)]
acc = sum(correct) / len(pred)
print(f"Accuracy: {acc}")
```

R Code

```r
# Split data into train and test
d_train = d %>% slice_sample(n=4000)
d_test = d %>% anti_join(d_train)



# Train model
compile(model, loss = "binary_crossentropy",
        optimizer = "adam", metrics = "accuracy")
fit(model, d_train$lemmata, d_train$value,
    epochs = 10, batch_size = 512,
    validation_split = 0.2)
# Validate against test data
eval=evaluate(model, d_test$lemmata, d_test$value)
print(glue("Accuracy: {eval['accuracy']}"))
```

Python Output

```
Train model
Epoch 1/5
32/32 [==============================] - 27s
   832ms/step - loss:
0.7204
Epoch 2/5
32/32 [==============================] - 25s
   782ms/step - loss:
0.4850
Epoch 3/5
32/32 [==============================] - 27s
   842ms/step - loss:
0.3335
Epoch 4/5
32/32 [==============================] - 25s
   786ms/step - loss:
0.2358
Epoch 5/5
32/32 [==============================] - 26s
   806ms/step - loss:
0.1632
Validate against test data
Accuracy: 0.5831180017226529
```

R Output

```
Accuracy: 0.4422911
```

Example 11.13 Deep Learning: Training and Testing the model.

11.5 Unsupervised Text Analysis: Topic Modeling

In Section 7.3, we discussed how clustering techniques can be used to find patterns in data, such as which cases or respondents are most similar. Similarly, especially in survey research it is common to use factor analysis to discover (or confirm) variables that form a scale.

In essence, the idea behind these techniques is similar: by understanding the regularities in the data (which cases or variables behave similarly), you can describe the relevant information in the data with fewer data points. Moreover, assuming that regularities capture interesting information and the deviations from these regularities are mostly uninteresting noise, these clusters of cases or variables can actually be substantively informative.

Since a document-term matrix (DTM) is “just” a matrix, you can also apply these clustering techniques to the DTM to find groups of words or documents. You can therefore use any of the techniques we described in Chapter 7, and in particular clustering techniques such as *k*-means clustering (see Section 7.3) to group documents that use similar words together.

It can be very instructive to do this, and we encourage you to play around with such techniques. However, in recent years, a set of models called *topic models* have become especially popular for the unsupervised analysis of texts. Very much like what you would do with other unsupervised techniques, also in topic modeling, you group words and documents into “topics”, consisting of words and documents that co-vary. If you see the word “agriculture” in a news article, there is a good chance you might find words such as “farm” or “cattle”, and there is a lower chance you will find a word like “soldier”. In other words, the words “agriculture” and “farm” generally occur in the same kind of documents, so they can be said to be part of the same topic. Similarly, two documents that share a lot of words are probably about the same topic, and if you know what topic a document is on (e.g., an agricultural topic), you are better able to guess what words might occur in that document (e.g., “cattle”).

Thus, we can formulate the goal of topic modeling as: given a corpus, find a set of *n* topics, consisting of specific words and/or documents, that minimize the mistakes we would make if we try to reconstruct the corpus from the topics. This is similar to regression where we try to find a line that minimizes the prediction error.

In early research on document clustering, a technique called Latent Semantic Analysis (LSA) essentially used a factor analysis technique called Singular Value Decomposition (see Section 7.3.3) on the DTM. This has yielded promising results in information retrieval (i.e., document search) and studying human memory and language use. However, it has a number of drawbacks including factor loadings that can be difficult to interpret substantively and is not a good way of dealing with words that can have multiple meanings (Landauer et al., 2013).

11.5.1 Latent Dirichlet Allocation (LDA)

The most widely used technique for topic modeling is Latent Dirichlet Allocation (LDA). Although the goal of LDA is the same as for clustering techniques, it starts from the other end with what is called a *generative model*. A generative model is a (simplified) formal model of how the data is assumed to have been generated. For example, if we would have a standard regression model predicting income based on age and education level, the implicit generative model is that to determine someone's income, you take their age and education level, multiply them both by their regression parameters, and then add the intercept and some random error. Of course, we know that's not actually how most companies determine wages, but it can be a useful starting point to analyze, e.g., labor market discrimination.

The generative model behind LDA works as follows. Assume that you are a journalist writing a 500 word news item. First, you would choose one or more *topics* to write about, for example 70% healthcare and 30% economy. Next, for each word in the item, you randomly pick one of these topics based on their respective weight. Finally, you pick a random word from the words associated with that topic, where again each word has a certain probability for that topic. For example, "hospital" might have a high probability for healthcare while "effectiveness" might have a lower probability but could still occur.

As said, we know (or at least strongly suspect) that this is not how journalists actually write their stories. However, this generative model helps understand the substantive interpretation of topics. Moreover, LDA is a *mixture model*, meaning it allows for each document to be about multiple topics, and for each word to occur in multiple topics. This matches with the fact that in many cases, our documents are in fact about multiple topics, from a news article about the economic effects of the COVID virus to an open survey answer containing multiple reasons for supporting a certain candidate. Additionally, since topic assignment is based on what other words occur in a document, the word "pupil" could be assigned either to a "biology" topic or to an "education" topic, depending on whether the document talks about eyes and lenses or about teachers and classrooms.

Figure 11.1 is a more formal notation of the same generative model. Starting from the left, for each document you pick a set of topics Θ. This set of topics is drawn from a *Dirichlet distribution* which itself has a parameter α (see note). Next, for each word you select a single topic z from the topics in that document. Finally, you pick an actual word w from the words in that topic, again controlled by a parameter β.

Now, if we know which words and documents are in which topics, we can start generating the documents in the corpus. In reality, of course, we have the reverse situation: we know the documents, and we want to know the topics. Thus, the task of LDA is to find the parameters that have the highest chance of generating these documents.

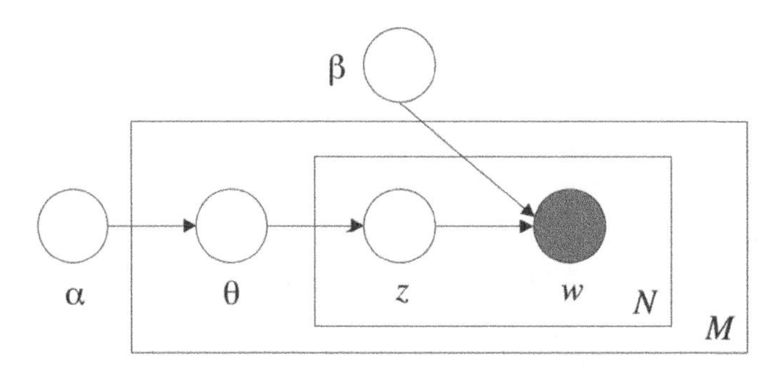

Figure 11.1 Latent Dirichlet Allocation in "Plate Model" notation (Blei et al., 2003, Fig. 1).

Since only the word frequencies are observed, this is a latent variable model where we want to find the most likely values for the (latent) topic z for each word in each document.

Unfortunately, there is no simple analytic solution to calculate these topic assignments like there is for OLS regression. Thus, like other more complicated statistical models such as multilevel regression, we need to use an iterative estimation that progressively optimizes the assignment to improve the fit until it converges.

An estimation method that is often used for LDA is Gibbs sampling. Simply put, this starts with a random assignment of topics to words. Then, in each iteration, it reconsiders each word and recomputes what likely topics for that word are given the other topics in that document and the topics in which that word occurs in other documents. Thus, if a document already contains a number of words placed in a certain topic, a new word is more likely to be placed in that topic as well. After enough iterations, this converges to a solution.

Note: The Dirichlet Distribution and its Hyperparameters

The Dirichlet distribution can be seen as a distribution over multinomial distributions, that is, every draw from a Dirichlet distribution results in a multinomial distribution. An easy way to visualize this is to see the Dirichlet distribution as a bag of dice. You draw a die from the bag, and each die is a distribution over the numbers one to six.

This distribution is controlled by a parameter called alpha (α), which is often called a *hyperparameter* because it is a parameter that controls how other parameters (the actual topic distributions) are estimated, similar to, e.g., the learning speed in many machine learning models. This alpha hyperparameter controls what kind of dice there are in the bag. A high alpha means that the dice are generally fair, i.e., give a uniform multinomial distribution. For topic models, this means that documents will in general contain an even spread of multiple topics. A low alpha means that each die is unfair in the sense of having a strong preference for some number(s), as if these numbers are weighted. You can then draw a die that prefers ones, or a die that prefers sixes. For topic models this means that each document tends to have one or two dominant topics. Finally, alpha can be symmetric (meaning dice are unfair, but randomly, so in the end each topic has the same chance) or asymmetric (they are still unfair, and now also favour some topics more than others). This would correspond to some topics being more likely to occur in all documents.

In our experience, most documents actually do have one or two dominant topics, and some topics are actually more prevalent across many documents then others – especially if you consider that procedural words and boilerplate also need to be fit into a topic unless they are filtered out beforehand. Thus, we would generally recommend a relatively low and asymmetric alpha, and in fact *gensim* provides an algorithm to find, based on the data, an alpha that corresponds to this recommendation (by specifying `alpha='auto'`). In R, we would recommend picking a lower alpha than the default value, probably around $\alpha=5/K$, and optionally try using an asymmetric alpha if you find some words that occur across multiple topics.

To get a more intuitive understanding of the effects of alpha, please see https://cssbook.net/lda for additional material and visualizations.

11.5.2 Fitting an LDA Model

Example 11.14 shows how you can fit an LDA model in Python or R. As example data, we use Obama's State of the Union Speeches using the corpus introduced in Chapter 10. Since such a speech generally touches on many different topics, we choose to first split by paragraph as these will be more semantically coherent (for Obama, at least). In R, we use the `corpus_reshape` function to split the paragraphs, while in Python we

Python Code

```python
url = "https://cssbook.net/d/sotu.csv"
sotu = pd.read_csv(url)
p_obama = (sotu[sotu.President == "Obama"]
           .text.str.split("\n\n").explode())


cv = CountVectorizer(min_df=.01,
                     stop_words="english")
dtm = cv.fit_transform(p_obama)
dtm
```

R Code

```r
url = "https://cssbook.net/d/sotu.csv"
sotu = read_csv(url)
p_obama = sotu %>%
  filter(President == "Obama") %>%
  corpus() %>%
  corpus_reshape("paragraphs")
dfm = p_obama %>%
  tokens(remove_punct=T) %>%
  dfm() %>%
  dfm_remove(stopwords("english")) %>%
  dfm_trim(min_docfreq=.01, docfreq_type = "prop")
dfm
```

Output (from R)

```
Document-feature matrix of: 738 documents, 746 features (97.0% sparse) and 6 docvars.
features
docs   speaker mr vice president members congress first united around come
 text1.1     1  1   1        1        1        1     1      1      1    0
 text1.2     0  0   0        0        0        0     0      0      0    1
 text1.3     0  0   0        0        0        0     0      0      0    0
 text1.4     0  0   0        0        0        0     0      1      0    0
 text1.5     0  0   0        0        0        0     0      0      0    0
 text1.6     0  0   0        0        0        0     0      0      0    0
[ reached max_ndoc ... 732 more documents, reached max_nfeat ... 736 more features ]
```

Python Code

```python
corpus = matutils.Sparse2Corpus(dtm,
    documents_columns=False)
vocab = dict(enumerate(cv.get_feature_names()))

lda = LdaModel(corpus, id2word=vocab, num_topics=10,
     random_state=123, alpha="asymmetric")
pd.DataFrame({f"Topic {n}":[w for (w,tw) in words]
     for (n, words) in
     lda.show_topics(formatted=False)})
```

R Code

```r
lda = dfm %>%
  convert(to = "topicmodels") %>%
  LDA(k=10, control=list(seed=123, alpha = 1/1:10))



terms(lda, 10)
```

R Output

Topic 1	Topic 2	Topic 3	Topic 4	Topic 5	Topic 6	Topic 7	Topic 8	Topic 9	Topic 10
college	care	us	can	years	energy	people	country	world	jobs
new	health	nation	make	first	change	get	time	american	year
workers	still	people	congress	economy	new	day	future	war	years
education	families	can	republicans	back	clean	now	american	security	new
help	americans	states	democrats	time	world	know	america	people	last
job	like	one	work	home	power	see	people	troops	tax
every	new	america	take	crisis	america	tax	done	us	million
kids	need	together	pay	financial	can	americans	work	america	$
small	must	united	cuts	two	economy	government	get	new	businesses
schools	job	president	banks	plan	research	american	now	nations	america

Example 11.14 LDA Topic Model of Obama's State of the Union speeches.

use *pandas*' `str.split`, which creates a list or paragraphs for each text, which we then convert into a paragraph per row using `explode`. Converting this to a DTM we get a reasonably sized matrix of 738 paragraphs and 746 unique words.

Next, we fit the actual LDA model using the package *gensim* (Python) and *topicmodels* (R). Before we can do this, we need to convert the DTM format into a format accepted by that package. For Python, this is done using the `Sparse2Corpus` helper function while in R this is done with the *quanteda* `convert` function. Then, we fit the model, asking for 10 topics to be identified in these paragraphs. There are three things to note in this line. First, we specify a *random seed* of 123 to make sure the analysis is replicable. Second, we specify an "asymmetric" of `1/1:10`, meaning the first topic has alpha 1, the second 0.5, etc. (in R). In Python, instead of using the default of `alpha='symmetric'`, we set `alpha='asymmetric'`, which uses the formula $\frac{1}{topic\_index + \sqrt{num\_topics}}$ to determine the priors. At the cost of a longer estimation time, we can even specify `alpha='auto'`, which will learn an asymmetric prior from the data. See the note on hyperparameters for more information. Third, for Python we also need to specify the vocabulary names since these are not included in the DTM.

The final line generates a data frame of top words per topic for first inspection (which in Python requires separating the words from their weights in a list comprehension and converting it to a data frame for easy viewing). As you can see, most topics are interpretable and somewhat coherent: for example, topic 1 seems to be about education and jobs, while topic 2 is health care. You also see that the word "job" occurs in multiple topics (presumably because unemployment was a pervasive concern during Obama's tenure). Also, some topics like topic 3 are relatively more difficult to interpret from this table. A possible reason for this is that not every paragraph actually has policy content. For example, the first paragraph of his first State of the Union was: *Madam Speaker, Mr. Vice President, Members of Congress, the First Lady of the United States – she's around here somewhere.* None of these words really fit a "topic" in the normal meaning of that term, but all of these words need to be assigned a topic in LDA. Thus, you often see "procedural" or "boilerplate" topics such as topic 3 occurring in LDA outputs.

Finally, note that we showed the R results here. As *gensim* uses a different estimation algorithm (and *scikit-learn* uses a different tokenizer and stopword list), results will not be identical, but should be mostly similar.

11.5.3 Analyzing Topic Model Results

Example 11.15 shows how you can combine the LDA results (topics per document) with the original document metadata. This could be your starting point for substantive analyses of the results, for example to investigate relations between topics or between, e.g., time or partisanship and topic use.

You can also use this to find specific documents for reading. For example, we noted above that topic 3 is difficult to interpret. As you can see in the table in Example 11.15 (which is sorted by value of topic 3), most of the high scoring documents are the first paragraph in each speech, which do indeed contain the "Madam speaker" boilerplate noted above. The other three documents are all calls for bipartisanship and support. As you can see from this example, carefully inspecting the top documents for each topic is very helpful for making sense of the results.

Python Code

```python
topics = pd.DataFrame(
    [dict(lda.get_document_topics(doc,
            minimum_probability=0.0))
     for doc in corpus])
meta = (sotu.iloc[p_obama.index]
        .drop(columns=["text"])
        .reset_index(drop=True))
tpd = pd.concat([meta, topics], axis=1)
tpd.head()
```

R Code

```r
topics = posterior(lda)$topics %>%
    as_tibble() %>%
    rename_all(~paste0("Topic_", .))
meta = docvars(p_obama) %>%
    select(President:Date) %>%
    add_column(doc_id=docnames(p_obama),.before=1)
tpd = bind_cols(meta, topics)
head(tpd)
```

Output

doc_id	President	Date	Topic_1	Topic_2	Topic_3	Topic_4	Topic_5	Topic_6	Topic_7
<chr>	<chr>	<date>	<dbl>	<dbl>	<dbl>	<dbl>	<dbl>	<dbl>	<dbl>
text7.73	Obama	2015-01-20	0.006225760	0.006215916	0.9440579	0.006217933	0.006207157	0.006213044	0.006217223
text5.1	Obama	2013-02-12	0.013922992	0.013920250	0.8746604	0.013945993	0.013920249	0.013922095	0.013926882
text4.1	Obama	2012-01-24	0.013928049	0.013941141	0.8745912	0.013940446	0.013929597	0.013926836	0.013949266
text6.1	Obama	2014-01-28	0.013928049	0.013941141	0.8745912	0.013940446	0.013929597	0.013926836	0.013949266
text2.12	Obama	2010-01-27	0.019553782	0.019579572	0.8239496	0.019602906	0.019529878	0.019576050	0.019597005
text2.95	Obama	2010-01-27	0.007127905	0.007119074	0.8126765	0.007138889	0.007124680	0.062978268	0.007120398

Python Code

```python
for docid in [622, 11, 322]:
    print(f"{docid}: {list(p_obama)[docid]}")
```

R Code

```r
for (id in c("text7.73", "text5.1", "text2.12")) {
  text = as.character(p_obama)[id]
  print(glue("{id}: {text}"))
}
```

R Output

```
text7.73: So the question for those of us here tonight is how we, all of us, can better reflect America's
   hopes. I've served in Congress
with many of you. I know many of you well. There are a lot of good people here on both sides of the aisle.
   And many of you have
told me that this isn't what you signed up for: arguing past each other on cable shows, the constant
   fundraising, always
looking over your shoulder at how the base will react to every decision.
text5.1: Mr. Speaker, Mr. Vice President, members of Congress, fellow citizens:
text2.12: And tonight, I'd like to talk about how together, we can deliver on that promise.
```

Example 11.15 Analyzing and inspecting LDA results.

11.5.4 Validating and Inspecting Topic Models

As we saw in the previous subsection, running a topic model is relatively easy. However, that doesn't mean that the resulting topic model will always be useful. As with all text analysis techniques, *validation* is the key to good analysis: are you measuring what you want to measure? And how do you know?

For topic modeling (and arguably for all text analysis), the first step after fitting a model is inspecting the results and establishing face validity. Top words per topic such as those listed above are a good place to start, but we would really encourage you to also look at the top documents per topic to better understand how words are used in context. Also, it is good to inspect the relationships between topics and look at documents that load high on multiple topics to understand the relationship.

If the only goal is to get an explorative understanding of the corpus, for example as a first step before doing a dictionary analysis or manual coding, just face validity is probably sufficient. For a more formal validation, however, it depends on the reason for using topic modeling.

If you are using topic modeling in a true unsupervised sense, i.e., without a predefined analytic schema in mind, it is difficult to assess whether the model measures what you want to measure – because the whole point is that you don't know what you want to measure. That said, however, you can have the general criteria that the model needs to achieve *coherence* and *interpretability*, meaning that words and documents that share a topic are also similar semantically.

In their excellent paper on the topic, Chang et al. (2009) propose two formal tasks to judge this using manual (or crowd) coding: in *word intrusion*, a coder is asked to pick the "odd one out" from a list where one other word is mixed in a group of topic words. In *topic intrusion*, the coder is presented with a document and a set of topics that occur in the document, and is asked to spot the one topic that was not present according to the model. In both tasks, if the coder is unable to identify the intruding word or topic, apparently the model does not fit our intuitive notion of "aboutness" or semantic similarity. Perhaps their most interesting finding is that goodness-of-fit measures like perplexity[1] are actually not good predictors of the interpretability of the resulting models.

If you are using topic models in a more confirmatory manner, that is, if you wish the topics to match some sort of predefined categorization, you should use regular gold standard techniques for validation: code a sufficiently large random sample of documents with your predefined categories, and test whether the LDA topics match those categories. In general, however, in such cases it is a better idea to use a dictionary or supervised analysis technique as topic models often do not exactly capture our categories. After all, unsupervised techniques mainly excel in bottom-up and explorative analyses (Section 11.1).

11.5.5 Beyond LDA

This chapter focused on regular or "vanilla" LDA topic modeling. Since the seminal publication, however, a large amount of variations and extensions on LDA have been proposed. These include, for example, *Dynamic Topic Models* (which incorporate time; Blei and Lafferty, 2006b), *Correlated Topic Models* (which explicitly model correlation between topics; Blei and Lafferty, 2006a). Although it is beyond the scope of this book to describe these models in detail, the interested reader is encouraged to learn more about these models.

Especially noteworthy are *Structural Topic Models* (R package stm; Roberts et al., 2014), which allow you to model covariates as topic or word predictors. This allows you, for example, to model topic shifts over time or different words for the same topic based on, e.g., Republican or Democrat presidents.

Python users should check out Hierarchical Topic Modeling (Griffiths et al., 2004). In hierarchical topic modeling, rather than the researcher specifying a fixed number of topics, the model returns a hierarchy of topics from few general topics to a large number of specific topics, allowing for a more flexible exploration and analysis of the data.

[1] Perplexity is a measure to compare and evaluate topic models using log-likelihood in order to estimate how well a model predicts a sample.

How many topics?

With topic modeling, the most important researcher choices are the number of topics and the value of alpha. These choices are called hyperparameters, since they determine how the model parameters (e.g. words per topic) are found.

There is no good theoretical solution to determine the "right" number of topics for a given corpus and research question. Thus, a sensible approach can be to ask the computer to try many models, and see which works best. Unfortunately, because this is an unsupervised (inductive) method, there is no single metric that determines how good a topic model is.

There are a number of such metrics proposed in the literature, of which we will introduce two. *Perplexity* is a score of how well the LDA model can fit (predict) the actual word distribution (or in other words: how "perplexed" the model is seeing the corpus). *Coherence* is a measure of how semantically coherent the topics are by checking how often the top token co-occurs in documents in each topic (Mimno et al., 2011).

Example 11.16 below shows how these can be calculated for a range of topic counts, and the same code could be used for trying different values of alpha. For both measures, lower values are better, and both essentially keep decreasing as you add more topics. What you are looking for is the inflection point (or "elbow point") where it goes from a steep decrease to a more gradual decrease. For coherence, this seems to be at 10 topics, while for perplexity this is at 20 topics.

There are two very important caveats to make here, however. First, these metrics are no substitute for human validation and the best model according to these metrics is not always the most interpretable or coherent model. In our experience, most metrics give a higher topic count that would be optimal from an interpretability perspective, but of course that also depends on how we operationalize interpretability. Nonetheless, these topic numbers are probably more indicative of a range of counts that should be inspected manually, rather than giving a definitive answer.

Second, the code below was written so it is easy to understand and quick to run. For real use in a research project, it is advised to include a broader range of topic counts and also vary the α. Moreover, it is smart to run each count multiple times so you get an indication of the variance as well as a single point (it is quite likely that the local minimum for coherence at k=10 is an outlier that will disappear if more runs are averaged). Finally, especially for a goodness-of-fit measure like perplexity it is better to split the data into a training and test set (see Section 11.4.1 for more details).

Python Code

```
result = []
for k in [5, 10, 15, 20, 25, 30]:
  m = LdaModel(corpus,num_topics=k,id2word=vocab,
        random_state=123, alpha="asymmetric")
  perplexity = m.log_perplexity(corpus)
  coherence=CoherenceModel(model=m,corpus=corpus,
      coherence="u_mass").get_coherence()
  result.append(dict(k=k, perplexity=perplexity,
                coherence=coherence))

result = pd.DataFrame(result)
result.plot(x="k", y=["perplexity", "coherence"])
plt.show()
```

R Code

```
results = list()
for (k in c(5, 10, 15, 20, 25, 30)) {
    alpha = 1/((1:k)+sqrt(k))
    dtm = convert(dfm, to="topicmodels")
    LDA(dtm,k=k,control=list(seed=99,alpha=alpha))
    results[[as.character(k)]] = data.frame(
      perplexity=perplexity(lda),
      coherence=mean(topic_coherence(lda, dtm)))
}
bind_rows(results, .id="k") %>%
  mutate(k=as.numeric(k)) %>%
  pivot_longer(-k) %>%
  ggplot() +
  geom_line(aes(x=k, y=value)) +
  xlab("Number of topics") +
  facet_grid(name ~ ., scales="free")
```

R Output

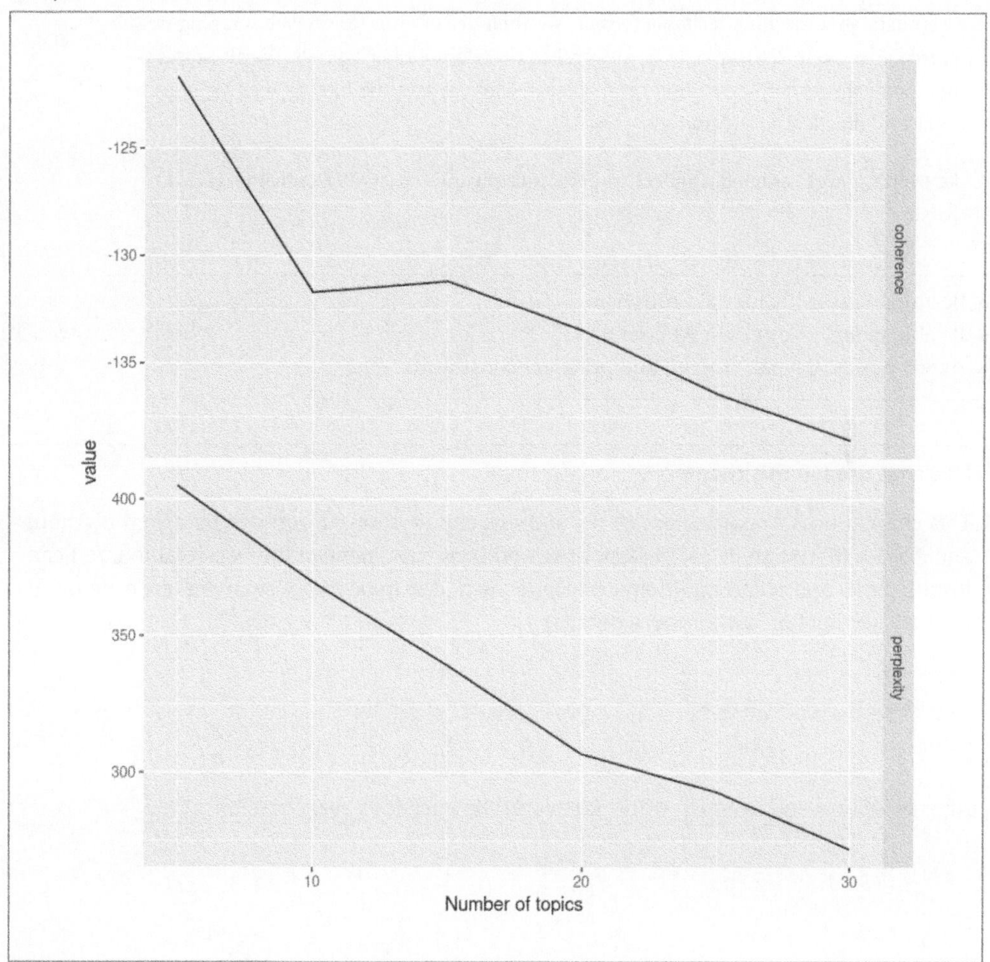

Example 11.16 Computing perplexity and coherence of topic models.

12

Scraping Online Data

Abstract

In this chapter, you'll learn how to retrieve your data from online sources. We first discuss the use of Application Programming Interfaces, so-called APIs, which allow you to retrieve data from social media platforms, but also government data or other forms of open data, in a machine-readable format. We then discuss how to do web scraping in a narrower sense to retrieve data from websites that do not offer an API. We also discuss how to deal with authentication mechanisms, cookies, and the like, as well as ethical, legal, and practical considerations.

Keywords web scraping, application programming interface (API), crawling, HTML parsing

- Be able to use APIs for data retrieval
- Be able to write your own web scraper
- Assess basics of legal, ethical, and practical constraints

Packages used in this chapter

This chapter uses in particular *httr* (R) and *requests* (Python) to retrieve data, *json* (Python) and *jsonlite* (R) to handle JSON responses, and *lxml* and *Selenium* for web scraping. You can install these and some additional packages (e.g., for geocoding) with the code below if needed (see Section 1.4 for more details):

Python Code

```
!pip3 install requests selenium
```

R Code

```
install.packages(c("tidyverse",
        "httr", "jsonlite", "glue",
        "data.table"))
```

After installing, you need to import (activate) the packages every session:

Computational Analysis of Communication: A Practical Introduction to the Analysis of Texts, Networks, and Images with Code Examples in Python and R, First Edition. Wouter van Atteveldt, Damian Trilling & Carlos Arcila Calderón.

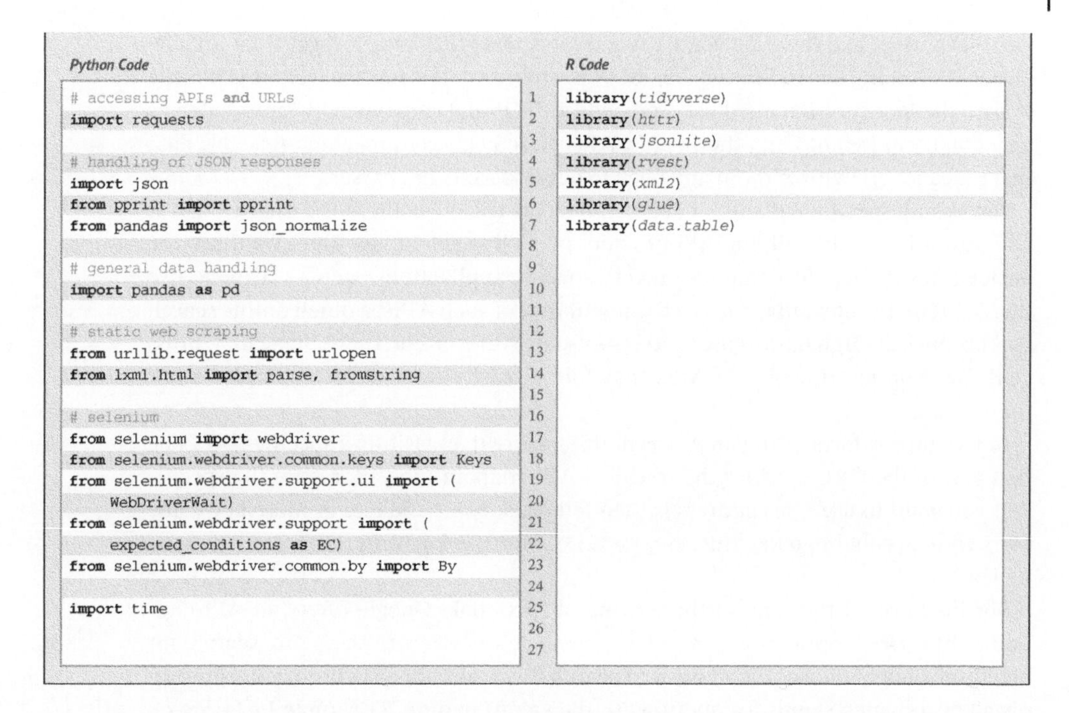

Python Code

```python
# accessing APIs and URLs
import requests

# handling of JSON responses
import json
from pprint import pprint
from pandas import json_normalize

# general data handling
import pandas as pd

# static web scraping
from urllib.request import urlopen
from lxml.html import parse, fromstring

# selenium
from selenium import webdriver
from selenium.webdriver.common.keys import Keys
from selenium.webdriver.support.ui import (
    WebDriverWait)
from selenium.webdriver.support import (
    expected_conditions as EC)
from selenium.webdriver.common.by import By

import time
```

R Code

```r
library(tidyverse)
library(httr)
library(jsonlite)
library(rvest)
library(xml2)
library(glue)
library(data.table)
```

12.1 Using Web APIs: From Open Resources to Twitter

Let's assume we want to retrieve data from some online service. This could be some social media platform, but could also be a government website, some *open data* platform or initiative, or sometimes a commercial organization that provides some online service. Of course, we could surf to their website, enter a search query, and somehow save the result. This would result in a lot of impracticalities, though. Most notably, websites are designed such that they are perfectly readable and understandable for humans, but the cues that are used often have no "meaning" for a computer program. As humans, we have no problem understanding which parts of a web page refer to the author of some item on a web page, what the numbers "2006" and "2008" mean, and on. But it is not trivial to think of a way to explain to a computer program how to identify variables like `author`, `title`, or `year` on a web page. We will learn how to do exactly that in Section 12.2. Writing such a *parser* is often necessary, but it is also error-prone and a detour, as we are trying to bring some information that has been optimized for human reading *back* to a more structured data structure.

Luckily, however, many online services not only have web interfaces optimized for human reading, but also offer another possibility to access the data they provide: an API (Application Programming Interface). The vast amount of contemporary web APIs work like this: you send a *request* to some URL, and you get back a JSON object. As you learned in Section 5.2, JSON is a nested data structure, very much like a Python dictionary or R named list (and, in fact, JSON data are typically represented as a dictionary in Python). In other words: APIs directly gives us machine-readable data that we can work with without any need to develop a custom parser.

Discussing specific APIs in a book can be a bit tricky, as there is a chance that it will be outdated: after all, the API provider may change it at any time. We therefore decided not to include a chapter on very specific applications such as "How to use the Twitter API" or similar – given the popularity of such APIs, a quick online search will produce enough up-to-date (and out-of-date) tutorials on these. Instead, we discuss the generic principles of APIs that should easily translate to examples other than ours.

In its simplest form, using an API is nothing more than visiting a specific URL. The first part of the URL specifies the so-called API endpoint: the address of the specific API you want to use. This address is then followed by a ? and one or more key-value pairs with an equal sign like this: `key=value`. Multiple key-value pairs are separated with a &.

For instance, at the time of the writing of this book, Google offers an API endpoint, `https://www.googleapis.com/books/v1/volumes`, to search for books on Google Books. If you want to search for books about Python, you can supply a key q (which stands for query) with the value "python" (Example 12.1). We do not need any specific software for this – we could, in fact, use a web browser as well. Popular packages that allow us to do it programatically are *httr* in combination with *jsonlite* (R) and *requests* (Python).

But how do we know which parameters (i.e., which key-value pairs) we can use? We need to look it up in the documentation of the API we are interested in (in this example https://developers.google.com/books/docs/v1/using). There is no other way of knowing that the key to submit a query is called q, and which other parameters can be specified.

In our example, we used a simple value to include in the request: the string "python". But what if we want to submit a string that contains, let's say, a space, or a character like & or ? which, as we have seen, have a special meaning in the request? In these cases, you need to "encode" your URL using a mechanism called URL encoding or percent encoding. You may have seen this earlier: a space, for instance, is represented by `%20`.

Python Code

```
r = requests.get("https://www.googleapis.com/"
                 "books/v1/volumes?q=python")
data = r.json()
print(data.keys()) # "items" seems most promising
pprint(data["items"][0]) # let's print the 1st one
```

R Code

```
url = str_c("https://www.googleapis.com/books/",
            "v1/volumes?q=python")
r = GET(url)
data = content(r, as="parsed")
print(names(data))
print(data$items[[1]])
```

Python Output (abridged)

```
dict_keys(['kind', 'totalItems', 'items'])
{'accessInfo': {'accessViewStatus': 'SAMPLE',
        ... ...
'etag': 'X93TTSzbXvk',
'id': 'wqeVv09Y6hIC',
'kind': 'books#volume',
'saleInfo': {'country': 'NL', 'isEbook': False, 'saleability': 'NOT_FOR_SALE'},
'searchInfo': {'textSnippet': 'A study of Delphic myths and their origins.'},
'selfLink': 'https://www.googleapis.com/books/v1/volumes/wqeVv09Y6hIC',
'volumeInfo': {'allowAnonLogging': False,
        'authors': ['Joseph Eddy Fontenrose'],
        ...
        ...
        'publishedDate': '1959',
        'publisher': 'Univ of California Press',
        'ratingsCount': 2,
        'readingModes': {'image': True, 'text': False},
        'subtitle': 'A Study of Delphic Myth and Its Origins',
        'title': 'Python'}}
```

Example 12.1 Retrieving JSON data from the Google Books API.

The data our request returns are nested data, and hence, they do not really "fit" in a tabular data frame. We could keep the data as they are (and then, for instance, just extract the key-value pairs that we are interested in), but – for the sake of getting a quick overview – let's flatten the data so that they can be represented in a data frame (Example 12.2). This works quite well here, but may be more problematic when the items have a widely varying structure. If that is the case, we probably would want to write a loop to iterate over the different items and extract the information we are interested in.

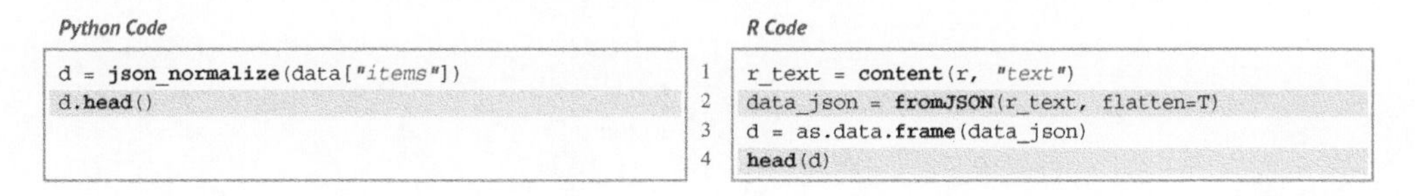

Python Code

```
d = json_normalize(data["items"])
d.head()
```

R Code

```
r_text = content(r, "text")
data_json = fromJSON(r_text, flatten=T)
d = as.data.frame(data_json)
head(d)
```

R Output

	kind	totalItems	items.kind	items.id	items.etag
	<chr>	<int>	<chr>	<chr>	<chr>
1	books#volumes	441	books#volume	wqeVv09Y6hIC	X93TTSzbXvk
2	books#volumes	441	books#volume	aJQILlLxRmAC	1x9G8zJhzJU
3	books#volumes	441	books#volume	6GzuBgAAQBAJ	2b3xQ+Fqzjc
4	books#volumes	441	books#volume	mCrrCAAAQBAJ	qJsyOcskT38
5	books#volumes	441	books#volume	carqdIdfVlYC	0vcIQokQKzc
6	books#volumes	441	books#volume	fzUCGtyg0MMC	RUkIPx/b4Wg

Example 12.2 Transforming the data into a data frame.

You may have realized that you did not get *all* results. This protects you from accidentally downloading a huge dataset (you may have underestimated the number of Python books available on the market), and saves the provider of the API a lot of bandwith. This does not mean that you cannot get more data. In fact, many APIs work with *pagination*: you first get the first "page" of results, then the next, and so on. Sometimes, the API response contains a specific key-value pair (sometimes called a "continuation key") that you can use to get the next results; sometimes, you can just say at which result you want to start (say, result number 11) and then get the next "page". You can then write a loop to retrieve as many results as you need (Example 12.3) – just make sure that you do not get stuck in an eternal loop. When you start playing around with APIs, make sure you do not cause unnecessary traffic, but limit the number of calls that are made (see also Section 12.4).

Many APIs work very much like the example we discussed, and you can adapt the logic above to many APIs once you have read their documentation. You would usually start by playing around with single requests, and then try to automate the process by means of a loop.

However, many APIs have restrictions regarding who can use them, how many requests can be made, and so on. For instance, you may need to limit the number of requests per minute by calling a `sleep` function within your loop to delay the execution of the next call. Or, you may need to authenticate yourself. In the example of the Google Books API, this will allow you to request more data (such as whether you own an (electronic) copy of the books you retrieved). In this case, the documentation outlines that you can simply pass an authentication token as a parameter with the URL. However, many APIs use more advanced authentication methods such as OAuth (see Section 12.3).

Lastly, for many APIs that are very popular with social scientists, specific wrapper packages exist (such as *tweepy* (Python) or *rtweet* (R) for downloading Twitter messages) which are a bit more user-friendly and handle things like authentication, pagination, respecting rate-limits, etc., for you.

Python Code

```python
allitems = []
i = 0
while True:
    r = requests.get("https://www.googleapis.com/"
        "books/v1/volumes?q=python&maxResults="
        f"40&startIndex={i}")
    data = r.json()
    try:
        allitems.extend(data["items"])
    except:
        print(f"Retrieved {len(allitems)},"
            "it seems like that's it")
        break
    i+=40
d = json_normalize(allitems)
```

R Code

```r
i = 0
j = 1
url = str_c("https://www.googleapis.com/books/",
            "v1/volumes?q=python&maxResults=40",
            "&startIndex={i}")
alldata = list()
while (TRUE) {
    r = GET(glue(url))
    r_text = content(r, "text")
    data_json = fromJSON(r_text, flatten=T)
    if (length(data_json$items)==0) {break}
    alldata[[j]] = as.data.frame(data_json)
    i = i + 40
    j = j + 1}
d = rbindlist(alldata, fill=TRUE)
```

Example 12.3 Full script including pagination.

12.2 Retrieving and Parsing Web Pages

Unfortunately, not all online services we may be interested in offer an API – in fact, it has even been suggested that computational researchers have arrived in an "post-API age" (Freelon, 2018), as API access for researchers has become increasingly restricted.

If data cannot be collected using an API (or a similar service, such as RSS feeds), we need to resort to web scraping. Before you start a web scraping project, make sure to ask the appropriate authorities for ethical and legal advice (see also Section 12.4).

Web scraping (sometimes also referred to as harvesting), in essence, boils down to automatically downloading web pages aimed at a human audience, and extracting meaningful information out of them. One could also say that we are reverse-engineering the way the information was published on the web. For instance, a news site may always use a specific formatting to denote the title of an article – and we would then use this to extract the title. This process is called "parsing", which in this context is just a fancy term for "extracting meaningful information".

When scraping data from the web, we can distinguish two different tasks: (1) downloading a (possibly large) number of webpages, and (2) parsing the content of the webpages. Often, both go hand in hand. For instance, the URL of the next page to be downloaded might actually be parsed from the content of the current page; or some overview page may contain the links and thus has to be parsed first in order to download subsequent pages.

We will first discuss how to parse a single HTML page (say, the page containing one specific product review, or one specific news article), and then describe how to "scale up" and repeat the process in a loop (to scrape, let's say, all reviews for the product; or all articles in a specific time frame).

12.2.1 Retrieving and Parsing an HTML Page

In order to parse an HTML file, you need to have a basic understanding of the structure of an HTML file. Open your web browser, visit a website of your choice (we suggest to use a simple page, such as https://cssbook.net/d/eat/index.html), and inspect its underlying HTML code (almost all browsers have a function called something like "view source", which enables you to do so).

You will see that there are some regular patterns in there. For example, you may see that each paragraph is enclosed with the tags <p> and </p>. Thinking back to Section 9.2, you may figure out that you could, for instance, use a regular expression to extract the text of the first paragraph. In fact, packages like *beautifulsoup* under the hood use regular expressions to do exactly that.

Writing your own set of regular expressions to parse an HTML page is usually not a good idea (but it can be a last resort when everything else fails). Chances are high that you will make a mistake or not handle some edge case correctly; and besides, it would be a bit like re-inventing the wheel. Packages like *rvest* (R), *beautifulsoup*, and *lxml* (both Python) already do this for you.

In order to use them, though, you need to have a basic understanding of what an HTML page looks like. Here is a simplified example:

```
<html>
<body>
<h1>This is a title</h1>
<div id="main">
<p>Some text with one <a href="test.html">link</a> </p>
<img src="plaatje.jpg">an image </img>
</div>
<div>
<p class="lead">Some more text </p>
<p>Even more...</p>
<p>And more.</p>
</div>
</body>
</html>
```

For now, it's not too important to understand the function of each specific tag (although it might help, for instance, to realize that a denotes a link, h1 a first-level heading, p a paragraph and div some kind of section).

What is important, though, is to realize that each tag is opened and closed (e.g., <p> is closed by </p>). Because tags can be nested, we can actually draw the code as a tree. In our example, this would look like this:

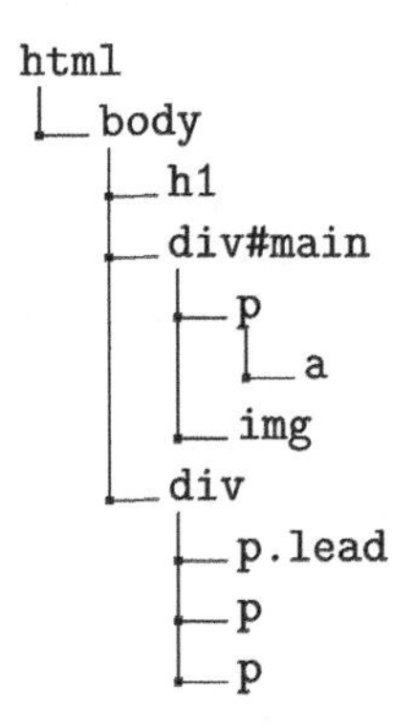

Additionally, tags can have *attributes*. For instance, the makers of a page with customer reviews may use attributes to specify what a section contains. They may write <p class="lead"> ... </p> to mark the lead paragraph of an article, and <a href=test.html"> ... </a< specifies the target of a hyperlink. Especially important here are the id and class attributes, which are often used by webpages to control the formatting. id (indicated with the hash sign # above) gives a unique ID to a single element, while class (indicated with a period) assigns a class label to one or more elements. This enables web sites to specify their layout and formatting using a technique called Cascading Style Sheets (CSS). For example, the web page could set the lead paragraph to be bold. The nice thing is that we can exploit this information to tell our parser where to find the elements we are interested in.

CSS Selectors. The easiest way to specify our parser to look for a specific element is to use a *CSS Selector*, which might be familiar to you if you have created web pages. For example, to find the lead paragraph(s) we specify p.lead. To find the node with id="body", we can specify #body. You can also use this to specify relations between nodes. For example, to find all paragraphs within the body element we would write #body p.

Table 12.1 gives an overview of the possibilities of CSS Select. In general, a CSS selector is a set of node specifiers (like h1, .lead or div#body), optionally with relation specifiers between them. So, #body p finds a p anywhere inside the id=body element, while #body > p requires the p to be directly contained inside the body (with no other nodes in between).

Table 12.1 Overview of CSS Select and XPath syntax.

Example	CSS Select	XPath
Basic tree navigation		
h1 anywhere in document	h1	//h1
h1 inside a body	body h1	//body//h1
h1 directly inside div	div > h1	//div/h1
Any node directly inside div	div > *	//div/*
p next to a h1	h1 ~ p	//h1/following-sibling::p
p next to a h1	h1 + p	//h1/following-sibling::p[1]
Node attributes		
<div id='x1'>	div#x1	//div[@id='x1']
any node with id x1	#x1	//*[@id='x1']
<div class='row'>	div.row	//div[@class='row']
any node with class row	.row	//*[@class='row']
a with href="#"	a[href="#"]	//a[@href="#"]
Advanced tree navigation		
a in a div with class 'meta' directly inside the main element	#main > div.meta a	//*[@id='main'] /div[@ class='meta']//a
First p in a div	div p:first-of-type	//div/p[1]
First child of a div	div :first-child	//div/*[1]
Second p in a div	div p:nth-of-type(2)	//div/p[2]
Second p in a div	div p:nth-of-type(2)	//div/p[2]
parent of the div with id x1	(not possible)	//div[@id='x1']/ parent::*

XPath. An alternative to CSS Selectors is *XPath*. Where CSS Selectors are directly based on HTML and CSS styling, XPath is a general way to describe nodes in XML (and HTML) documents. The general form of XPath is similar to CSS Select: a sequence of node descriptors (such as `h1` or `*[@id='body']`). Contrary to CSS Select, you always have to specify the relationship, where // means any direct or indirect descendant and / means a direct child. If the relationship is not a child or descendant relationship (but for example a sibling or parent), you specify the *axis* with e.g. `//a/parent::p` meaning an `a` anywhere in the document (`//a`) which has a direct parent (`/parent::`) that is a `p`.

A second difference with CSS Selectors is that the *class* and *id* attributes are not given special treatment, but can be used with the general `[@attribute='value']` pattern. Thus, to get the lead paragraph you would specify `//p[@class='lead']`.

The advantage of XPath is that it is a very powerful tool. Everything that you can describe with a CSS Selector can also be described with an XPath pattern, but there are some things that CSS Selectors cannot describe such as parents. On the other hand, XPath patterns can be a bit harder to write, read, and debug. You can choose to use either tool, and you can even mix and match them in a single script, but our general recommendation is to use CSS Selectors unless you need to use the specific abilities of XPath.

Example 12.4 shows how to use XPATHs and CSS selectors to parse an HTML page. To fully understand it, open https://cssbook.net/d/eat/index.html in a browser and look at its source code (all modern browsers have a function "View page source" or similar), or – more comfortable – right-click on an element you are interested in (such as a restaurant name) and select "Inspect element" or similar. This will give you a user-friendly view of the HTML code.

Of course, Example 12.4 only parses one possible element of interest: the restaurant names. Try to retrieve other elements as well!

Python Code

```python
tree=parse(urlopen(
    "https://cssbook.net/d/eat/index.html"))

# get the restaurant names via XPATH
print([e.text_content().strip() for e in
        tree.xpath("//h3")])

# get the restaurant names via CSS Selector
print([e.text_content().strip() for e in
        tree.getroot().cssselect("h3")])
```

R Code

```r
url = "https://cssbook.net/d/eat/index.html"
page = html(url)

# get the restaurant names via XPATH
page %>% html_nodes(xpath="//h3") %>% html_text()

# get the restaurant names via CSS Selector
page %>% html_nodes("h3") %>% html_text()
```

R Output

```
['Pizzeria Roma', 'Trattoria Napoli', 'Curry King']
['Pizzeria Roma', 'Trattoria Napoli', 'Curry King']
```

Example 12.4 Parsing websites using XPATHs or CSS selectors.

> **Do you care about children?**
>
> Regardless of whether you use XPATHS or CSS Selectors to specify which part of the page you are interested in, it is often the case that there are other elements within it. Depending on whether you want to also retrieve the text of these elements or not, you have to use different approaches (see Example 12.5).

Notably, you may want to parse links. In HTML, links use a specific tag, a. These tags have an attribute, href, which contains the link itself. Example 12.6 shows how, after selecting the a tags, we can access these attributes.

Python Code

```
# three ways of extracting text
print("Appending `/text()` to the XPATH gives you"
      "exactly the text that is in the element"
      "itself, including line-breaks that happen"
      "to be in the source code:" )
print(tree.xpath(
    "//div[@class='restaurant']/text()"))

print("\nUsing the `text` property of the"
      "elements in the list of elements that are"
      "matched by the XPATH expression gives you"
      "the text of the elements themselves"
      "without the line breaks:")
print([e.text for e in tree.xpath(
      "//div[@class='restaurant']")])

print("\nUsing the `text_content()` method"
      "instead returns the text of the element"
      "*and the text of its children*:")
print([e.text_content() for e in tree.xpath(
      "//div[@class= 'restaurant']")])

print("\nThe same but using CSS Selectors (note"
      "the .getroot() method, because the"
      "selectors can only be applied to HTML"
      "elements, not to DOM trees):")
print([e.text_content() for e in
      tree.getroot().cssselect(". restaurant")])
```

R Code

```
url = "http://cssbook.net/d/eat/index.html")
page = read_html(url)

glue("Appending `/text()` to the XPATH gives you\\
exactly the text that is in the element itself,\\
including line-breaks that happen to be in the\\
source code:" )
page %>% html_nodes(xpath=
      "//div[@class='restaurant']/text()")

glue("\nUsing the `html_text` function instead\\
returns the text of the element *and the text\\
of its children*:")
page %>% html_nodes(xpath=
      "//div[@class='restaurant']") %>% html_text()

glue("\The same but using CSS Selectors:")
page %>% html_nodes(".restaurant") %>% html_text()
```

Python Output

```
Appending "/text()" to the XPATH gives you exactly the text that is in the element itself, including
line-breaks that happen to be in the source code:
[' ', '\n ', '\n ', '\n ', ' ', '\n ', '\n ', '\n ', ' ', '\n ', '\n ', '\n ']
Using the "text" property of the elements in the list of elements that are matched by the XPATH expression
gives you the text of the elements themselves without the line breaks:
[' ', ' ', ' ']
Using the "text_content()" method instead returns the text of the element *and the text of its children*:
[' Pizzeria Roma \n Here you can get ... ... \n Read the full review here\n ', ' Trattoria Napoli \n Another
restaurant ... ... \n Read the full review here\n ', ' Curry King \n Some description. \n Read the full review
here\n ']
The same but using CSS Selectors (note the .getroot() method, because the selectors can only be applied to
HTML elements, not to DOM trees):
[' Pizzeria Roma \n Here you can get ... ... \n Read the full review here\n ', ' Trattoria Napoli \n Another
restaurant ... ... \n Read the full review here\n ', ' Curry King \n Some description. \n Read the full review
here\n ']
```

Example 12.5 Getting the text of an HTML element versus getting the text of the element and its children.

Python Code

```python
linkelements = tree.xpath("//a")
linktexts = [e.text for e in linkelements]
links = [e.attrib["href"] for e in linkelements]

print(linktexts)
print(links)
```

R Code

```r
page %>%
  html_nodes(xpath="//a") %>%
  html_text()
page %>%
  html_nodes(xpath="//a") %>%
  html_attr("href")
```

Python Output

```
['here', 'here', 'here']
['review0001.html', 'review0002.html', 'review0003.html']
```

Example 12.6 Parsing link texts and links.

Pretending to be a specific browser

When *lxml*, *rvest*, or your web browser download an HTML page, they send a so-called HTTP request. This request contains the URL, but also some meta-data, such as a so-called user-agent string. This string specifies the name and version of the browser. Some sites may block specific user agents (such as, for instance, the ones that *lxml* or *rvest* use); and sometimes, they deliver different content for different browsers. By using a more powerful module for downloading the HTML code (such as *requests* or *httr*) before parsing it, you can specify your own user-agent string and thus pretend to be a specific browser. If you do a web search, you will quickly find long lists with popular strings. In Example 12.7, we rewrote Example 12.4 such that a custom user-agent can be specified.

Python Code

```python
headers = {"User-Agent": "Mozilla/5.0 (Windows "
    "NT 10.0; Win64; x64; rv:60.0) "
    "Gecko/20100101 Firefox/60.0"}

htmlsource = requests.get(
    "https://cssbook.net/d/eat/index.html",
    headers = headers).text
tree = fromstring(htmlsource)
print([e.text_content().strip() for e in
       tree.xpath("//h3")])
```

R Code

```r
r = GET("http://cssbook.net/d/eat/index.html",
    user_agent=str_c("Mozilla/5.0 (Windows NT "
    "10.0; Win64; x64; rv:60.0) Gecko/20100101 ",
    "Firefox/60.0"))
page = read_html(r)
page %>% html_nodes(xpath="//h3") %>% html_text()
```

Python Output

```
['Pizzeria Roma', 'Trattoria Napoli', 'Curry King']
```

Example 12.7 Specifying a user agent to pretend to be a specific browser.

12.2.2 Crawling Websites

Once we have mastered parsing a single HTML page, it is time to scale up. Only rarely are we interested in parsing a single page. In most cases, we want to use an HTML page as a starting point, parse it, follow a link to some other interesting page, parse it as well, and so on. There are some dedicated frameworks for this such as *scrapy*, but in our experience, it may be more of a burden to learn that framework than to just implement your crawler yourself.

Staying with the example of a restaurant review website, we might be interested in retrieving all restaurants from a specific city, and for all of these restaurants, all available reviews.

Our approach, thus, could look as follows:

1. Retrieve the overview page.
2. Parse the names of the restaurants and the corresponding links.
3. Loop over all the links, retrieve the corresponding pages.
4. On each of these pages, parse the interesting content (i.e., the reviews, ratings, and so on).

So, what if there are multiple overview pages (or multiple pages with reviews)? Basically, there are two possibilities: the first possibility is to look for the link to the next page, parse it, download the next page, and so on. The second possibility exploits the fact that often, URLs are very systematic: for instance, the first page of restaurants might have a URL such as http://myreviewsite.com/amsterdam/restaurants.html?page=1. If this is the case, we can simply construct a list with all possible URLs (Example 12.8). Afterwards, we would just loop over this list and retrieve all the pages (a bit like how we approached Example 12.3 in Section 12.1).

However, often, things are not as straightforward, and we need to find the correct links on a page that we have been parsing – that's why we *crawl* through the website.

Python Code

```
baseurl="https://reviews.com/?page="
tenpages = [f"{baseurl}{i+1}" for i in range(10)]
print(tenpages)
```

R Code

```
baseurl="https://reviews.com/?page="
tenpages = glue("{baseurl}{1:10}")
print(tenpages)
```

Python Output

```
[http://myreviewsite.com/amsterdam/hotels.html?page=1', 'http://myreviewsite.com/amsterdam/hotels.
    html?page=2', 'http://myreviewsite.com/amsterdam/hotels.html?page=3 ', 'http://myreviewsite.com/
    amsterdam/hotels.html?page=4', 'http://myreviewsite.com/amsterdam/hotels.html?page=5 ', 'http://
    myreviewsite.com/amsterdam/hotels.html?page=6', 'http://myreviewsite.com/amsterdam/hotels.
    html?page=7 ', 'http://myreviewsite.com/amsterdam/hotels.html?page=8', 'http://myreviewsite.com/
    amsterdam/hotels.html?page=9 ', 'http://myreviewsite.com/amsterdam/hotels.html?page=10']
```

Example 12.8 Generating a list of URLs that follow the same pattern.

Writing a good crawler can take some time, and they will look very differently for different pages. The best advice is to build them up step-by-step. Carefully inspect the website you are interested in. Take a sheet of paper, draw its structure, and try to find out which pages you need to parse, and how you can get from one page to the next. Also think about how the data that you want to extract should be organized.

We will illustrate this process using our mock-up review website https://cssbook.net/eat/. First, have a look at the site and try to understand its structure.

You will see that it has an overview page, index.html, with the names of all restaurants and, per restaurant, a link to a page with reviews. Click on these links, and note your observations, such as:

- the pages have different numbers of reviews;
- each review consists of an author name, a review text, and a rating;
- some, but not all, pages have a link saying "Get older reviews"
- ...

If you combine what you just learned about extracting text and links from HTML pages with your knowledge about control structures like loops and conditional statements (Section 3.2), you can now write your own crawler.

Writing a scraper is a craft, and there are several ways of achieving your goal. You probably want to develop your scraper in steps: first write a function to parse the overview page, then a function to parse the review pages, then try to combine all elements into one script. Before you read on, try to write such a scraper.

To show you one possible solution, we implemented a scraper in Python that crawls and parses all reviews for all restaurants (Example 12.9), which we describe in detail below.

First, we need to get a list of all restaurants and the links to their reviews. That's what is done in the function *get_restaurants*. This is actually the first thing we do (see line 39).

We now want to loop over these links and retrieve the reviews. We decided to use a *generator* (Section 3.2): instead of writing a function that collects *all* reviews in a list first, we let the function yield each review immediately – and then append that review to a file. This has a big advantage: if our scraper fails (for instance, due to a time out, a block, or a programming error), then we have already saved the reviews we got so far.

We loop over the links to the restaurants (line 43) and call the function *get_reviews* (line 45). Each review it returns (the review is a dict) gets the name of the restaurant as an extra key, and then gets written to a file which contains one JSON-object per line (also known as a jsonlines-file).

The function `get_reviews` takes a link to a review page as input and yields reviews. If we knew all pages with reviews already, then we would not need the while loop statement in line 15 and the lines 31–36. However, as we have seen, some review pages contain a link to older reviews. We therefore use a loop that runs forever (that is what `while True:` does), *unless* it encounters a `break` statement (line 36). An inspection of the HTML code shows that these links have a `span` tag with the attribute `class="backbutton"`. We therefore check if such a button exists (line 31), and if so, we get its `href` attribute (i.e., the link itself), overwrite the `url` variable with it, and then go back to line 15, the beginning of the loop, so that we can download and parse this next URL. This goes on until such a link is no longer found.

12.2.3 Dynamic Web Pages

You may have realized that all our scraping efforts until now proceeded in two steps: we retrieved (downloaded) the HTML source of a web page and then parsed it. However, modern websites more and more frequently are dynamic rather than static. For example, after being loaded, they load additional content, or what is displayed changes based on what the user does. Frequently, some JavaScript is run within the user's browser to do that. However, we do not have a browser here. The HTML code we downloaded may contain some instructions for the browser that some code needs to be run, but in the absence of a browser, our Python or R script cannot do this.

As a first test to check out whether this is a concern, you can simply check whether the HTML code in your browser is the same as that you would get if you downloaded

Python code

```
1 BASEURL = "https://cssbook.net/d/eat/"
2
3 def get_restaurants(url):
4     """takes the URL of an overview page as input
5     returns a list of (name, link) tuples"""
6     tree = parse(urlopen(url))
7     names = [e.text.strip() for e in
8       tree.xpath("//div[@class='restaurant']/h3")]
9     links = [e.attrib["href"] for e in
10      tree.xpath("//div[@class='restaurant']//a")]
11    return list(zip(names, links))
12
13 def get_reviews(url):
14    """yields reviews on the specified page"""
15    while True:
16        print(f"Downloading {url}...")
17        tree = parse(urlopen(url))
18        names = [e.text.strip() for e in
19            tree.xpath("//div[@class='review']/h3")]
20        texts = [e.text.strip() for e in
21            tree.xpath("//div[@class='review']/p")]
22        ratings = [e.text.strip() for e in tree.xpath(
23            "//div[@class="review"]/div[@class='rating']")]
24        for u, txt, rat in zip(names, texts, ratings):
25            review = {}
26            review["username"] = u.replace("wrote:","")
27            review["reviewtext"] = txt
28            review["rating"] = rat
29            yield review
30        bb= tree.xpath("//span[@class='backbutton']")) > 0:
31        if bb
32            print("Processing next page")
33            url = BASEURL+bb[0].attrib["href"]
34          else:
35            print("No more pages found.")
36            break
37
38 print("Retrieving all restaurants...")
39 links = get_restaurants(BASEURL+"index.html")
40 print(links)
41
42 with open("reviews.json", mode = "w") as f:
43     for restaurant, link in links:
44         print(f"Processing {restaurant}...")
45         for r in get_reviews(BASEURL+link):
46             r["restaurant"] = restaurant
47             f.write(json.dumps(r))
48             f.write("\n")
49
50 # You can process the results with pandas
51 # (using lines=True since it"s one json per line)
52 df = pd.read_json("reviews.json", lines=True)
53 print(df)53
```

Python Output

```
Retrieving all restaurants and their links...
[('Pizzeria Roma', 'review0001.html'), ('Trattoria Napoli', 'review0002.html'), ('Curry King',
        'review0003.html')]
Processing reviews for Pizzeria Roma...
Downloading and parsing review page http://cssbook.net/d/restaurants/review0001.html...
No more pages found.
Processing reviews for Trattoria Napoli...
Downloading and parsing review page http://cssbook.net/d/restaurants/review0002.html...
No more pages found.
Processing reviews for Curry King...
Downloading and parsing review page http://cssbook.net/d/restaurants/review0003.html...
Found page with older reviews! I'll process that one next
Downloading and parsing review page http://cssbook.net/d/restaurants/review0003-1.html...
Found page with older reviews! I'll process that one next
Downloading and parsing review page http://cssbook.net/d/restaurants/review0003-2.html
No more pages found.
        username                   reviewtext rating      restaurant
0    gourmet2536 The best thing to do is orderi 7.0/10 Pizzeria Roma
1       foodie12    The worst food I ever had! 1.0/10 Pizzeria Roma
2  mrsdiningout If nothing else is open, you c 6.5/10 Trattoria Napoli
3       foodie12        Best Italian in town! 8.6/10  Trattoria Napoli
4          smith                    Love it! 9.0/10        Curry King
5       foodie12                    Superb! 9.2/10        Curry King
6     dontlikeit As expected, I didn't like it 4.0/10     Curry King
7      otherguy Try the yoghurt curry! 7.7/10             Curry King
8          tasty We went here for dinner once a 7.0/10    Curry King
9           anna I have mixed feeling about thi 6.2/10    Curry King
10          hans                Not much to say 5.0/10    Curry King
11       bee1983                I am a huge fan! 10/10    Curry King
12        rhebjf The service is good, the food 6.5/10     Curry King
13 foodcritic555        Once and never again!. 1.0/10     Curry King
```

Example 12.9 Crawling a website.

Python Code

```python
with open("test.html", mode="w") as fo:
    fo.write(htmlsource)
```

R Code

```r
fileConn<-file("test.html")
writeLines(content(r, as = "text"), fileConn)
close(fileConn)
```

Example 12.10 Dumping the HTML source to a file.

it with R or Python. After having retrieved the page (Example 12.7), you simply dump it to a file (Example 12.10) and open this file in your browser to verify that you indeed downloaded what you intended to download (and not, for instance, a login page, a cookie wall, or an error message).

If this test shows that the data you are interested in is indeed not part of the HTML code you can retrieve with R or Python, and use the following checklist to find

1. Does using a different user-agent string (see above) solve the issue?
2. Is the issue due to some cookie that needs to be accepted or you need to log in (see below)?
3. Is a different page delivered for different browsers, devices, display settings, etc.?

If all of this does not help, or if you already know for sure that the content you are interested in is dynamically fetched via JavaScript or similar, you can use *Selenium* to literally start a browser and extract the content you are interested in from there. Selenium has been designed for testing web sites and allows you to automate clicks in a browser window, and also supports CSS selectors and xpaths to specify parts of the web page.

Using Selenium may require some additional setup on your computer, which may depend on your operating system and the software versions you are using – check out the usual online sources for guidance if needed. It is possible to use Selenium through R using *Rselenium*. However, doing so can be quite a hassle and requires, running a separate Selenium server, for instance, using Docker. If you opt to use Selenium for web scraping, your safest bet is probably to follow an online tutorial and/or to dive into the documentation. To give you a first impression of the general working, Example 12.11 shows you how to (at the time of writing of this book) open Firefox, surf to DuckDuckGo, search for Tintin by entering that string and pressing the return key, click on the first link containing that string, and take a screenshot of the result.

Losing your head

If you want to run long-lasting scraping processes using Selenium in the background (or on a server without a graphical user interface), you may want to look into what is called a "headless" browser. For instance, Selenium can start Firefox in "headless" mode, which means that it will run without making any connection to a graphical interface. Of course, that also means that you cannot watch Selenium scrape, which may make debugging more difficult. You could opt for developing your scraper first using a normal browser, and then changing it to use a headless browser once everything works.

Python Code

```
1 driver = webdriver.Firefox()
2 driver.implicitly_wait(10)
3 driver.get("https://www.duckduckgo.com")
4 element = driver.find_element_by_name("q")
5 # also check out other options such as
6 # .find_element_by_xpath
7 # or .find_element_by_css_selector
8 element.send_keys("TinTin")
9 element.send_keys(Keys.RETURN)
10 try:
11     driver.find_element_by_css_selector(
12        "#links a").click()
13    # let"s be cautious and wait 10 seconds
14    # so that everything is loaded
15    time.sleep(10)
16    driver.save_screenshot("screenshotTinTin.png")
17 finally:
18    # whatever happens, close the browser
19    driver.quit()
```

Example 12.11 Using Selenium in Python to open a browser window, input text, click on a link, and take a screenshot.

12.3 Authentication, Cookies, and Sessions

12.3.1 Authentication and APIs

When we introduced APIs in Section 12.1, we used the example of an API where you did not need to authenticate yourself. As we have seen, using such an API is as simple as sending an HTTP request to an endpoint and getting a response (usually, a JSON object) back. And indeed, there are plenty of interesting APIs (think for instance of open government APIs) that work this way.

While this has obvious advantages for you, it also has some serious downsides from the perspective of the API provider as well as from a security and privacy standpoint. The more confidential the data is, the more likely it is that the API provider needs to know who you are in order to determine which data you are allowed to retrieve; and even if the data are not confidential, authentication may be used to limit the number of requests that an individual can make in a given time frame.

In its most simple form, you just need to provide a unique key that identifies you as a user. For instance, Example 12.12 shows how such a key can be passed along as an HTTP header, essentially as additional information next to the URL that you want to retrieve (see also Section 12.3.2). The example shows a call to an endpoint of a commercial API for natural language processing to inform how many requests we have made today.

As you see, using an API that requires authentication by passing a key as an HTTP header is hardly more complicated than using APIs that do not require authentication such as outlined in Section 12.1. However, many APIs use more complex protocols for authentication.

The most popular one is called OAuth, and it is used by many APIs provided by major players such as Google, Facebook, Twitter, Github, LinkedIn, etc. Here, you have a client ID and a client secret (sometimes also called consumer key and consumer secret, or API key and API secret) and an access token with associated access token secret. The first pair authenticates you as a user, the second pair authenticates the specific "app" (i.e., your script). Once authenticated, your script can then interact

Python Code

```python
requests.get("https://api.textrazor.com/account/",
headers={"x-textrazor-key": "SECRET"}).json()
```

R Code

```r
r = GET("https://api.textrazor.com/account/",
    add_headers('x- textrazor-key'= "SECRET"))
print(content(r, "text"))
```

Output

```
{'ok': True,
 'response': {'planDailyRequestsIncluded': 15000,
  'requestsUsedToday': 10734,
  'plan': 'TR_FRIEND',
  'concurrentRequestLimit': 2,
  'concurrentRequestsUsed': 1}}
```

Example 12.12 Passing a key as HTTP request header to authenticate at an API endpoint.

with the API. While it is possible to directly work with OAuth HTTP requests using *requests_oauthlib* (Python) or *httr* (R), chances of having to do so are relatively low, unless you plan on really developing your own app or even your own API: for all popular API's, so-called wrappers, packages that provide a simpler interface to the API, are available on pypi and CRAN. Still, all of these require to have at least a consumer key and a consumer secret. The access token sometimes is generated via a web interface where you manage your account (e.g., in the case of Twitter), or can be acquired by your script itself, which then will redirect the user to a website in which they are asked to authenticate the app. The nice thing about this is that it only needs to happen once: once your app is authenticated, it can keep making requests.

12.3.2 Authentication and Webpages

In this section, we briefly discuss different approaches for dealing with websites where you need to log on, accept something (e.g., a so-called cookie wall), or have to otherwise authenticate yourself. One approach can be the use of a web testing framework like Selenium (see Section 12.2.3): you let your script literally open a browser and, for instance, fill in your login information.

However, sometimes that's not necessary and we can still use simpler and more efficient webscraping without invoking a browser. As we have already seen in Section 12.2.1, when making an HTTP request, we can transmit additional information, such as the so-called user-agent string. In a similar way, we can pass other information, such as cookies.

In the developer tools of your browser (which we already used to determine XPath and CSS selectors), you can look up which cookies a specific website has placed. For instance, you could inspect all cookies *before* you logged on (or passed a cookie wall) and again inspect them afterwards to determine what has changed. With this kind of reverse-engineering, you can find out what cookies you need to manually set.

In Example 12.13, we illustrate this for a specific page (at the time of writing of our book). Here, by inspecting the cookies in Firefox, we found out that clicking "Accept" on the cookie wall landing page caused a cookie with the name cpc and the value 10 to

Python Code

```
1 URL = "https://www.geenstijl.nl/5160019/page"
2
3 # circumvent cookie wall by setting a specific
4 # cookie: the key-value pair (cpc: 10)
5 client = requests.session()
6 r = client.get(URL)
7 cookies = client.cookies.items()
8 cookies.append(("cpc", "10"))
9 response = client.get(URL,cookies=dict(cookies))
10 # end circumvention
11
12 tree = fromstring(response.text)
13 allcomments = [e.text_content().strip() for e in
14               tree.cssselect(".cmt-content")]
15 print(f"There are {len(allcomments)} comments.")
```

Python Output

```
There are 72 comments
```

Example 12.13 Explicitly setting a cookie to circumvent a cookie wall.

be set. To set those cookies in our scraper, the easiest way is to retrieve that page first and store the cookies sent by the server. In Example 12.13, we therefore start a *session* and try to download the page. We know that this will only show us the cookie wall – but it will also generate the necessary cookies. We then store these cookies, and add the cookie that we want to be set (`cpc=10`) to this cookie jar. Now, we have all cookies that we need for future requests. They will stay there for the whole session.

If we only want to get a single page, we may not need to start a session to remember all the cookies, and we can just directly pass the single cookie we care about to a request instead (Example 12.14).

Python Code

```
1 r = requests.get(URL,cookies={"cpc": "10"})
2 tree = fromstring(r.text)
3 allcomments = [e.text_content().strip() for e in
4               tree.cssselect(".cmt-content")]
5 print(f"There are {len(allcomments)} comments.")
```

Example 12.14 Shorter version of Example 12.13 for single requests.

12.4 Ethical, Legal, and Practical Considerations

Web scraping is a powerful tool, but it needs to be handled responsibly. Between the white area of sites that explicitly consented to creating a copy of their data (for instance, by using a creative commons license) and the black area of an exact copy of copyrighted material and redistributing it as it is, there is a large gray area where it is less clear what is acceptable and what is not.

There is a tension between legitimate interests of the operators of web sites and the producers of content on the one hand, and the societal interest of studying online communication on the other hand. Which interest prevails may differ on a case-to-case basis. For instance, when using APIs as described in Section 12.1, in most cases, you have to consent to the terms of service (TOS) of the API provider.

For example, Twitter's TOS allow you to redistribute the numerical tweet ids, but not the tweets themselves, and therefore, it is common to share such lists of ids with fellow researchers instead of the "real" Twitter datasets. Of course, this is not optimal from a reproducibility point of view: if another researcher has to retrieve the tweets again based on their ids, then this is not only cumbersome, but most likely also leads to a slightly different dataset, because tweets may have been deleted in the meantime. At the same time, it is a compromise most people can live with.

Other social media platforms have closed their APIs or tightened the restrictions a lot, making it impossible to study many pressing research questions. Therefore, some have even called researchers to neglect these TOS, because "in some circumstances the benefits to society from breaching the terms of service outweigh the detriments to the platform itself" (Bruns, 2019, p. 1561). Others acknowledge the problem, but doubt that this is a good solution (Puschmann, 2019). In general, one needs to distinguish between the act of collecting the data and sharing the data. For instance, in many jurisdictions, there are legal exemptions for collecting data for scientific purposes, but that does not mean that they can be re-distributed as they are (Van Atteveldt et al., 2019).

This chapter can by no means replace the consultation of a legal expert and/or an ethics board, but we would like to offer some strategies to minimize potential problems.

Be nice. Of course, you could send hundreds of requests per minute (or second) to a website and try to download everything that they have ever published. However, this causes unnecessary load on their servers (and you would probably get blocked). If, on the other hand, you carefully think about what you really need to download, and include a lot of waiting times (for instance, using `sys.sleep` (R) or `time.sleep` (Python) so that your script essentially does the same as could be done by hiring a couple of student assistants to copy-paste the data manually, then problems are much less likely to arise.

Collaborate. Another way to minimize traffic and server load is to collaborate more. A concerted effort with multiple researchers may lead to less duplicate data and in the end probably an even better, re-usable dataset.

Be extra careful with personal data. Both from an ethical and a legal point of view, the situation changes drastically as soon as personal data are involved. Especially since the General Data Protection Regulation (GDPR) regulations took effect in the European Union, collecting and processing such data requires a lot of additional precaution and is usually subject to explicit consent. It is clearly infeasible to ask every Twitter user for consent to process their tweet and doing so is probably covered by research exceptions, the general advice is to store as little personal data as possible and only what is absolutely needed. Most likely, you need to have a data management plan in place, and should get appropriate advice from your legal department. Therefore, think carefully whether you really need, for instance, the user names of the authors of reviews you are going to scrape, or whether the text alone suffices.

Once all ethical and legal concerns are sorted out and you have made sure that you have written a scraper in such a way that it does not cause unnecessary traffic and load on the servers from which you are scraping, and after doing some test runs, it is time to think about how to actually run it on a larger scale. You may already have figured that you probably do not want to run your scraper from a Jupyter Notebook that is constantly open in your browser on your personal laptop. Also here, we would like to offer some suggestions.

Consider using a database. Imagine the following scenario: your scraper visits hundreds of websites, collects its results in a list or in a data frame, and after hours of running suddenly crashes – maybe because some element that you were sure must exist on each page, exists only on 999 out of 1000 pages, because a connection timed out, or any other error. Your data is lost, you need to start again (not only annoying, but also undesirable from a traffic minimization point of view). A better strategy may be to immediately write the data for each page to a file. But then, you need to handle a potentially huge number of files later on. A much better approach, especially if you plan to run your scraper repeatedly over a long period of time, is to consider the use of a database in which you dump the results immediately after a page has been scraped (see Section 15.1).

Run your script from the command line. Store your scraper as a .py or .R script and run it from your terminal (your command line) by typing `python myscript.py` or `R myscript.R` rather than using an IDE such as Spyder or R Studio or a Jupyter Notebook. You may want to have your script print a lot of status information (for instance, which page it is currently scraping), so that you can watch what it is doing. If you want to, you can have your computer run this script in regular intervals (e.g., once an hour). On Linux and MacOS, for instance, you can use a so-called *cron* job to automate this.

Run your script on a server. If your scraper runs for longer than a couple of hours, you may not want to run it on your laptop, especially if your Internet connection is not stable. Instead, you may consider using a server. As we will explain in Section 15.2, it is quite affordable to set up a Linux VM on a cloud computing platform (and next to commercial services, in some countries and institutions there are free services for academics). You can then use tools like *nohup* or *screen* to run your script on the background, even if you are no longer connected to the server (see Section 15.2).

13

Network Data

Abstract

Especially social media data, but also other types of data can often be represented as networks. This chapter introduces *igraph* (R and Python) and *networkx* (Python) to showcase how to deal with such data, perform Social Network Analysis (SNA) and represent it visually.

Keywords graphs, social network analysis

- Understand how can networks be represented and visualized
- Conduct basic description of networks
- Perform Social Network Analysis

Packages used in this chapter

This chapter uses functions from the package *igraph* in R and the package *networkx* in Python. In Python we will also use the *python-louvain* packages which introduces the Louvain clustering functions in *community*. You can install these packages with the code below if needed (see Section 1.4 for more details):

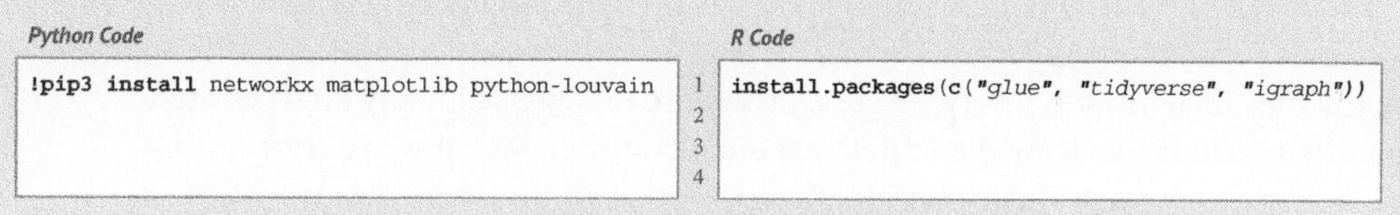

Python Code

```
!pip3 install networkx matplotlib python-louvain
```

R Code

```
install.packages(c("glue", "tidyverse", "igraph"))
```

After installing, you need to import (activate) the packages every session:

Python Code

```
import urllib.request
%matplotlib inline
import matplotlib.pyplot as plt
import networkx as nx
import networkx.algorithms.community as nxcom
import community # from python-louvain package
```

R Code

```
library(glue)
library(igraph)
library(tidyverse)
```

Computational Analysis of Communication: A Practical Introduction to the Analysis of Texts, Networks, and Images with Code Examples in Python and R, First Edition. Wouter van Atteveldt, Damian Trilling & Carlos Arcila Calderón.

13.1 Representing and Visualizing Networks

How can networks help us to understand and represent social problems? How can we use social media as a source for small and large-scale network analysis? In the computational analysis of communication these questions become highly relevant given the huge amount of social media data produced every minute on the Internet. In fact, although graph theory and SNA were already being used during the last two decades of the 20th century, we can say that the widespread adoption of the Internet and especially social networking services such as Twitter and Facebook really unleashed their potential. Firstly, computers made it easier to compute graph measures and visualize their general and communal structures. Secondly, the emergence of a big spectrum of social media network sites (i.e. Facebook, Twitter, Sina Weibo, Instagram, Linkedin, etc.) produced an unprecedented number of online social interactions, which still is certainly an excellent arena to apply this framework. Thus, the use of social media as a source for network analysis has become one of the most exciting and promising areas in the field of computational social science.

This section presents a brief overview of graph structures (nodes and edges) and types (directed, weighted, etc.), together with their representation in R and Python. We also include visual representations and basic graph analysis.

A graph is a structure derived from a set of elements and their relationships. The element could be a neuron, a person, an organization, a street, or even a message, and the relationship could be a synapse, a trip, a commercial agreement, a drive connection or a content transmission. This is a different way to represent, model and analyze the world: instead of having rows and columns as in a typical data frame, in a graph we have *nodes* (components) and *edges* (relations). The mathematical representation of a graph $G = (V, E)$ is based on a set of nodes (also called vertices):

$$v_1, v_2, \ldots v_n$$

and the edges or pair of nodes:

$$(v_1, v_2), (v_1, v_3), (v_2, v_3) \ldots (v_m, v_n) \in E$$

As you may imagine, it is a very versatile procedure to represent many kinds of situations that include social, media or political interactions. In fact, if we go back to 1934 we can see how graph theory (originally established in the 18th century) was first applied to the representation of social interactions (Moreno, 1934) in order to measure the attraction and repulsion of individuals of a social group[1].

The network approach in social sciences has an enormous potential to model and predict *social actions*. There is empirical evidence that we can successfully apply this framework to explain distinct phenomena such as political opinions, obesity, and happiness, given the influence of our friends (or even of the friends of our friends) over

---

[1] See also the mathematical problem of the *Seven Bridges of Königsberg*, formulated by Leonhard Euler in 1736, that is considered the basis of graph theory. Inspired by a city divided by a river and connected by several bridges, the problem consisted of walking through the whole city crossing each bridge exactly once.

our behaviour (Christakis and Fowler, 2009). The network created by this sophisticated structure of human and social connections is an ideal scenario to understand how close we are to each other in terms of degrees of separation (Watts, 2004) in small (such as a school) and large-scale (such as a global pandemic) social dynamics. Moreover, the network approach can help us to track the propagation either of a virus in epidemiology, or a fake news story in political and social sciences, such as in the work by Vosoughi et al. (2018).

Now, let us show you how to create and visualize network structures in R and Python. As we mentioned above, the structure of a graph is based on nodes and edges, which are the fundamental components of any network. Suppose that we want to model the social network of five American politicians in 2017 (Donald Trump, Bernie Sanders, Hillary Clinton, Barack Obama and John McCain), based on their *imaginary* connections on Facebook (friending) and Twitter (following)[2]. Technically, the base of any graph is a list of edges (written as pair of nodes that indicate the relationships) and a list of nodes (some nodes might be isolated without any connection!). For instance, the friendship on Facebook between two politicians would normally be expressed as two strings separated by comma (e.g. "Hillary Clinton", "Donald Trump"). In Example 13.1 we use libraries *igraph* (R)[3] and *networkx* (Python) to create from scratch a simple graph with five nodes and four edges, using the above-mentioned structure of pairs of nodes (notice that we only include the edges while the vertices are automatically generated).

In both cases we generated a graph object `g1` which contains the structure of the network and different attributes (such as `number_of_nodes()` in *networkx*). You

Python Code

```
edges = [("Hillary Clinton", "Donald Trump"),
         ("Bernie Sanders","Hillary Clinton"),
         ("Hillary Clinton", "Barack Obama"),
         ("John McCain", "Donald Trump"),
         ("Barack Obama", "Bernie Sanders")]
g1 = nx.Graph()
g1.add_edges_from(edges)
print("Imaginary Facebook network of 5 "
      "American politicians")
print("Nodes:", g1.number_of_nodes(),
      "Edges: ", g1.number_of_edges())
print(g1.edges)
```

R Code

```
edges=c("Hillary Clinton", "Donald Trump",
        "Bernie Sanders","Hillary Clinton",
        "Hillary Clinton", "Barack Obama",
        "John McCain", "Donald Trump",
        "Barack Obama", "Bernie Sanders")
g1 = make_graph(edges, directed = FALSE)
glue("Imaginary Facebook network of 5 ",
"American politicians")
g1
```

Python Output

```
Imaginary Facebook network of 5 American
    politicians
Nodes: 5 Edges: 5
[('Hillary Clinton', 'Donald Trump'), ('Hillary
    Clinton', 'Bernie Sanders'), ('Hillary
    Clinton', 'Barack Obama'), ('Donald Trump',
    'John McCain'), ('Bernie Sanders', 'Barack
    Obama')]
```

R Output

```
IGRAPH c0819d5 UN-- 5 5 --
+ attr: name (v/c)
+ edges from c0819d5 (vertex names):
[1] Hillary Clinton--Donald Trump Hillary
     Clinton--Bernie Sanders
[3] Hillary Clinton--Barack Obama Donald Trump
     --John McCain
[5] Bernie Sanders --Barack Obama
```

Example 13.1 Creating a graph from scratch.

[2] The connections among these politicians on Facebook and Twitter in the examples are of course purely fictional and were created *ad hoc* to illustrate small social networks.

[3] You can use this library in Python with the adapted package *python-igraph*.

can add/remove nodes and edges to/from this initial graph, or even modify the names of the vertices. One of the most useful functions is the visualization of the network (`plot` in *igraph* and `draw` or `draw_networkx` in *networkx*). Example 13.2 shows a basic visualization of the imaginary network of friendships of five American politicians on Facebook.

Using network terminology, either nodes or edges can be *adjacent* or not. In the figure we can say that nodes representing Donald Trump and John McCain are adjacent because they are connected by an edge that depicts their friendship. Moreover, the edges representing the friendships between John McCain and Donald Trump, and Hillary Clinton and Donald Trump, are also adjacent because they share one node (Donald Trump).

Now that you know the relevant terminology and basics of working with graphs, you might be wondering: what if I want to do the same with Twitter? Can I represent the relationships between users in the very same way as Facebook? Well, when you

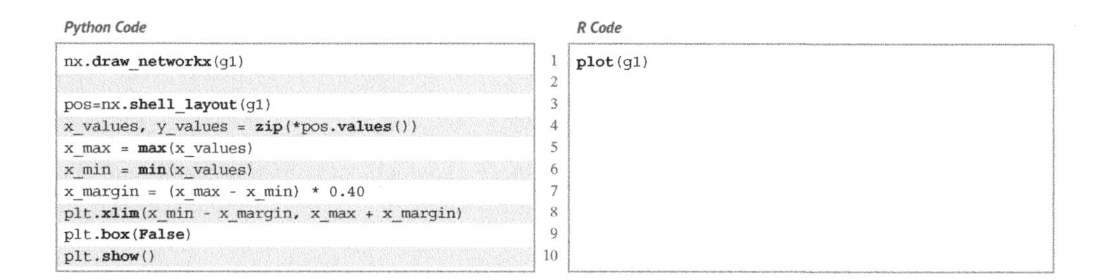

Python Code

```python
nx.draw_networkx(g1)

pos=nx.shell_layout(g1)
x_values, y_values = zip(*pos.values())
x_max = max(x_values)
x_min = min(x_values)
x_margin = (x_max - x_min) * 0.40
plt.xlim(x_min - x_margin, x_max + x_margin)
plt.box(False)
plt.show()
```

R Code

```r
plot(g1)
```

R output

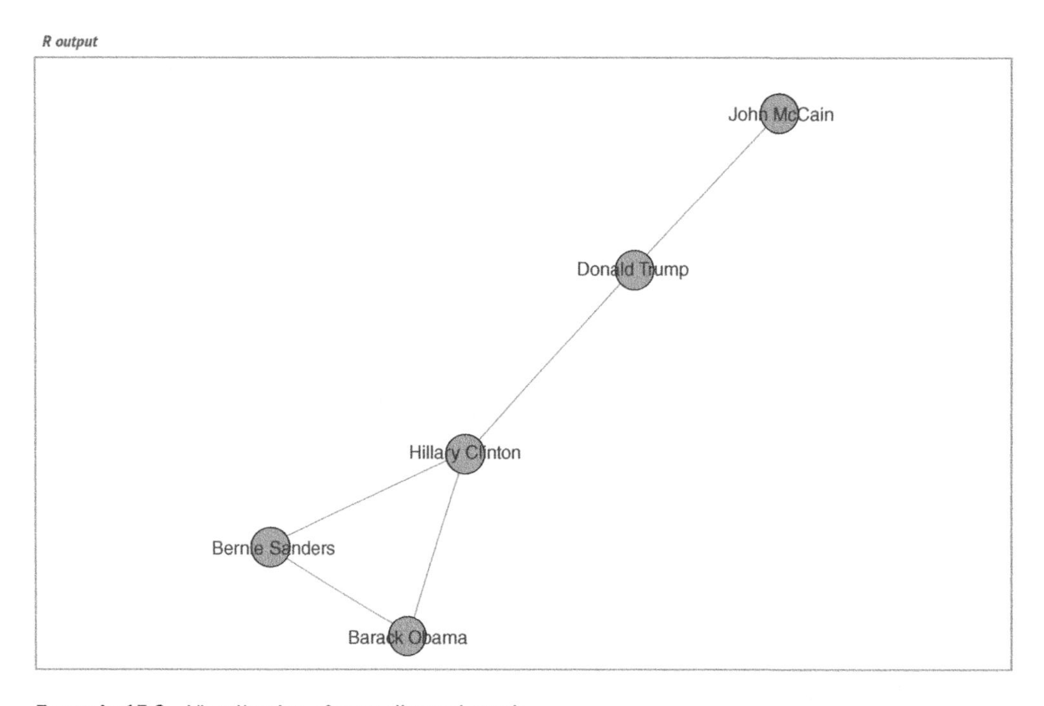

Example 13.2 Visualization of an undirected graph.

model networks it is extremely important that you have a clear definition of what you mean with nodes and edges, in order to maintain a coherent interpretation of the graph. In both, Facebook and Twitter, the nodes represent the users, but the edges might not be the same. In Facebook, an edge represents the friendship between two users and this link *has no direction* (once a user accepts a friend request, both users become friends). In the case of Twitter, an edge could represent various relationships. For example, it could mean that two users follow each other, or that one user is following another user, but not the other way around! In the latter case, the edge *has a direction*, which you can establish in the graph. When you give directions to the edges you are creating a *directed graph*. In Example 13.3 the directions are declared with the order of the pair of nodes: the first position is for the "from" and the second for the "to". In *igraph* (R) we set the argument `directed` of the function `make_graph` to `TRUE`. In *networkx* (Python), you use the class `DiGraph` instead of `Graph` to create the object `g2`.

In the new graph the edges represent the action of following a user on Twitter. The first declared edge indicates that Hillary Clinton follows Donald Trump, but does not indicate the opposite. In order to provide the directed graph with more *arrows* we included in `g2` two new edges (Obama following Clinton and Clinton following Sanders), so we can have a couple of reciprocal relationships besides the unidirectional ones. You can visualize the directed graph in Example 13.4 and see how the edges now contain useful arrows.

The edges and nodes of our graph can also have weights and features or attributes. When the edges have specific values that depict a feature of every pair of nodes (i.e., the distance between two cities) we say that we have a *weighted graph*. This type of graph is extremely useful for creating a more accurate representation of a network. For example, in our hypothetical network of American politicians on Twitter (`g2`) we

Python Code

```
edges += [("Hillary Clinton", "Bernie Sanders"),
          ("Barack Obama","Hillary Clinton")]
g2 = nx.DiGraph()
g2.add_edges_from(edges)
print("Imaginary Twitter network of 5 "
      "American politicians")
print("Nodes:", g2.number_of_nodes(),
      "Edges: ", g2.number_of_edges())
print(g2.edges)
```

R Code

```
edges = c(edges,
          "Hillary Clinton", "Bernie Sanders",
          "Barack Obama","Hillary Clinton")
g2 = make_graph(edges, directed = TRUE)
glue("Imaginary Facebook network of 5 ",
"American politicians")
print(g2)
```

Python Output

```
Imaginary Twitter network of 5 American
      politicians
Nodes: 5 Edges: 7
[('Hillary Clinton', 'Donald Trump'), ('Hillary
      Clinton', 'Barack Obama'), ('Hillary
      Clinton', 'Bernie Sanders'), ('Bernie
      Sanders', 'Hillary Clinton'), ('Barack
      Obama', 'Bernie Sanders'), ('Barack Obama',
      'Hillary Clinton'), ('John McCain', 'Donald
      Trump')]
```

R Output

```
Imaginary Twitter network of 5 American
      politicians
IGRAPH d9fd45f DN-- 5 7 --
+ attr: name (v/c)
+ edges from d9fd45f (vertex names):
[1] Hillary Clinton->Donald Trump Bernie Sanders
      ->Hillary Clinton
[3] Hillary Clinton->Barack Obama John McCain
      ->Donald Trump
[5] Barack Obama ->Bernie Sanders Hillary
      Clinton->Bernie Sanders
[7] Barack Obama ->Hillary Clinton
```

Example 13.3 Creating a directed graph.

Python Code

```
nx.draw_networkx(g2)

pos=nx.shell_layout(g2)
x_values, y_values = zip(*pos.values())
x_max = max(x_values)
x_min = min(x_values)
x_margin = (x_max - x_min) * 0.40
plt.xlim(x_min - x_margin, x_max + x_margin)
plt.box(False)
plt.show()
```

R Code

```
plot(g2)
```

R Output

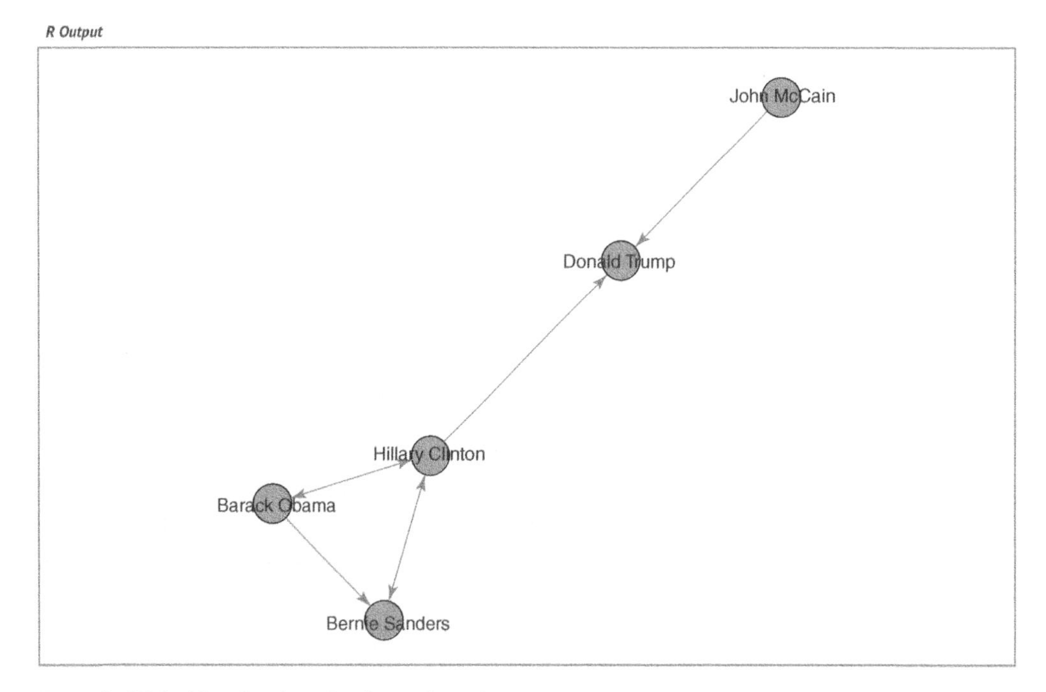

Example 13.4 Visualization of a directed graph.

can assign weights to the edges by including the number of likes that each politician has given to the followed user. This value can serve as a measure of the distance between the nodes (i.e., the higher the number of likes the shorter the social distance). In Example 13.5 we include the weights for each edge: Clinton has given five likes to Trumps' tweets, Sanders 20 to Clinton's messages, and so on. In the plot you can see how the sizes of the lines between the nodes change as a function of the weights.

You can include more properties of the components of your graph. Imagine you want to use the number of followers of each politician to determine the size of the nodes, or the gender of the user to establish a color. In Example 13.6 we added the variable *followers* to each of the nodes and asked the packages to plot the network using this value as the size parameter (in fact we multiplied the values by 0.001 to make it realistic on the screen, but you could also normalize these values when needed). We also included the variable *party* that was later recoded in a new one called *color* in order to represent

Python Code

```python
edges_w = [("Hillary Clinton", "Donald Trump", 5),
    ("Bernie Sanders","Hillary Clinton", 20),
    ("Hillary Clinton", "Barack Obama", 30),
    ("John McCain", "Donald Trump", 40),
    ("Barack Obama", "Hillary Clinton", 50),
    ("Hillary Clinton", "Bernie Sanders", 10),
    ("Barack Obama", "Bernie Sanders", 15)]
g2 = nx.DiGraph()
g2.add_weighted_edges_from(edges_w)

edge_labels=dict([((u,v,),d["weight"]) for
                u,v,d in g2.edges(data=True)])

nx.draw_networkx_edge_labels(g2,pos,
                    edge_labels=edge_labels)
nx.draw_networkx(g2, pos)

pos=nx.spring_layout(g2)
x_values, y_values = zip(*pos.values())
x_max = max(x_values)
x_min = min(x_values)
x_margin = (x_max - x_min) * 0.40
plt.xlim(x_min - x_margin, x_max + x_margin)
plt.box(False)
plt.show()
```

R Code

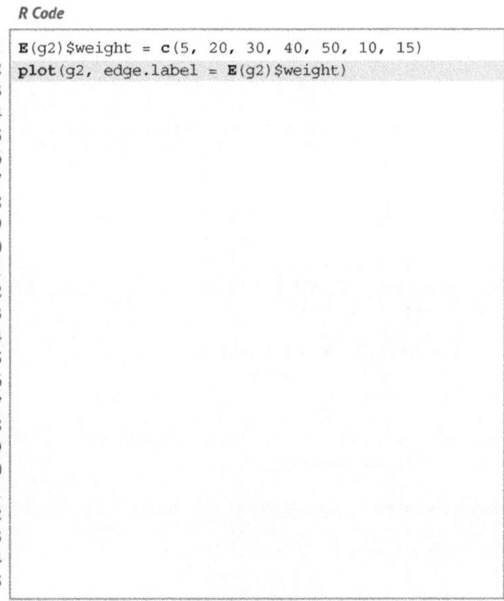

```r
E(g2)$weight = c(5, 20, 30, 40, 50, 10, 15)
plot(g2, edge.label = E(g2)$weight)
```

R Output

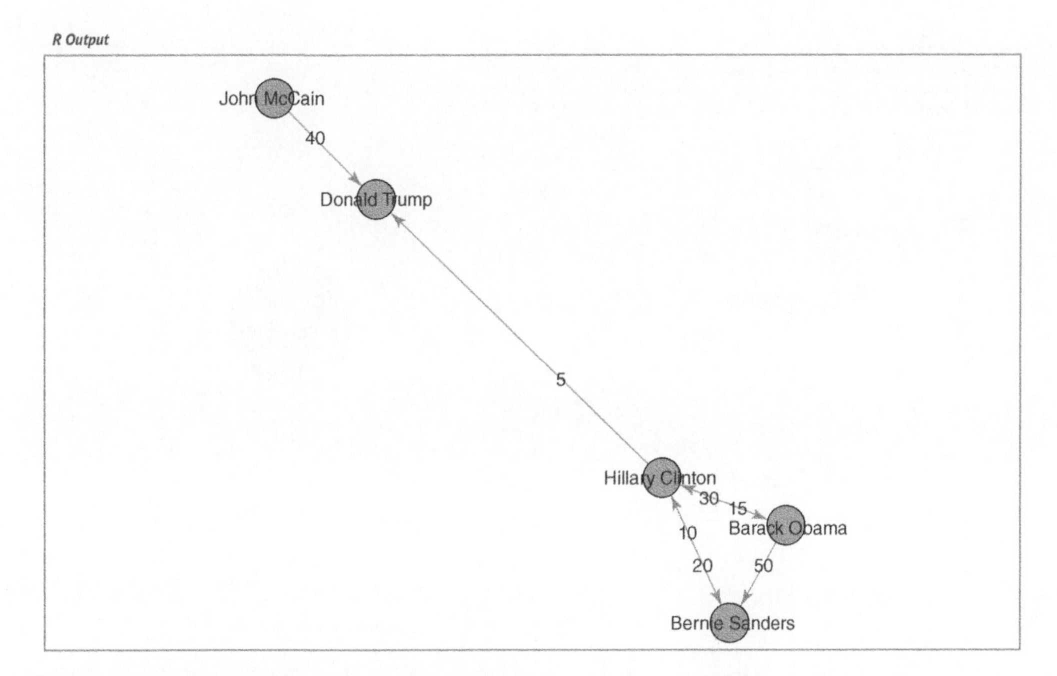

Example 13.5 Visualization of a weighted graph.

Republicans with red and Democrats with blue. You may need to add other features to the nodes or edges, but with this example you have an overview of what you can do.

We can mention a third type of graphs: the *induced subgraphs*, which are in fact subsets of nodes and edges of a bigger graph. We can represent these subsets as $G' = (V', E')$. In Example 13.7 we extract two induced subgraphs from our original

Python Code

```python
attrs = {"Hillary Clinton": {"followers": 100000,
                             "party": "Democrat"},
         "Donald Trump": {"followers": 200000,
                          "party": "Republican"},
         "Bernie Sanders": {"followers": 50000,
                            "party": "Democrat"},
         "Barack Obama": {"followers": 500000,
                          "party": "Democrat"},
         "John McCain": {"followers": 40000,
                         "party": "Republican"} }
nx.set_node_attributes(g2, attrs)
size = nx.get_node_attributes(g2, "followers")
size = list(size.values())

colors= nx.get_node_attributes(g2, "party")
colors = list(colors.values())
colors = [w.replace("Democrat", "blue") for
          w in colors]
colors = [w.replace("Republican", "red") for
          w in colors]

nx.draw_networkx_edge_labels(g2,pos,
                             edge_labels=edge_labels)
nx.draw_networkx(g2, pos, node_size=
                 [x * 0.002 for x in size],
                 node_color=colors)

pos=nx.spring_layout(g2)
x_values, y_values = zip(*pos.values())
x_max = max(x_values)
x_min = min(x_values)
x_margin = (x_max - x_min) * 0.40
plt.xlim(x_min - x_margin, x_max + x_margin)
plt.box(False)
plt.show()
```

R Code

```r
V(g2)$followers = c(100000, 200000,
                    50000, 500000, 40000)
V(g2)$party = c("Democrat", "Republican",
                "Democrat", "Democrat", "Republican")
V(g2)$color = V(g2)$party
V(g2)$color = gsub("Democrat", "blue",
                   V(g2)$color)
V(g2)$color = gsub("Republican", "red",
                   V(g2)$color)
plot(g2, edge.label = E(g2)$weight,
     vertex.size = V(g2)$followers*0.0001)
```

R Output

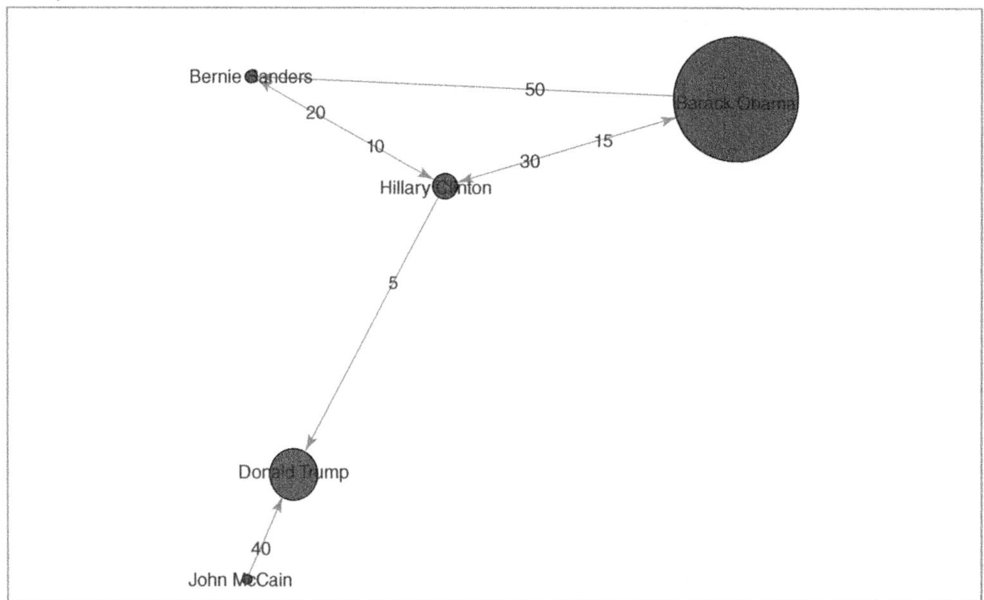

Example 13.6 Visualization of a weighted graph including vertex sizes.

Python Code

```python
g3 = g1.subgraph(["Hillary Clinton",
                  "Bernie Sanders",
                  "Barack Obama"])
print("Democrats:")
print("Nodes:", g3.number_of_nodes(),
      "Edges: ", g3.number_of_edges())
print(g3.edges)

g4 = g1.subgraph(["Donald Trump", "John McCain"])
print("Republicans:")
print("Nodes:", g4.number_of_nodes(),
      "Edges: ", g4.number_of_edges())
print(g4.edges)
```

R Code

```r
g3 = induced_subgraph(g1, c(1,3,4))
glue("Democrats subgraph")
print(g3)
g4 = induced_subgraph(g1, c(2,5))
glue("Republicans subgraph")
print(g4)
```

Python Output

```
Democrats:
Nodes: 3 Edges: 3
[('Hillary Clinton', 'Bernie Sanders'), ('Hillary
      Clinton', ' Barack Obama'), ('Bernie
      Sanders', 'Barack Obama')]
Republicans:
Nodes: 2 Edges: 1
[('John McCain', 'Donald Trump')]
```

R Output

```
Democrats subgraph
IGRAPH 5a711f9 UN-- 3 3 --
+ attr: name (v/c)
+ edges from 5a711f9 (vertex names):
[1] Hillary Clinton--Bernie Sanders Hillary
       Clinton--Barack Obama
[3] Bernie Sanders --Barack Obama
Republicans subgraph
IGRAPH 67fde4a UN-- 2 1 --
+ attr: name (v/c)
+ edge from 67fde4a (vertex names):
[1] Donald Trump--John McCain
```

Example 13.7 Induced subgraphs for Democrats and Republicans.

network of American politicians on Facebook (`g1`): the first (`g3`) is built with the edges that contain only Democrat nodes, and the second (`g4`) with edges formed by Republican nodes. There is also a special case of an induced subgraph, called a *clique*, which is an independent or complete subset of an undirected graph (each node of the clique must be connected to the rest of the nodes of the subgraph).

Keep in mind that in network visualization you can always configure the size, shape and color of your nodes or edges. It is out of the scope of this book to go into more technical details, but you can always check the online documentation of the recommended libraries.

So far we have created networks from scratch, but most of the time you will have to create a graph from an existing data file. This means that you will need an input data file with the graph structure, and some functions to load them as objects onto your workspace in R or Python. You can import graph data from different specific formats (such as Graph Modeling Language (GML), GraphML, JSON, etc.), but one popular and standardized procedure is to obtain the data from a text file containing a list of edges or a matrix. In Example 13.8 we illustrate how to read graph data in *igraph* and *networkx* using a simple adjacency list that corresponds to our original imaginary Twitter network of American politicians (`g2`).

13.2 Social Network Analysis

This section gives an overview of the existing measures to conduct Social Network Analysis (SNA). Among other functions, we explain how to examine paths and reachability, how to calculate centrality measures (degree, closeness, betweenness, eigenvector) to quantify the importance of a node in a graph, and how to detect communities in the graph using clustering.

Python Code

```python
url="https://cssbook.net/d/poltwit.csv"
fn, _headers = urllib.request.urlretrieve(url)
g2 = nx.read_adjlist(fn, create_using=nx.DiGraph,
                     delimiter=",")
print("Nodes:", g2.number_of_nodes(),
      "Edges: ", g2.number_of_edges())
```

R Code

```r
edges = read_csv(
"https://cssbook.net/d/poltwit.csv",
    col_names=FALSE)
g2 = graph_from_data_frame(d=edges)
glue("Nodes: ", gorder(g2),
    " Edges: ", gsize(g2))
plot(g2)
```

Python Output

```
Nodes: 5 Edges: 7
```

R Output

```
Nodes: 5 Edges: 7
```

Example 13.8 Reading a graph from a file.

13.2.1 Paths and Reachability

The first idea that comes to mind when analyzing a graph is to understand how their nodes are connected. When multiple edges create a network we can observe how the vertices constitute one or many paths that can be described. In this sense, a `sequence` between node x and node y is a path where each node is `adjacent` to the previous. In the imaginary social network of friendship of American politicians contained in the undirected graph `g1`, we can determine the sequences or simple paths between any pair of politicians. As shown in Example 13.9 we can use the function `all_simple_paths` contained in both *igraph* (R) and *networkx* (Python), to obtain the two possible routes between Barack Obama and John McCain. The shortest path includes the nodes Hillary Clinton and Donald Trump; and the longer includes Sanders, Clinton, and Trump.

One specific type of path is the one in which the initial node is the same than the final node. This closed path is called a *circuit*. To understand this concept let us recover the inducted subgraph of Democrat politicians (`g3`) in which we only have three nodes. If you plot this graph, as we do in Example 13.10, you can clearly visualize how a circuit works.

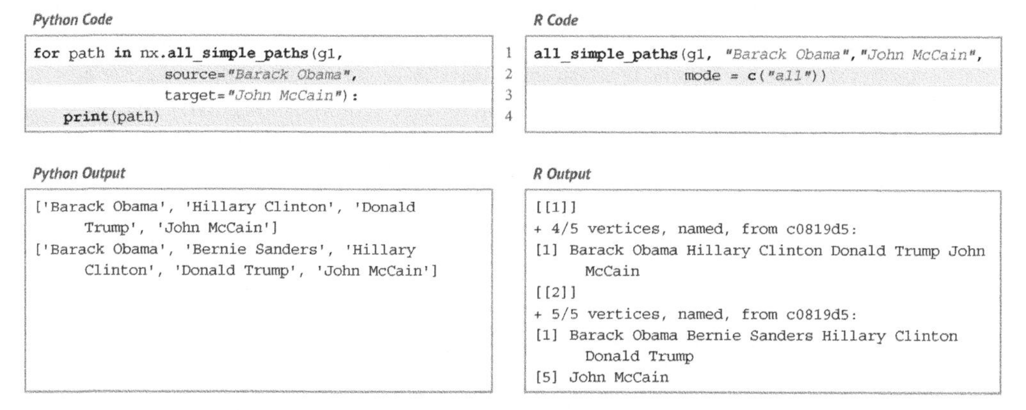

Python Code

```python
for path in nx.all_simple_paths(g1,
                source="Barack Obama",
                target="John McCain"):
    print(path)
```

R Code

```r
all_simple_paths(g1, "Barack Obama","John McCain",
                 mode = c("all"))
```

Python Output

```
['Barack Obama', 'Hillary Clinton', 'Donald
    Trump', 'John McCain']
['Barack Obama', 'Bernie Sanders', 'Hillary
    Clinton', 'Donald Trump', 'John McCain']
```

R Output

```
[[1]]
+ 4/5 vertices, named, from c0819d5:
[1] Barack Obama Hillary Clinton Donald Trump John
    McCain
[[2]]
+ 5/5 vertices, named, from c0819d5:
[1] Barack Obama Bernie Sanders Hillary Clinton
    Donald Trump
[5] John McCain
```

Example 13.9 Possible paths between two nodes in the imaginary Facebook network of American politicians.

In SNA it is extremely important to be able to describe the possible paths since they help us to estimate the reachability of the vertices. For instance, if we go back to our original graph of American politicians on Facebook (`g1`) visualized in Example 13.2, we can see that Sanders is reachable from McCain because there is a path between them (McCain–Trump–Clinton–Sanders). Moreover, we observe that this social network is fully *connected* because you can reach any given node from any other node in the graph. But it might not always be that way. Imagine that we remove the friendship of Clinton and Trump by deleting that specific edge. As you can observe in Example 13.11, when we create and visualize the graph `g6` without this edge we can see that the network is not longer fully connected and it has two *components*. Technically speaking, we would say for example that the subgraph of Republicans is a connected component of the network of American politicians, given that this connected subgraph is part of the bigger graph while not connected to it.

When analyzing paths and reachability you may be interested in knowing the distances in your graph. One common question is what is the average path length of a social network, or in other words, what is the average of the shortest distance between

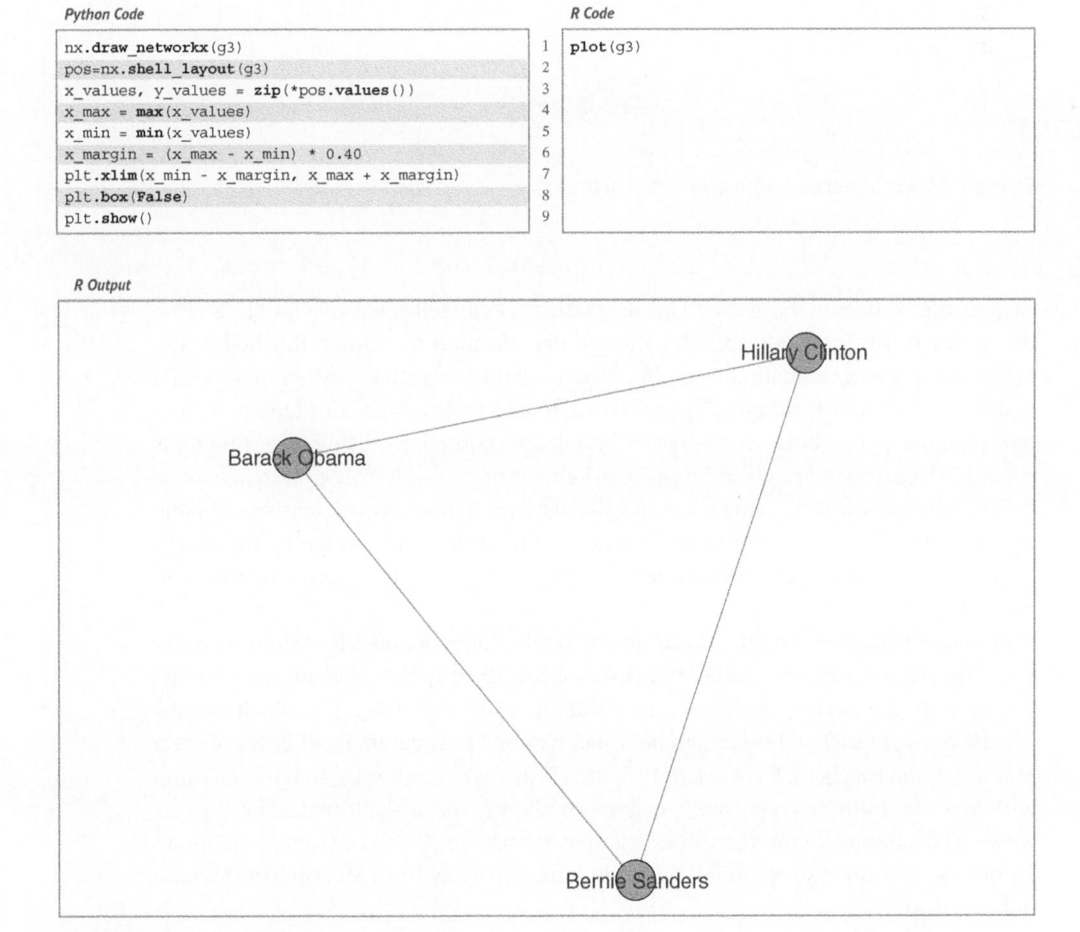

Python Code

```
nx.draw_networkx(g3)
pos=nx.shell_layout(g3)
x_values, y_values = zip(*pos.values())
x_max = max(x_values)
x_min = min(x_values)
x_margin = (x_max - x_min) * 0.40
plt.xlim(x_min - x_margin, x_max + x_margin)
plt.box(False)
plt.show()
```

R Code

```
plot(g3)
```

R Output

Example 13.10 Visualization of a circuit.

Python Code

```
#Remove the friendship between Clinton and Trump
g6 = g1.copy()
g6.remove_edge("Hillary Clinton","Donald Trump")
nx.draw_networkx(g6)
pos=nx.shell_layout(g6)
x_values, y_values = zip(*pos.values())
x_max = max(x_values)
x_min = min(x_values)
x_margin = (x_max - x_min) * 0.40
plt.xlim(x_min - x_margin, x_max + x_margin)
plt.box(False)
plt.show()
```

R Code

```
# Remove the friendship between Clinton and Trump
g6 = delete.edges(g1, E(g1, P=
        c("Hillary Clinton","Donald Trump")))
plot(g6)
```

R Output

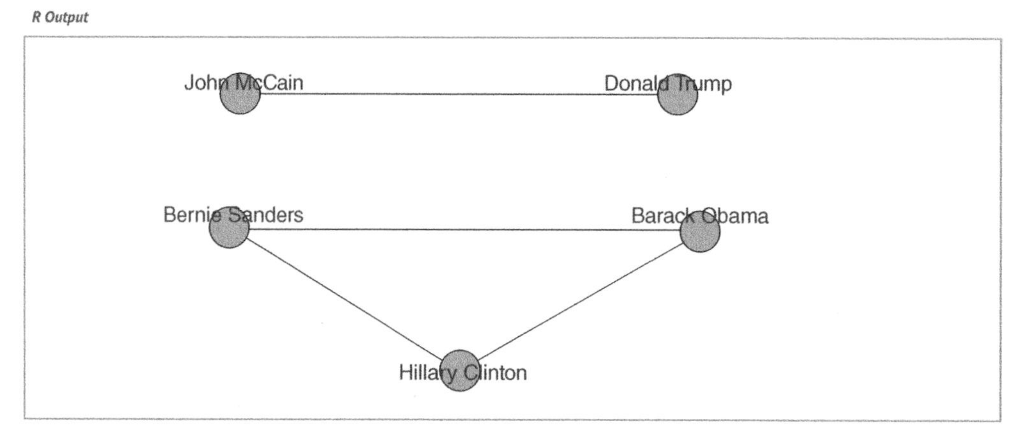

Example 13.11 A network with two components..

each pair of vertices in the graph? This *mean distance* can tell you a lot about how close the nodes in the network are: the shorter the distance the closer the nodes are. Moreover, you can estimate the specific distance (shortest path) between two specific nodes. As shown in Example 13.12 we can estimate the average path length (1.7) in our imaginary Facebook network of American politicians using the functions `mean_distance` in *igraph* and `average_shortest_path_length` in *networkx*. In this example we also estimate the specific distance in the network between Obama and McCain (3) using the function `distances` in *igraph* and estimating the length (`len`) of the shortest path (first result of `shortest_simple_paths` minus 1) in *networkx*.

In terms of distance, we can also wonder what the edges or nodes that share a border with any given vertex are. In the first case, we can identify the *incident edges* that go out or into one vertex. As shown in Example 13.13, by using the the functions `incident` in *igraph* and `edges` in *networkx* we can easily get incident edges of John McCain in the Facebook Network (`g1`), which is just one single edge that joins Trump with McCain. In the second case, we can also identify its adjacent nodes, or in other words its neighbors. In the very same example, we use `neighbors` (same function in *igraph* and *networkx*) to obtain all the nodes one step away from McCain (in this case only Trump).

Python Code

```
print("Average path length in Facebook network: ",
      nx.average_shortest_path_length(g1))
paths = list(nx.shortest_simple_paths(g1,
              "Barack Obama", "John McCain"))
print("Distance between Obama and McCain",
      len(paths[0])-1)
```

R Code

```
glue("Average path length in Facebook network: ",
     mean_distance(g1, directed = T))
glue("Distance between Obama and McCain",
     "in Facebook network: ",
     distances(g1, v="Barack Obama",
               to="John McCain", weights=NA))
```

Python Output

```
Average path length in Facebook network: 1.7
Distance between Obama and McCain in Facebook
     network: 3
```

R Output

```
Average path length in Facebook network: 1.7
Distance between Obama and McCain in Facebook
     network: 3
```

Example 13.12 Estimating distances in the network.

Python Code

```
print("Incident edges of John McCain:",
      g1.edges("John McCain"))
print("Neighbors of John McCain",
      [n for n in g1.neighbors("John McCain")])
```

R Code

```
#mode: all, out, in
glue("Incident edges of John McCain in",
     "Facebook Network:")
incident(g1, V(g1)["John McCain"], mode="all")
glue("Neighbors of John McCain in",
     "Facebook Network:")
neighbors(g1, V(g1)["John McCain"], mode="all")
```

Python Output

```
Incident edges of John McCain in Facebook Network:
     [('John McCain', 'Donald Trump')]
Neighbors of John McCain in Facebook Network:
     ['Donald Trump']
```

R Output

```
Incident edges of John McCain in Facebook Network:
+ 1/5 edge from c0819d5 (vertex names):
[1] Donald Trump--John McCain
Neighbors of John McCain in Facebook Network:
+ 1/5 vertex, named, from c0819d5:
[1] Donald Trump
```

Example 13.13 Incident edges and neighbors of J. McCain the imaginary Facebook Network.

There are some other interesting descriptors of social networks. One of the most common measures is the *density* of the graph, which accounts for the proportion of edges relative to all possible ties in the network. In simpler words, the density tells us from 0 to 1 how much connected the nodes of a graph are. This can be estimated for both undirected and directed graphs. Using the functions `edge_density` in *igraph* and `density` in *networkx* we obtain a density of 0.5 (middle level) in the imaginary Facebook network of American politicians (undirected graph) and 0.35 in the Twitter network (directed graph).

In undirected graphs we can also measure *transitivity* (also known as *clustering coefficient*) and *diameter*. The first is a key property of social networks that refers to the ratio of triangles over the total amount of connected triples. It is to say that we wonder how likely it is that two nodes are connected if they share a mutual neighbor. Applying the function `transitivity` (included in *igraph* and *networkx*) to `g1` we can see that this tendency is 0.5 in the Facebook network (there is a 50% probability that two politicians are friends when they have a common contact). The second descriptor, the diameter, depicts the length of the network in terms of the longest geodesic distance[4]. We use the function `reciprocity` (included in *igraph* and *networkx*) in the Facebook network and get a diameter of 3, which you can also check if you go back to the visualization of `g1` in Example 13.2.

Additionally, in directed graphs we can calculate the *reciprocity*, which is just the proportion of reciprocal ties in a social network and can be computed with the function `diameter` (included in *igraph* and *networkx*). For the imaginary Twitter network (directed graph) we get a reciprocity of 0.57 (which is not bad for a Twitter graph where important people usually have much more followers than follows!).

In Example 13.14 we show how to estimate these four measures in R and Python. Notice that in some of the network descriptors you have to decide whether or not to include the edge weights for computation (in the provided examples we did not take these weights into account).

13.2.2 Centrality Measures

Now let us move to *centrality measures*. Centrality is probably the most common, popular or known measure in the analysis of social networks because it gives you a clear idea of the importance of any of the nodes within a graph. Using its measures you can pose many questions such as which is the most central person in a network of friends on Facebook, who can be considered an opinion leader on Twitter or who is an influencer on Instagram. Moreover, knowing the specific importance of every node of the network can help us to visualize or label only certain vertices that overpass a previously determined threshold, or to use the color or size to distinguish the most central nodes from the others. There are four typical centrality measures: *degree*, *closeness*, *eigenvector* and *betweenness*.

Python Code

```
print("Density in Facebook network: ",
      nx.density(g1))
print("Density in Twitter network: ",
      nx.density(g2))
print("Transitivity in Facebook network: ",
      nx.transitivity(g1))
print("Diameter in Facebook network: ",
      nx.diameter(g1, e=None, usebounds=False))
print("Reciprocity in Twitter network: ",
      nx.reciprocity(g2))
```

R Code

```
glue("Density in Facebook network: ",
     edge_density(g1))
glue("Density in Twitter network: ",
     ecount(g2)/(vcount(g2)*(vcount(g2)-1)) )
glue("Transitivity in Facebook network: ",
     transitivity(g1), type="global")
glue("Diameter in Facebook network: ",
     diameter(g1, directed = F, weights = NA))
glue("Reciprocity in Twitter network: ",
     reciprocity(g2))
```

Python Output

```
Density in Facebook network: 0.5
Density in Twitter network: 0.35
Transitivity in Facebook network: 0.5
Diameter in Facebook network: 3
Reciprocity in Twitter network:
    0.5714285714285714
```

R Output

```
Density in Facebook network: 0.5
Density in Twitter network: 0.35
Transitivity in Facebook network: 0.5
Diameter in Facebook network: 3
Reciprocity in Twitter network: 0.571428571428571
```

Example 13.14 Incident edges and neighbors of J. McCain the imaginary Facebook Network.

[4] The *geodesic distance* is the shortest number of edges between two vertices

The *degree* of a node refers to the number of ties of that vertex, or in other words, to the number of edges that are incident to that node. This definition is constant for undirected graphs in which the directions of the links are not declared. In the case of directed graphs, you will have three options to measure the degree. First, you can think of the number of edges pointing *into* a node, which we call *indegree*; second, we have the number of edges pointing `out` of a node, or *outdegree*. In addition, we could also have the total number of edges pointing in and out any node. Degree, as well as other measures of centrality mentioned below, can be expressed in absolute numbers, but we can also *normalize*[5] these measures for better interpretation and comparison. We will prefer this latter approach in our examples, which is also the default option in many SNA packages.

We can then estimate the degree of two of our example networks. In Example 13.15, we first estimate the degree of each of the five American politicians in the imaginary Facebook network, which is an undirected graph; and then the total degree in the Twitter network, which is a directed graph. For both cases, we use the functions `degree` in *igraph* (R) and `degree_centrality` in *networkx* (Python). We later compute the `in` and `out` degree for the Twitter network. Using *igraph* we again used the function `degree` but now adjust the parameter `mode` to `in` or `out`, respectively. Using *networkx*, we employ the functions `in_degree_centrality` and `out_degree_centrality`.

There are three other types of centrality measures. *Closeness centrality* refers to the geodesic distance of a node to the rest of nodes in the graph. Specifically, it indicates how close a node is to the others by taking the length of the shortest paths between the vertices. *Eigenvector centrality* takes into account the importance of the surrounding nodes and computes the centrality of a vertex based on the centrality of its neighbors. In technical words, the measure is proportional to the sum of connection centralities. Finally, *betweenness centrality* indicates to what extent the node is in the paths that connect many other nodes. Mathematically it is computed as the sum of the fraction of every pair of (shortest) paths that go through the analyzed node.

Python Code

```
print("Degree centrality of Facebook"
      "network (undirected): \n",
      nx.degree_centrality(g1))
print("Degree centrality of Twitter"
      "network (directed): \n",
      nx.degree_centrality(g2))
print("In degree centrality of Twitter"
      "network (directed): \n",
      nx.in_degree_centrality(g2))
print("Out degree centrality of Twitter"
      "network (directed): \n",
      nx.out_degree_centrality(g2))
```

R Code

```
1  print("Degree centrality")
2  print("Facebook network (undirected):" )
3  print(degree(g1, normalized = T))
4  print("Degree centrality")
5  print("Twitter network (directed):" )
6  print(degree(g2, normalized = T, mode="all"))
7  print("In degree centrality")
8  print("Twitter network (directed):" )
9  print(degree(g2, normalized = T, mode="in"))
10 print("Out degree centrality")
11 print("Twitter network (directed):" )
12 print(degree(g2, normalized = T, mode="out"))
```

[5] The approach is to divide by the maximum possible number of vertices (*N*), or by *N*-1. We may also estimate the `weighted degree` of a node, which is the same degree but ponderated by the weight of the edges.

Python Output

```
Degree centrality of the Facebook network
      (undirected):
 {'Hillary Clinton': 0.75, 'Donald Trump': 0.5,
      'Bernie Sanders': 0.5, 'Barack Obama': 0.5,
      'John McCain': 0.25}
Degree centrality of the Twitter network
      (directed):
 {'Hillary Clinton': 1.25, 'Donald Trump': 0.5,
      'Bernie Sanders': 0.75, 'Barack Obama':
      0.75, 'John McCain': 0.25}
In degree centrality of the Twitter network
      (directed):
 {'Hillary Clinton': 0.5, 'Donald Trump': 0.5,
      'Bernie Sanders': 0.5, 'Barack Obama': 0.25,
      'John McCain': 0.0}
Out degree centrality of the Twitter network
      (directed):
 {'Hillary Clinton': 0.75, 'Donald Trump': 0.0,
      'Bernie Sanders': 0.25, 'Barack Obama': 0.5,
      'John McCain': 0.25}
```

R Output

```
[1] "Degree centrality of the Facebook network
      (undirected):"
Hillary Clinton Donald Trump Bernie Sanders Barack
      Obama John     McCain
    0.75     0.50     0.50     0.50     0.25
[1] "Degree centrality of the Twitter network
      (directed):"
Hillary Clinton Bernie Sanders John McCain Barack
      Obama      Donald Trump
    1.25     0.75     0.25     0.75     0.50
[1] "In degree centrality of the Twitter network
      (directed):"
Hillary Clinton Bernie Sanders John McCain Barack
      Obama      Donald Trump
    0.50     0.50     0.00     0.25     0.50
[1] "Out degree centrality of the Twitter network
      (directed):"
Hillary Clinton Bernie Sanders John McCain Barack
      Obama      Donald Trump
    0.75     0.25     0.25     0.50     0.00
```

Example 13.15 Computing degree centralities in undirected and directed graphs.

As shown in Example 13.16, we can obtain these three measures from undirected graphs using the functions `closeness`, `eigen_centrality` and `betweenness` in *igraph*, and `closeness_centrality`, `eigenvector_centrality` and `betweenness_centrality` in *networkx*. If we take a look to the centrality measures for every politician of the imaginary Facebook network we see that Clinton seems to be a very important and central node of the graph, just coinciding with the above-mentioned findings based on the degree. It is not a rule that we obtain the very same trend in each of the centrality measures but it is likely that they have similar results although they are looking for different dimensions of the same construct.

We can use these centrality measures in many ways. For example, you can take the degree centrality as a parameter of the node size and labeling when plotting the network. This may be of great utility since the reader can visually identify the most important nodes of the network while minimizing the visual impact of those that are less central. In Example 13.17 we decided to specify the size of the nodes (parameters `vertex.size` in *igraph* and `node_size` in *networkx*) with the degree centrality of each of the American politicians in the Twitter network (directed graph) contained in `g2`. We also used the degree centrality to filter the labels in the graph, and then included only those that overpassed a threshold of 0.5 (parameters `vertex.label` in *igraph* and `labels` in *networkx*). These two simple parameters of the plot give you a fair image of the potential of the centrality measures to describe and understand your social network.

13.2.3 Clustering and Community Detection

One of the greatest potentials of SNA is the ability to identify how nodes are interconnected and thus define *communities* within a graph. This is to say that most of the time

Python Code

```python
print("Closeness centrality of Facebook"
      "network (undirected): \n",
      nx.closeness_centrality(g1))
print("Eigenvector centrality of Facebook"
      "network (undirected): \n",
      nx.eigenvector_centrality(g1))
print("Betweenness centrality of Facebook"
      "network (undirected): \n",
      nx.betweenness_centrality(g1))
```

R Code

```r
print("Closeness centrality")
print("Facebook network (undirected):")
print(closeness(g1, normalized = T))
print("Eigenvector centrality")
print("Facebook network (undirected):")
print(eigen_centrality(g1, scale=F)$vector)
print("Betweenness centrality")
print("Facebook network (undirected):")
print(betweenness(g1, normalized = T))
```

Python Output

```
Closeness centrality of the Facebook network (undirected):
{'Hillary Clinton': 0.8, 'Donald Trump': 0.6666666666666666, 'Bernie Sanders': 0.5714285714285714, 'Barack
      Obama': 0.5714285714285714, 'John McCain': 0.4444444444444444}
Eigenvector centrality of the Facebook network (undirected):
{'Hillary Clinton': 0.6037035301706529, 'Donald Trump': 0.34248744909850964, 'Bernie Sanders':
      0.49715259845254134, 'Barack Obama': 0.49715259845254134, 'John McCain': 0.15467056143060928}
Betweenness centrality of the Facebook network (undirected):
{'Hillary Clinton': 0.6666666666666666, 'Donald Trump': 0.5, 'Bernie Sanders': 0.0, 'Barack Obama': 0.0,
      'John McCain': 0.0}
```

Example 13.16 Estimations of closeness, eigenvector and betweenness centralities.

the nodes and edges in our network are not distributed homogeneously, but they tend to form clusters that can later be interpreted. In a social network you can think for example of the principle of *homophily*, which is the tendency of human beings to associate and interact with similar individuals; or you can think of extrinsic factors (e.g., economical or legal) that may generate the cohesion of small groups of citizens that belong to a wider social structure. While it is of course difficult to make strong claims regarding the underlying causes, we can use different computational approaches to model and detect possible communities that emerge from social networks and even interpret and label those groups. The creation of clusters as an unsupervised machine learning technique was introduced in Section 7.3 for structured data and in Section 11.5 for text analysis (topic modeling). We will use some similar unsupervised approaches for community detection in social networks.

Many social and communication questions may arise when clustering a network. The identification of subgroups can tell us how diverse and fragmented a network is, or how the behaviour of a specific community relates to other groups and to the entire graph. Moreover, the concentration of edges in some nodes of the graph would let us know about the social structure of the networks which in turn would mean a better understanding of its inner dynamic. It is true that the computational analyst will need more than the provided algorithms when labeling the groups to understand the communities, which means that you must become familiar with the way the graph has been built and what the nodes, edges or weights represent.

A first step towards an analysis of subgroups within a network is to find the available complete subgraphs in an undirected graph. As we briefly explained at the end of Section 13.1, these independent subgraphs are called `cliques` and

Python Code

```
size = list(nx.degree_centrality(g2).values())
size = [x * 1000 for x in size]
labels_filtered = {k: v for k, v in
    nx.degree_centrality(g2).items() if v > 0.5 }
labels = {}
for k, v in labels_filtered.items():
    labels[k] = k

nx.draw_networkx(g2, node_size= size,
                 labels=labels)

pos=nx.shell_layout(g2)
x_values, y_values = zip(*pos.values())
x_max = max(x_values)
x_min = min(x_values)
x_margin = (x_max - x_min) * 0.40
plt.xlim(x_min - x_margin, x_max + x_margin)
plt.box(False)
plt.show()
```

R Code

```
plot(g2, vertex.label.cex = 2,
    vertex.size= degree(g2, normalized = T)*40,
    vertex.label = ifelse(degree(g2,
        normalized = T) > 0.5, V(g2)$name, NA))
```

R Output

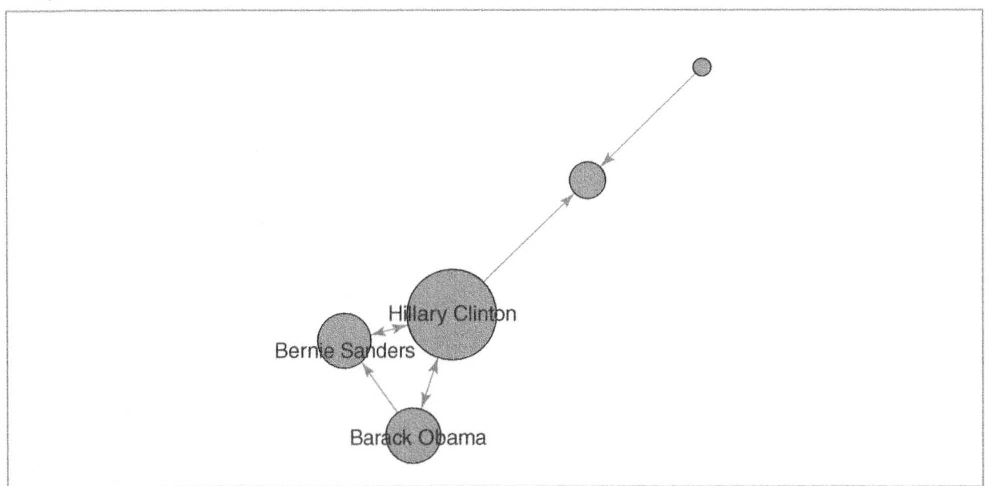

Example 13.17 Using the degree centrality to change the size and labels of the nodes.

refer to subgroups where every vertex is connected to every other vertex. We can find the `maximal cliques` (a clique is maximal when it cannot be extended to a bigger clique) in the imaginary undirected graph of American politicians on Facebook (`g1`) by using the functions `max_cliques` in *igraph* (Eppstein et al., 2010) and `max_cliques` in *networkx* (Cazals and Karande, 2008). As you can see in Example 13.18, we obtain a total of three subgraphs, one representing the Democrats, another the Republicans, and one more the connector of the two parties (Clinton–Trump).

Now, in order to properly detect communities we will apply some common algorithms to obtain the most likely subgroups in a social network. The first of these models is the so called *edge-between* or Girvan–Newman algorithm (Newman and Girvan, 2004).

Python Code

```
print("Number f cliques: ",
      nx.graph_number_of_cliques(g1))
print("Cliques: ", list(nx.find_cliques(g1)))
```

R Code

```
glue("Number of cliques: {clique_num(g1)}")
max_cliques(g1)
```

Python Output

```
Number of cliques: 3
Cliques: [['Hillary Clinton', 'Bernie Sanders',
      'Barack Obama'], ['Hillary Clinton', 'Donald
      Trump'], ['John McCain', ' Donald Trump']]
```

R Output

```
Number of cliques: 3
[[1]]
+ 2/5 vertices, named, from c0819d5:
[1] Donald Trump John McCain

[[2]]
+ 2/5 vertices, named, from c0819d5:
[1] Donald Trump Hillary Clinton

[[3]]
+ 3/5 vertices, named, from c0819d5:
[1] Hillary Clinton Bernie Sanders Barack Obama
```

Example 13.18 Finding all the maximal cliques in an undirected graph.

This algorithm is based on divisive hierarchical clustering (explained in Section 7.3) by breaking down the graph into pieces and iteratively removing edges from the original one. Specifically, the Girvan–Newman approach uses the betweenness centrality measure to remove the most central edge at each iteration. You can easily visualize this splitting process in a dendogram, as we do in Example 13.19, where we estimated `cluster1` to detect possible communities in the Facebook network. We used the functions `cluster_edge_betweenness` in *igraph* and `girvan_newman` in *networkx*.

When you look at the figure you will notice that the final leaves correspond to the nodes (the politicians) and then you have different partition levels (1 to 4), which in fact are different cluster possibilities. In edge-betweenness clustering, the big question is which partition level to choose, or in other words, which of the community division is better. The concept of *modularity* arises as a good measure (–1 to 1) to evaluate how good the division is (technically it's measured as the fraction of edges that fall within any given groups, let's say group 1 and group 2, minus the expected number of edges within those groups distributed at random). Thus, we can choose which of the four proposed divisions is the best based on the highest value of their modularities: the higher the modularity the more dense the connections *within* the community and the more sparse the connections *across* communities. In the case of `cluster_edge_betweenness` in *igraph* it automatically estimates that the best division (on modularity) is the first one with two communities.

With community detection algorithms we can then estimate the length (number of suggested clusters), membership (to which cluster belongs each node) and modularity (how good is the clustering). In the case of *igraph* in R we apply the functions `length` (base), `membership` and `modularity` over the produced clustering object (i.e., `cluster1`). In the case of *networkx* in Python we first have to specify that we want to use the first component of the divisions (out of four) using the function `next`. Then, we can apply the functions `len` (base) and `modularity` to get the descriptors, and print the fist division (stored as `communities1`) to obtain the membership (see Example 13.20).

Python Code

```python
cluster1 = nxcom.girvan_newman(g1)
```

R Code

```r
cluster1 = cluster_edge_betweenness(g1)
dendPlot(cluster1, mode="hclust")
```

R Output

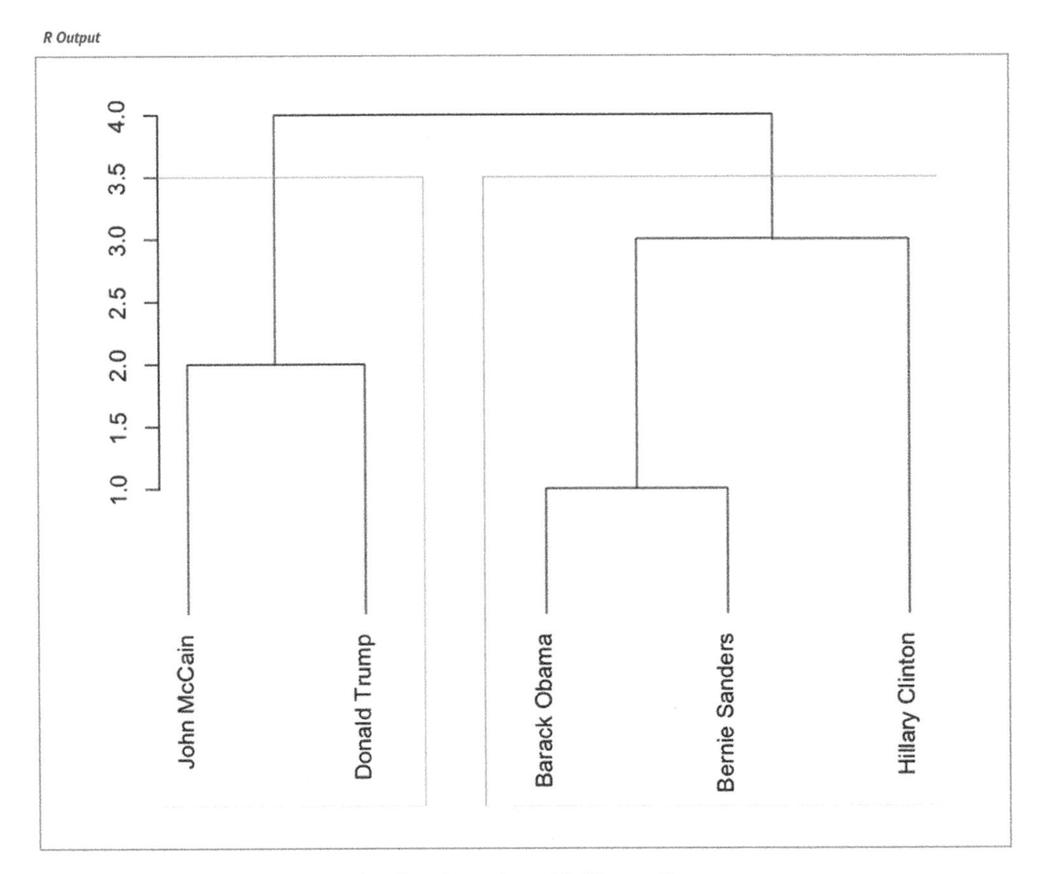

Example 13.19 Dendrogram to visualize clustering with Girvan–Newman.

Python Code

```python
c1 = next(cluster1)
print(f"Girvan-Neuman:\nLength {len(c1)}")
print(f"Modularity: "
      f"{nxcom.modularity(g1, c1):.2f}")
print(f"Membership: {c1}")
```

R Code

```r
print("Girvan-Neuman")
print(glue("Length: {length(cluster1)}"))
print(glue("Modularity: {modularity(cluster1)}"))

print(membership(cluster1))
```

R Output

```
Girvan-Neuman:
Length 2
Modularity: 0.22
Membership: ({'Bernie Sanders', 'Hillary Clinton', 'Barack Obama'}, {'John McCain', 'Donald Trump'})
```

Example 13.20 Community detection with Girvan–Newman.

We can estimate the communities for our network using many other more clustering algorithms, such as the *Louvain algorithm*, the *Propagating Label algorithm*, and *Greedy Optimization*, among others. Similar to Girvan–Newman, the Louvain algorithm uses the measure of modularity to obtain a multi-level optimization (Blondel et al., 2008) and its goal is to obtain optimized clusters which minimize the number of edges between the communities and maximize the number of edges within the same community. For its part, the *Greedy Optimization* algorithm is also based on the modularity indicator (Clauset et al., 2004). It does not consider the edges' weights and works by initially setting each vertex in its own community and then joining two communities to increase modularity until obtaining the maximum modularity. Finally, the Propagating Label algorithm – which takes into account edges weights – initializes each node with a unique label and then iteratively each vertex adopts the label of its neighbors until all nodes have the most common label of their neighbors (Raghavan et al., 2007). The process can be conducted asynchronously (as done in our example), synchronously or semi-synchronously (it might produce different results).

In Example 13.21 we use `cluster_louvain`, `cluster_fast_greedy` and `cluster_label_prop` in *igrapgh* (R) and `best_partition`, `greedy_modularity_communities` and `asyn_lpa_communities` in *networkx* (Python). You can see that the results are quite similar[6] and it is pretty clear that there are two communities in the Facebook network: Democrats and Republicans!

Python Code

```
print("Louvain")
cluster2 = community.best_partition(g1)
print("Length: "
      f"{float(len(set(cluster2.values())))}")
print("Modularity: "
      f"{community.modularity(cluster2, g1):.2f}")
print(f"Membership: {cluster2}")

print("\nGreedy optimization")
cluster3 = nxcom.greedy_modularity_communities(g1)
c3 = sorted(cluster3, key=len, reverse=True)
print(f"Length {len(c3)}")
print("Modularity: "
      f"{nxcom.modularity(g1, c3):.2f}")
print(f"Membership: {c3}")

print("\nPropagating label: ")
cluster4 = nxcom.asyn_lpa_communities(g1)
c4 = sorted(cluster4, key=len, reverse=True)
print("Length: ", len(c4))
print("Modularity:"
      f"{nxcom.modularity(g1, c4):.2f}")
print("Membership: ", c4)
```

R Code

```
print("Louvain:")
cluster2 = cluster_louvain(g1)
print(glue("Length: {length(cluster2)}"))
print(glue("Modularity: {modularity(cluster2)}"))
print(membership(cluster2))

print("Greedy optimization:")
cluster3 = cluster_fast_greedy(g1)
print(glue("Length: {length(cluster3)}"))
print(glue("Modularity: {modularity(cluster3)}"))
print(membership(cluster3))

print("Label propagation:")
cluster4 = cluster_label_prop(g1)
print(glue("Length {length(cluster4)}"))
print(glue("Modularity: {modularity(cluster4)}"))
print(membership(cluster4))
```

[6] This similarity is because our example network is extremely small. In larger networks, the results might not be that similar.

Python output

```
Louvain
Length: 2.0
Modularity: 0.22
Membership: {'Hillary Clinton': 0, 'Donald Trump': 1, 'Bernie Sanders': 0, 'Barack Obama': 0, 'John
     McCain': 1}

Greedy optimization
Length 2
Modularity: 0.22
Membership: [frozenset({'Bernie Sanders', 'Hillary Clinton', 'Barack Obama'}), frozenset({'John McCain',
     'Donald Trump'})]

Propagating label:
Length: 1
Modularity: -0.00
Membership: [{'Hillary Clinton', 'John McCain', 'Barack Obama', 'Bernie Sanders', 'Donald Trump'}]
```

Example 13.21 Community detection with Louvain Propagating Label and Greedy Optimization.

We can plot each of those clusters for better visualization of the communities. In Example 13.22 we generate the plots with the Greedy Optimization algorithm in R and the Louvain algorithm in Python, and we get two identical results.

There are more ways to obtain subgraphs of your network (such as the `K-core decomposition`) or to evaluate the homophily of your graph (using the indicator of `assortativity` that measures the degree to which the nodes associate to similar vertices). In fact, there are many other measures and techniques you can use to conduct SNA that we have deliberately omitted in this section for reasons of space, but we have covered the most important aspects and procedures you need to know to initiate yourself in the computational analysis of networks.

So far we have seen how to conduct SNA over "artificial" graphs for the sake of simplicity. However, the representation and analysis of "real world" networks will normally be more challenging because of their size or their complexity. To conclude this chapter we will show you how to apply some of the explained concepts to real data.

Using the Twitter API (see Section 12.1), we retrieved the names of the first 100 followers of the five most important politicians in Spain by 2017 (Mariano Rajoy, Pedro Sánchez, Albert Rivera, Alberto Garzón and Pablo Iglesias). With this information we produced an undirected graph[7] of the "friends" of these Spanish politicians in order to understand how these leaders where connected through their followers. In Example 13.23 we load the data into a graph object `g_friends` that contains the 500 edges of the network. As we may imagine the five mentioned politicians were normally the most central nodes, but if we look at the degree, betweenness and closeness centralities we can easily get some of the relevant nodes of the Twitter network: CEARefugio, elenballesteros or Unidadpopular. These accounts deserve special attention since they contribute to the connection of the main leaders of that country. In fact, if we conduct clustering analysis using Louvain algorithm we will find a high modularity (0.77, which indicates that the clusters are well separated) and not surprisingly five clusters.

When we visualize the clusters in the network (Example 13.24) using the degree centrality for the size of the node, we can locate the five politicians in the center of the

[7] We deliberately omitted the directions of the edges given their impossible reciprocity.

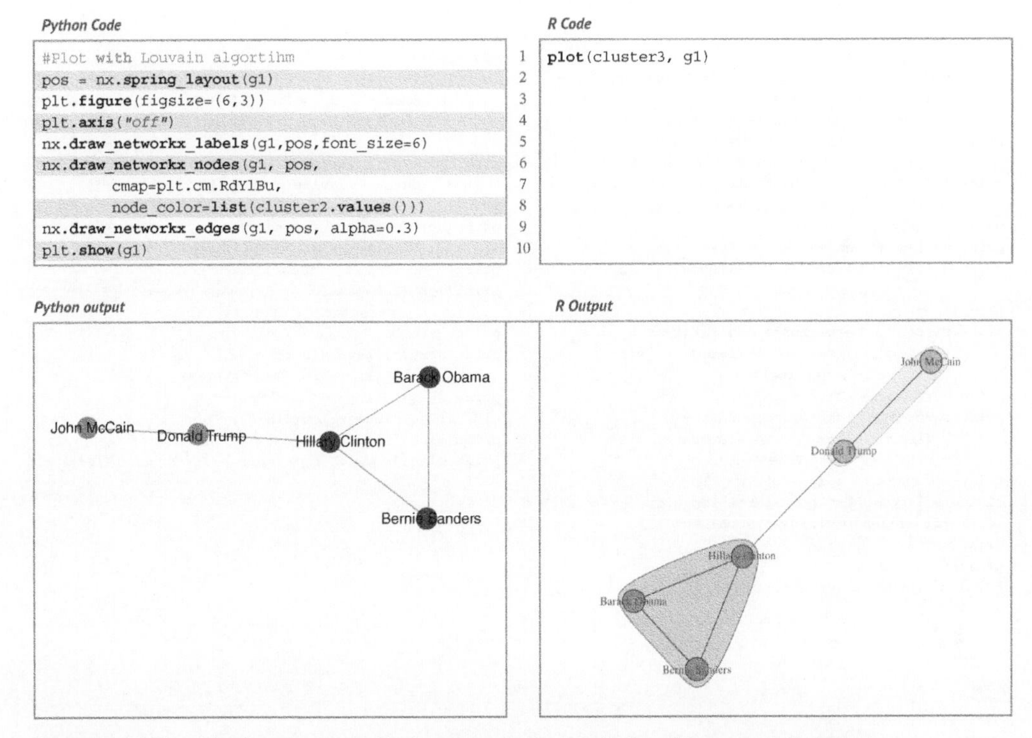


```
#Plot with Louvain algortihm
pos = nx.spring_layout(g1)
plt.figure(figsize=(6,3))
plt.axis("off")
nx.draw_networkx_labels(g1,pos,font_size=6)
nx.draw_networkx_nodes(g1, pos,
        cmap=plt.cm.RdYlBu,
        node_color=list(cluster2.values()))
nx.draw_networkx_edges(g1, pos, alpha=0.3)
plt.show(g1)
```

```
plot(cluster3, g1)
```

Example 13.22 Plotting clusters with Greedy optimization in R and Louvain in Python.

clusters (depicted with different colors). More interesting, we can see that even when some users follow two of the political leaders, they are just assigned to one of the clusters. This the case of the node joining Garzón and Sánchez who is assigned to the Sánchez's cluster, or the node joining Garzón and Rajoy who is assigned to Rajoy's cluster. In the plot you can also see two more interesting facts. First, we can see a triangle that groups Sánchez, Garzón and Iglesias, which are leaders of the left-wing parties in Spain. Second, some pair of politicians (such as Iglesias–Garzón or Sánchez–Rivera) share more friends than the other possible pairs.

Python Code

```python
url = "https://cssbook.net/d/friends3.csv"
fn, _headers = urllib.request.urlretrieve(url)
g_friends = nx.read_adjlist(fn,
                  create_using=nx.Graph, delimiter=";")
print("Nodes:", g_friends.number_of_nodes(),
      "Edges: ", g_friends.number_of_edges())
print("Nodes:", g_friends.number_of_nodes(),
      "Edges: ", g_friends.number_of_edges())
print("Degree centrality: ")
print(sorted(nx.degree_centrality(
      g_friends).items(), key=lambda x: x[1],
              reverse=True)[0:9])
print("Betweenness centrality: ")
print(sorted(nx.betweenness_centrality(
      g_friends).items(), key=lambda x: x[1],
              reverse=True)[0:9])
print("Closeness centrality: ")
print(sorted(nx.closeness_centrality(
      g_friends).items(), key=lambda x: x[1],
              reverse=True)[0:9])
print("Clustering with Louvain: ")
cluster5 = community.best_partition(g_friends)
size = float(len(set(cluster5.values())))
print("Length: ", size)
print("Modularity: "
f"{community modularity(cluster5, g_friends):.2f}")
```

R Code

```r
edges = read_delim(
    "https://cssbook.net/d/friends3.csv",
    col_names=FALSE, delim=";")
g_friends = graph_from_data_frame(d=edges,
                                   directed = FALSE)
glue("Nodes: ", gorder(g_friends),
     " Edges: ", gsize(g_friends))
print("Degree centrality:")
print(sort(degree(g_friends, normalized = T),
           decreasing = TRUE)[1:10])
print("Betweenness centrality:")
print(sort(betweenness(g_friends, normalized = T),
           decreasing = TRUE)[1:10])
print("Closeness centrality:")
print(sort(closeness(g_friends, normalized = T),
           decreasing = TRUE)[1:10])
print("Clustering with Louvain:")
cluster5 = cluster_louvain(g_friends)
print(glue("Length: {length(cluster5)}"))
print(glue("Modularity: {modularity(cluster5)}"))
```

Output

```
Nodes: 491 Edges: 500
Nodes: 491 Edges: 500
Degree centrality:
[('Pablo_Iglesias_', 0.20612244897959187), ('Albert_Rivera', 0.20408163265306123), ('agarzon',
        0.20408163265306123), ('sanchezcastejon', 0.20408163265306123), ('marianorajoy',
     0.20408163265306123), ('CEARefugio', 0.004081632653061225), ('VictorLapuente', 0.004081632653061225),
     ('javierfernandez', 0.004081632653061225), ('mas_demo', 0.004081632653061225)]
Betweenness centrality:
[('sanchezcastejon', 0.4847681328312369), ('agarzon', 0.44044055921356695), ('Albert_Rivera',
     0.4226327682769384), ('marianorajoy', 0.384657568548892), ('Pablo_Iglesias_', 0.35788117228958527),
     ('elenballesteros', 0.19595159798490358), ('Unidadpopular__', 0.13749254219017618), ('kanciller',
     0.08787927105192601), ('JuanfranGuevara', 0.08787927105192601)]
Closeness centrality:
[('sanchezcastejon', 0.3592375366568915), ('agarzon', 0.33653846153846156), ('Pablo_Iglesias_',
     0.3356164383561644), ('Unidadpopular__', 0.3353867214236824), ('CEARefugio', 0.3349282296650718),
     ('VictorLapuente', 0.3349282296650718), ('javierfernandez', 0.3349282296650718), ('mas_demo',
     0.3349282296650718), ('elenballesteros', 0.31511254019292606)]
Clustering with Louvain:
Length: 5.0
Modularity: 0.77
```

Example 13.23 Loading and analyzing a real network of Spanish politicians and their followers on Twitter.

Python Code

```python
pos = nx.spring_layout(g_friends)
plt.figure(figsize=(10,10))
plt.axis("off")

size = list(nx.degree_centrality(
     g_friends).values())
size = [x * 7000 for x in size]
labels_filtered = {k: v for k, v in
     nx.degree_centrality(g_friends).items() if
                 v > 0.005 }
labels = {}
for k, v in labels_filtered.items():
     labels[k] = k

nx.draw_networkx_labels(g_friends,
     pos,font_size=10,
     labels=labels)
nx.draw_networkx_nodes(g_friends,
     pos, node_size= size, cmap=plt.cm.RdYlBu,
     node_color=list(cluster5.values()))
nx.draw_networkx_edges(g_friends, pos, alpha=0.5)
plt.show(g_friends)
```

R Code

```r
plot(cluster5, g_friends, vertex.label.cex = 2,
   vertex.size=degree(g_friends, normalized=T)*40,
   vertex.label = ifelse(degree(g_friends,
                          normalized=T) > 0.005,
                          V(g_friends)$name, NA))
```

Output: Python

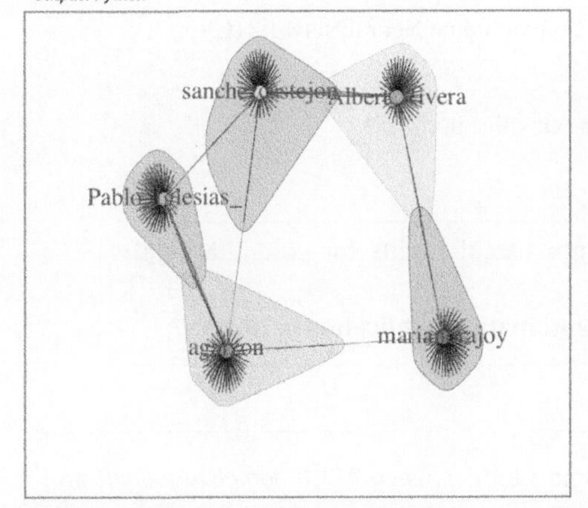

Output: R

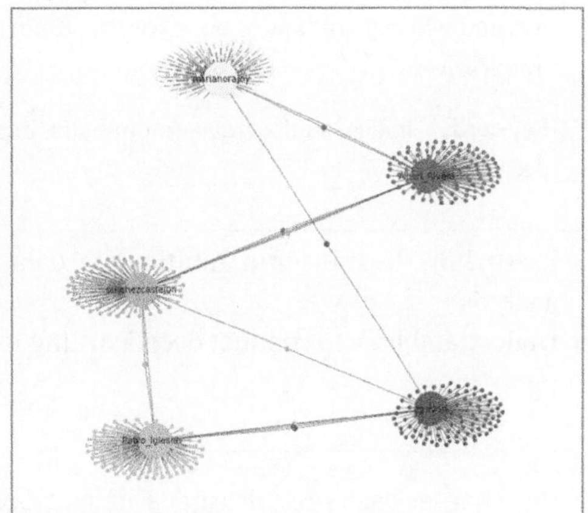

Example 13.24 Visualizing the network of Spanish politicians and their followers on Twitter and plotting its clusters.

14

Multimedia Data

Abstract

Digitally collected data often do not only contain texts, but also audio, images, and videos. Instead of using only textual features as we did in previous chapters, we can also use pixel values to analyze images. First, we will see how to use existing libraries, commercial services or APIs to conduct multimedia analysis (such as optical character recognition, speech-to-text or object recognition). Then we will show how to store, represent, and convert image data in order to use it as an input in our computational analysis. We will focus on image analysis using machine learning classification techniques based on deep learning, and will explain how to build (or fine-tune) a Convolutional Neural Network (CNN) by ourselves.

Keywords image, audio, video, multimedia, image classification, deep learning

- Learn how to transform multimedia data into useful inputs for computational analysis
- Understand how to conduct deep learning to automatic classification of images

This chapter uses *tesseract* (generic) and Google Cloud Speech API (*googleLanguageR* and *google-cloud-language* in Python) to convert images or audio files into text, respectively. We will use *PIL* (Python) and *imagemagic* (generic) to convert pictures as inputs; and *Tensorflow* and *keras* (both in Python and R) to build and fine-tune CNNs.

You can install these and other auxiliary packages with the code below if needed (see Section 1.4 for more details):

Python Code

```
!pip3 install Pillow requests numpy sklearn
!pip3 install tensorflow keras
```

R Code

```
install.packages(c("magick","glue","lsa",
    "tidyverse","dslabs","randomForest","caret",
    "tensorflow","keras"))
```

After installing, you need to import (activate) the packages every session:

Computational Analysis of Communication: A Practical Introduction to the Analysis of Texts, Networks, and Images with Code Examples in Python and R, First Edition. Wouter van Atteveldt, Damian Trilling & Carlos Arcila Calderón.

Python Code

```python
import matplotlib.pyplot as plt
from PIL import Image
import requests
import numpy as np

import sklearn
from sklearn.datasets import fetch_openml
from sklearn.metrics.pairwise import (
    cosine_similarity)
from sklearn.ensemble import (
    RandomForestClassifier)
from sklearn.metrics import accuracy_score

import tensorflow as tf
import keras
from keras.applications import resnet50
```

R Code

```r
library(magick)
library(lsa)
library(tidyverse)
library(dslabs)
library(randomForest)
library(caret)
library(tensorflow)
library(keras)
```

14.1 Beyond Text Analysis: Images, Audio and Video

A book about the *computational analysis of communication* would be incomplete without a chapter dedicated to analyzing visual data. In fact, if you think of the possible contents derived from social, cultural and political dynamics in the current digital landscape, you will realize that written content is only a limited slice of the bigger cake. Humans produce much more oral content than text messages, and are more agile in deciphering sounds and visual content. Digitalization of social and political life, as well as the explosion of self-generated digital content in the web and social media, have provoked an unprecedented amount of multimedia content that deserve to be included in many types of research.

Just imagine a collection of digital recorded radio stations, or the enormous amount of pictures produced every day on Instagram, or even the millions of videos of social interest uploaded on Youtube. These are definitely goldmines for social researchers who traditionally used manual techniques to analyze just a very small portion of this multimedia content. However, it is also true that computational techniques to analyze audio, images or video are still little developed in social sciences given the difficulty of application for non-computational practitioners and the novelty of the discoveries in fields such as computer vision.

This section gives a brief overview of different formats of multimedia files. We explain how to generate useful inputs into our pipeline to perform computational analysis.

You are probably already familiar with digital formats of images (.jpg, .bmp, .gif, etc.), audio (.mp3, .wav, .wma, flac, etc.) or video (.avi, .mov, .wmv, .flv, etc.), which is the very first step to use these contents as input. However, similar to the case of texts you will need to do some preprocessing to put these formats into good shape and get a proper mathematical representation of the content.

In the case of audio, there are many useful computational approaches to do research over these contents: from voice recognition, audio sentiment analysis or sound classification, to automatic generation of music. Recent advances in the field of artificial intelligence have created a prosperous and diversified field with multiple academic and commercial applications. Nevertheless, computational social scientists can obtain great insights just by using specific applications such as speech-to-text transformation and then apply text analytics (already explained in Chapters 9, 10 and 11) to the results. As you will see in Section 14.2, there are some useful libraries in R and Python to use pre-trained models to transcribe voice in different languages.

Even when this approach is quite limited (just a small portion of the audio analytics world) and constrained (we will not address how to create the models), it will show how a specific, simple and powerful application of the automatic analysis of audio inputs can help answering many social questions (i.e., what are the topics of a natural conversation, what are the sentiments expressed in the scripts of radio news pieces, or which actors are named in oral speeches of any political party). In fact, automated analysis of audio can enable new research questions, different from those typically applied to text analysis. This is the case of the research by Knox and Lucas (2021), who used a computational approach over audio data from the Supreme Court Oral Arguments (407 arguments and 153 hours of audio, comprising over 66 000 justice utterances and 44 million moments) to demonstrate that some crucial information such as the skepticism of legal arguments was transmitted by vocal delivery (i.e., speech tone), something indecipherable to text analysis. Or we could also mention the work by Dietrich et al. (2019) who computationally analyzed the vocal pitch of more than 70 000 Congressional floor audio speeches and found that female members of the Congress spoke with greater *emotional intensity* when talking about women.

On the other hand, applying computational methods to video input is probably the most challenging task in spite of the recent and promising advances in computer vision. For the sake of space, we will not cover specific video analytics in this chapter, but it is important to let you know that most of the computational analysis of video is based on the inspection of image and audio contents. With this standard approach you need to specify which key frames you are going to extract from the video (for example take a still image every 1000 frames) and then apply computer vision techniques (such as object detection) to those independent images. Check for example version 3 of the object detection architecture *You Only Look Once Take* (YOLOv3)[1] created by Redmon and Farhadi (2018), which uses a pre-trained Convolutional Neural Network (CNN) (see Section 14.4) to locate objects within the video (Figure 14.1). To answer many social science questions you might complement this frame-to-frame image analysis with an analysis of audio features. In any case, this approach will not cover some interesting aspects of the video such as the camera frame shots and movements, or the editing techniques, which certainly give more content information.

[1] https://pjreddie.com/darknet/yolo/

Figure 14.1 A screen shot of a real-time video analyzed by YOLOv3 on its website https://pjreddie.com/darknet/yolo/ *Source:* https://pjreddie.com/darknet/yolo/

14.2 Using Existing Libraries and APIs

In the following sections we will show you how to deal with multimedia contents from scratch, with special attention to image classification using state-of-the-art libraries. However, it might be a good idea to begin by using existing libraries that directly implement multimedia analyses or by connecting to commercial services to deploy classification tasks remotely using their APIs. There is a vast variety of available libraries and APIs, which we cannot cover in this book, but we will briefly mention some of them that may be useful in the computational analysis of communication.

One example in the field of visual analytics is the *optical character recognition* (OCR). It is true that you can train your own models to deploy multi-class classification and predict every letter, number or symbol in an image, but it will be a task that will take you a lot of effort. Instead, there are specialized libraries in both R and Python such as *tesseract* that deploy this task in seconds with high accuracy. It is still possible that you will have to apply some pre-processing to the input images in order to get them in good shape. This means that you may need to use packages such as *PIL* or *Magick* to improve the quality of the image by cropping it or by reducing the background noise. In the case of PDF files you will have to convert them first into images and then apply OCR.

In the case of more complex audio and image documents you can use more sophisticated services provided by private companies (e.g., Google, Amazon, Microsoft, etc.). These commercial services have already deployed their own machine learning models with very good results. Sometimes you can even customize some of their models, but as a rule their internal features and configuration are not transparent to the user. Moreover, these services offer friendly APIs and normally free quota to deploy your first exercises.

To work with audio files, many social researchers might need to convert long conversations, radio programs, or interviews to plain text. For this propose, *Google Cloud* offers the service *Speech-to-Text*[2] that remotely transcribes the audio to a text format supporting multiple languages (more than 125!). With this service you can remotely use the advanced deep learning models created by Google Platform from your own local computer (you must have an account and connect with the proper packages such as *googleLanguageR* or *google-cloud-language* in Python).

If you apply either OCR to images or Speech-to-Text recognition to audio content you will have juicy plain text to conduct NLP, sentiment analysis, topic modelling, among other techniques (see Chapter 11). Thus, it is very likely that you will have to combine different libraries and services to perform a complete computational pipeline, even jumping from R to Python, and vice versa!

Finally, we would like to mention the existence of the commercial services of *autotaggers*, such as Google's Cloud Vision, Microsoft's Computer Vision or Amazon's Recognition. For example, if you connect to the services of Amazon's Recognition you can not only detect and classify images, but also conduct sentiment analysis over faces or predict sensitive contents within the images. As in the case of Google Cloud, you will have to obtain commercially sold credentials to be able to connect to Amazon's Recognition API (although you get a free initial "quota" of API access calls before you are required to pay for usage). This approach has two main advantages. The first is the access to a very well trained and validated model (continuously re-trained) over millions of images and with the participation of thousands of coders. The second is the scalability because you can store and analyze images at scale at a very good speed using cloud computing services.

As an example, you can use Amazon's Recognition to detect objects in a news photograph of refugees in a lifeboat (Figure 14.2) and you will obtain a set of accurate labels: *Clothing* (99.95%), *Apparel* (99.95%), *Human* (99.70%), *Person* (99.70%), *Life*

Figure 14.2 A photograph of refugees on a lifeboat, used as an input for Amazon's Recognition API. The commercial service detects in the picture classes such as clothing, apparel, human, person, life jacket or vest.

[2] https://cloud.google.com/speech-to-text

jacket (99.43%) and *Vest* (99.43%). With a lower confidence you will also find labels such as *Coat* (67.39%) and *People* (66.78%). This example also highlights the need for validation, and the difficulty of grasping complex concepts in automated analyses: while all of these labels are arguably correct, it is safe to say that they fail to actually grasp the essence of the picture and the social context. One may even go as far as saying that – knowing the picture is about refugees – some of these labels, were they given by a human to describe the picture, would sound pretty cynical.

In Section 14.4 we will use this very same image (stored as `myimg2_RGB`) to detect objects using a classification model trained with an open-access database of images (ImageNet). You will find that there are some different predictions in both methods, but especially that the time to conduct the classification is shorter in the commercial service, since we don't have to train or choose a model. As you may imagine, you can neither modify the commercial models nor have access to their internal details, which is a strong limitation if you want to build your own customized classification system.

14.3 Storing, Representing, and Converting Images

In this section we will focus on learning how to store, represent, and convert images for further computational analysis. For a more exhaustive discussion of the computational analysis of images, see Williams et al. (2020).

To perform basic image manipulation we have to: (i) load images and transform their shape when it is necessary (by cropping or resizing), and (ii) create a mathematical representation of the image (normally derived from its size, colors and pixel intensity) such as a three-dimensional matrix (x, y, color channel) or a flattened vector. You have some useful libraries in Python and R (*pil* and *imagemagik*, respectively) to conduct research in these initial stages, but you will also find that more advanced libraries in computer vision will include functions or modules for pre-processing images. At this point you can work either locally or remotely, but keep in mind that images can be heavy files and if you are working with thousands of files you will probably need to store or process them in the cloud (see Section 15.2).

You can load any image as an object into your workspace as we show in Example 14.1. In this case we load two pictures of refugees published by mainstream media in Europe (see Amores et al., 2019), one is a JPG and the other is a PNG file. For this basic loading step we used the `open` function of the `Image` module in *pil* and `image_read` function in *imagemagik*. The JPG image file is a 805 × 453 picture with the color model *RGB* and the PNG is a 1540 × 978 picture with the color model *RGBA*. As you may notice the two objects have different formats, sizes and color models, which means that there is little analysis you can do if you don't create a standard mathematical representation of both.

The good news when working with digital images is that the concept of `pixel` (picture element) will help you to understand the basic mathematical representation behind computational analysis of images. A rectangular grid of pixels is represented by a dot matrix which in turn generates a `bitmap image` or `raster graphic`. The dot matrix data structure is a basic but powerful representation of the images since we can conduct multiple simple and advanced operations with the matrices. Specifically, each dot in the matrix is a number that contains information about the intensity of each pixel (that commonly ranges from 0 to 255) also known as bit or color depth (Figure 14.3). This means that the numerical representation of a pixel can have

Python Code

```python
myimg1 = Image.open(requests.get(
    "https://cssbook.net/d/259_3_32_15.jpg",
    stream=True).raw)
myimg2 = Image.open(requests.get(
    "https://cssbook.net/d/298_5_52_15.png",
    stream=True).raw)
print(myimg1)
print(myimg2)
```

R Code

```r
myimg1 = image_read(
    "https://cssbook.net/d/259_3_32_15.jpg")
myimg2 = image_read(
    "https://cssbook.net/d/298_5_52_15.png")
rbind(image_info(myimg1), image_info(myimg2))
```

Python Output

```
<PIL.JpegImagePlugin.JpegImageFile image
        mode=RGB size=805x453 at 0x104CA5B90>
<PIL.PngImagePlugin.PngImageFile image mode=RGBA
        size=1540x978 at 0x104D1FC90>
```

R Output

```
format width height colorspace matte filesize
     density
1 JPEG 805 453 sRGB FALSE 75275 72x72
2 PNG 1540 978 sRGB TRUE 2752059 57x57
```

Example 14.1 Loading JPG and PNG pictures as objects.

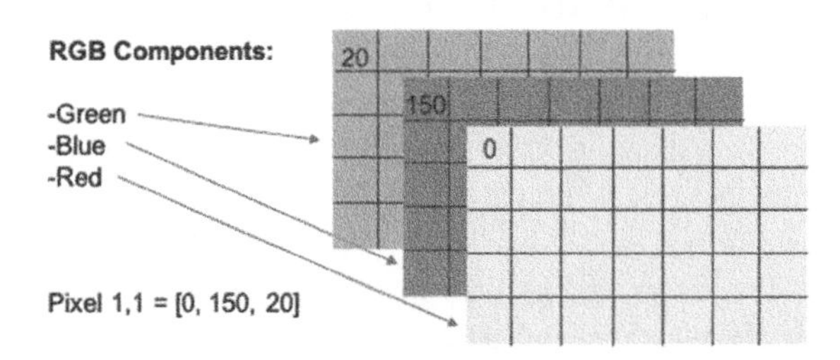

Figure 14.3 Representation of the matrix data structure of a RGB image in which each pixel contains information for the intensity of each color component.

256 different values, 0 being the darkest tone of a given color and 255 the lightest. Keep in mind that if you divide the pixel values by 255 you will have a 0–1 scale to represent the intensity.

In a black-and-white picture we will only have one color (gray-scale), with the darker points representing the black and the lighter ones the white. The mathematical representation will be a single matrix or a two-dimensional array in which the number of rows and columns will correspond to the dimensions of the image. For instance in a 224×224 black-and-white picture we will have 50 176 integers (0–255 scales) representing each pixel intensity.

In Example 14.2 we convert our original JPG picture to gray-scale and then create an object with the mathematical representation (a 453×805 matrix).

By contrast, color images will have multiple color channels that depend on the color model you chose. One standard color model is the three-channel RGB (*red*, *green* and *blue*), but you can find other variations in the chosen colors and the number of channels such as: RYB (*red*, *yellow* and *blue*), RGBA (*red*, *green*, *blue* and *alpha*[3]) or CMYK (*cyan*, *magneta*, *yellow* and *key*[4]). Importantly, while schemes used for printing such as

[3] Alpha refers to the opacity of each pixel.

[4] Key refers to *black*.

Python Code

```
myimg1_L = myimg1.convert("L")
print(type(myimg1_L))
myimg1_L_array = np.array(myimg1_L)
print(type(myimg1_L_array))
print(myimg1_L_array.shape)
```

R Code

```
myimg1_L = image_convert(myimg1,
                         colorspace = "gray")
print(class(myimg1_L))
myimg1_L_array = as.integer(myimg1_L[[1]])
print(class(myimg1_L_array))
print(dim(myimg1_L_array))
```

Python Output

```
<class' PIL.Image.Image'>
<class' numpy.ndarray'>
(453, 805)
```

R Output

```
[1] "magick-image"
[1] "array"
[1] 978 1540 1
```

Example 14.2 Converting images to gray-scale and creating a two-dimensional array.

CMYK are *substractive* (setting all colors to their highest value results in black, setting them to their lowest value results in white), schemes used for computer and television screens (such as RGB) are *additive*: setting all of the colors to their maximal value results in white (pretty much the opposite as what you got with your paintbox in primary school).

We will mostly use RGB in this book since it is the most used representation in the state-of-the-art literature in computer vision given that normally these color channels yield more accurate models. RGB's mathematical representation will be a three-dimensional matrix or a collection of three two-dimensional arrays (one for each color) as we showed in Figure 14.3. Then an RGB 224 × 224 picture will have 50 176 pixel intensities for each of the three colors, or in other words a total of 150 528 integers!

Now, in Example 14.3 we convert our original JPG file to a RGB object and then create a new object with the mathematical representation (a 453 × 805 × 3 matrix).

Instead of pixels, there are other ways to store digital images. One of them is the *vector graphics*, with formats such as .ai, .eps, .svg or .drw. Differently to bitmap images, they don't have a grid of dots but a set of *paths* (lines, triangles, square, curvy shapes, etc.) that have a start and end point, so simple and complex images are created with paths. The great advantage of this format is that images do not get "pixelated" when you enlarge them because the paths can easily be transformed while remaining smooth. However, to obtain the standard mathematical representation of images you can convert the vector graphics to raster graphics (the way back is a bit more difficult and often only possible by approximation).

Sometimes you need to convert your image to a specific size. For example, in the case of image classification this is a very important step since all the input images of the model must have the same size. For this reason, one of the most common tasks in the preprocessing stage is to change the dimensions of the image in order to adjust width and height to a specific size. In Example 14.4 we use the `resize` method provided by *pil* and the `image_scale` function in *imagemagik* to reduce the first of our original pictures in RGB (`myimg1_RGB`) to 25% . Notice that we first obtain the original dimensions of the photograph (i.e. `myimg_RGB.width` or `image_info(myimg_RGB)['width'][[1]]`) and then multiply it by 0.25 in order to obtain the new size which is the argument required by the functions.

Python Code

```python
myimg1_RGB = myimg1.convert("RGB")
print(type(myimg1_RGB))
myimg1_RGB_array = np.array(myimg1_RGB)
print(type(myimg1_RGB_array))
print(myimg1_RGB_array.shape)
```

R Code

```r
myimg1_RGB = image_convert(myimg1,
                           colorspace = "RGB")
print(class(myimg1_RGB))
myimg1_RGB_array = as.integer(myimg1_RGB[[1]])
print(class(myimg1_RGB_array))
print(dim(myimg1_RGB_array))
```

Python Output

```
<class' PIL.Image.Image'>
<class' numpy.ndarray'>
(453, 805, 3)
```

R Output

```
[1] "magick-image"
[1] "array" [1]
453 805 3
```

Example 14.3 Converting images to RGB color model and creating three two-dimensional arrays.

Python Code

```python
#Resize and visalize myimg1. Reduce to 25%
myimg1_RGB_25 = myimg1_RGB.resize(
    (int(myimg1_RGB.width * 0.25),
     int(myimg1_RGB.height * 0.25)))
myimg1_RGB_25
```

R Code

```r
#Resize and visalize myimg1. Reduce to 25%
myimg1_RGB_25 = image_scale(myimg1_RGB,
          image_info(myimg1_RGB)["width"][[1]]*0.25)
plot(myimg1_RGB_25)
```

Output: Python *Output: R*

Example 14.4 Resize to 25% and visualize a picture.

Now, using the same functions of the latter example, we specify in Example 14.5 how to resize the same picture to 224 × 244, which is one of the standard dimensions in computer vision.

You may have noticed that the new image has now the correct width and height but that it looks deformed. The reason is that the original picture was not squared and our order was to force it to fit into a 224 × 224 square, losing its original aspect. There are different alternatives to solving this issue, but probably the most extended is to *crop* the original image to create a squared picture. As you can see in Example 14.6 we can create a function that first determines the orientation of the picture (vertical versus horizontal) and then cut the margins (up and down if it is vertical; and left and right if it is horizontal) to create a square. After applying this ad hoc function `crop` to the original image we can resize again to obtain a non-distorted 224 × 224 image.

Python Code

```python
#Resize to 224 x 224
myimg1_RGB_224 = myimg1_RGB.resize(
    (224,224))
myimg1_RGB_224
```

R Code

```r
#Resize and visalize myimg1. Resize to 224 x 224
#! indicates to resize width and height exactly
myimg1_RGB_224 = image_scale(myimg1_RGB,
                             "!224x!224")
plot(myimg1_RGB_224)
```

Output: Python

Output: R

Example 14.5 Resize to 224 x 224 and visualize a picture.

Python Code

```python
#Crop and resize to 224 x 224

#Adapted from Webb, Casas & Wilkerson (2020)
def crop(img):
    height = img.height
    width = img.width
    hw_dif = abs(height - width)
    hw_halfdif = hw_dif / 2
    crop_leftright = width > height
    if crop_leftright:
        y0 = 0
        y1 = height
        x0 = 0 + hw_halfdif
        x1 = width - hw_halfdif
    else:
        y0 = 0 + hw_halfdif
        y1 = height - hw_halfdif
        x0 = 0
        x1 = width
    return img.crop((x0, y0, x1, y1))

myimg1_RGB_crop = crop(myimg1_RGB)
myimg1_RGB_crop_224 = myimg1_RGB_crop.resize(
     (224,224))
myimg1_RGB_crop_224
```

R Code

```r
#Crop and resize to 224 x 224
#Create function
crop = function(img) {
  width = image_info(img)["width"][[1]]
  height = image_info(img)["height"][[1]]
  if (width > height) {
      return (image_crop(img,
sprintf("%dx%d+%d", height,
      height, (width-height)/2)))
   }  else {
      return (image_crop(img,
sprintf("%sx%s+%s+%s", width,
      width, (width-width), (height-width)/2)))
      }
  }

myimg1_RGB_crop = crop(myimg1_RGB)
myimg1_RGB_crop_224 = image_scale(myimg1_RGB_crop,
                                  "!224x!224")
plot(myimg1_RGB_crop_224)
```

Output: Python

Output: R

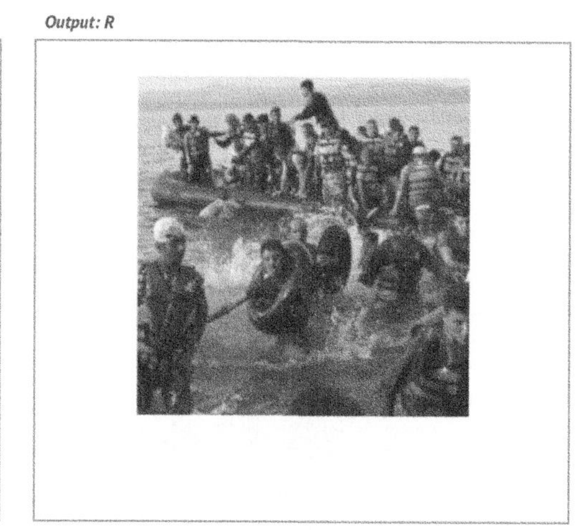

Example 14.6 Function to crop the image to create a square and the resize the picture.

Of course you are now losing part of the picture information, so you may think of other alternatives such as filling a couple of sides with blank pixels (or `padding`) in order to create the square by adding information instead of removing it.

You can also adjust the orientation of the image, flip it, or change its background, among other commands. These techniques might be useful for creating extra images in order to enlarge the training set in image classification (see Section 14.4). This is called *data augmentation* and consists of duplicating the initial examples on which the model was trained and altering them so that the algorithm can be more robust and generalize better. In Example 14.7 we used the `rotate` method in *pil* and `image_rotate` function in *imagemagik* to rotate 45 degrees the above resized image myimg1_RGB_`224` to see how easily we can get an alternative picture with similar information to include in an augmented training set.

Finally, the numerical representation of visual content can help us to *compare* pictures in order to find similar or even duplicate images. Let's take the case of RGB images which in Example 14.3 we showed how to transform to a three two-dimensional array. If we now convert the three-dimensional matrix of the image into a flattened vector we can use this simpler numerical representation to estimate similarities. Specifically, as we do in Example 14.8, we can take the vectors of two *flattened images* of resized 15 × 15 images to ease computation (`img_vect1` and `img_vect2`) and use the *cosine similarity* to estimate how akin those images are. We stacked the two vectors in a matrix and then used the `cosine_similarity` function of the `metrics` module of the *sklearn* package in Python and the `cosine` function of the *lsa* package in R.

As you can see in the resulting matrix when the images are compared with themselves (that would be the case of an exact duplicate) they obtain a value of 1. Similar images would obtain values under 1 but still close to it, while dissimilar images would obtain low values.

Python Code

```
#Rotate 45 degrees
myimg1_RGB_224_rot=myimg1_RGB_224.rotate(-45)
myimg1_RGB_224_rot
```

R Code

```
1 #Rotate 45 degrees
2 myimg1_RGB_224_rot = image_rotate(
3     myimg1_RGB_224, 45)
4 plot(myimg1_RGB_224_rot)
```

Output: Python

Output: R

Example 14.7 Rotating a picture 45 degrees.

Python Code

```
#Create two 15x15 small images to compare

#image1
myimg1_RGB_crop_15 = myimg1_RGB_crop_224.resize(
    (15,15))
#image2
myimg2_RGB = myimg2.convert("RGB")
myimg2_RGB_array = np.array(myimg2_RGB)
myimg2_RGB_crop = crop(myimg2_RGB)
myimg2_RGB_crop_224 = myimg2_RGB_crop.resize(
    (224,224))
myimg2_RGB_crop_15 = myimg2_RGB_crop_224.resize(
    (15,15))

img_vect1 = np.array(myimg1_RGB_crop_15).flatten()
img_vect2 = np.array(myimg2_RGB_crop_15).flatten()

matrix = np.row_stack((img_vect1, img_vect2))

sim_mat = cosine_similarity(matrix)
sim_mat
```

R Code

```
1  #Create two 15x15 small images to compare
2
3  #image1
4  myimg1_RGB_crop_15 = image_scale(
5      myimg1_RGB_crop_224, 15)
6  img_vect1 = as.integer(myimg1_RGB_crop_15[[1]])
7  img_vect1 = as.vector(img_vect1)
8
9  #image2
10 myimg2_RGB = image_convert(myimg2,
11                            colorspace = "RGB")
12 myimg2_RGB_crop = crop(myimg2_RGB)
13 myimg2_RGB_crop_15 = image_scale(
14     myimg2_RGB_crop, 15)
15 img_vect2 = as.integer(myimg2_RGB_crop_15[[1]])
16 #drop the extra channel for comparision
17 img_vect2 = img_vect2[,,-4]
18 img_vect2 = as.vector(img_vect2)
19
20 matrix = cbind(img_vect1, img_vect2)
21
22 cosine(img_vect1, img_vect2)
23 cosine(matrix)
```

Python Output

```
array([[1. , 0.75303996],
[0.75303996, 1. ]])
```

R Output

```
[,1]
[1,] 0.8994653
          img_vect1 img_vect2
img_vect1 1.0000000 0.8994653
img_vect2 0.8994653 1.0000000
```

Example 14.8 Comparing two flattened vectors to detect similarities between images.

14.4 Image Classification

The implementation of computational image classification can help to answer many scientific questions, from testing traditional hypotheses to opening new fields of interest in social science research. Just think about the potential of detecting at scale *who* appears in news photographs or what are the facial *emotions* expressed in the profiles of a social network. Moreover, imagine you can automatically label whether an image contains a certain action or not. For example, this is the case of Williams et al. (2020) who conducted a binary classification of pictures related to the *Black Lives Matter* movement in order to model if a picture was a protest or not, which can help to understand the extent to which the media covered a relevant social and political issue.

There are many other excellent examples of how you can adopt image classification tasks to answer specific research questions in social sciences such as those of Horiuchi et al. (2012) who detected smiles in images of politicians to estimate the effects of facial appearance on election outcomes; or the work by Peng (2018) who used automated recognition of facial traits in American politicians to investigate the bias of media portrayals.

In this section, we will learn how to conduct computational image classification which is probably the most extended computer vision application in communication and social sciences (see Table 14.1 for some terminology). We will first discuss how to apply a *shallow* algorithm and then a deep-learning approach, given a labelled data set.

Technically, in an image classification task we train a model with examples (e.g., a corpus of pictures with labels) in order to predict the category of any given new sample. It is the same logic used in supervised text classification explained in Section 11.4 but using images instead of texts. For example, if we show many pictures of cats and houses the algorithm would learn the constant features in each and will tell you with some degree of confidence if a new picture contains either a cat or a house. It is the same with letters, numbers, objects or faces, and you can apply either binary or multi-class classification. Just think when your vehicle registration plate is recognized by a camera or when your face is automatically labelled in pictures posted on Facebook.

Beyond image classification we have other specific tasks in computer vision such as *object detection* or *semantic segmentation* (Figure 14.4): To conduct object detection we first have to locate all the possible objects contained in a picture by predicting a bounding box (i.e., the four points corresponding to the vertical and horizontal coordinates of the center of the object), which is normally a regression task. Once the bounding boxes are placed around the objects, we must apply multi-class classification as explained earlier. In the case of semantic segmentation, instead of classifying objects, we classify each pixel of the image according to the class of the object the pixel belongs to, which means that different objects of the same class might not be distinguished. See Géron (2019) for a more detailed explanation and graphical examples of object detection versus image segmentation.

It is beyond the scope of this book to address the implementation of object detection or semantic segmentation, but we will focus on how to conduct basic image classification in state-of-the-art libraries in R and Python. As you may have imagined we will

Table 14.1 Some computer vision concepts used in computational analysis of communication.

Computer vision lingo	Definition
bitmap	Format to store digital images using a rectangular grid of points of colors. Also called "raster image".
pixel	Stands for "picture element" and is the smallest point of a bitmap image.
color model	Mathematical representation of colors in a picture. The standard in computer vision is RGB, but there are others such as RYB, RGBA or CMYK.
vector graphic	Format to store digital images using lines and curves formed by points.
data augmentation	Technique to increase the training set of images by creating new ones base on the modification of some of the originals (cropping, rotating, etc.)
image classification	Machine learning task to predict a class of an image based on a model. State-of-the-art image classification is conducted with Convolutional Neural Networks (CNN). Related tasks are object detection and image segmentation.
activation function	Parameter of a CNN that defines the output of a layer given the inputs of the previous layer. Some usual activation functions in image classification are sigmoid, softmax, or RELU.
loss function	Parameter of a CNN which accounts for the difference between the prediction and the target variable (confidence in the prediction). A common one common in image classification is the cross entropy loss.
optimization	Parameter of a CNN that updates weights and biases in order to reduce the error. Some common optimizers in image classification are Stochastic Gradient Descent and ADAM.
transfer learning	Using trained layers of other CNN architectures to fine tune a new model investing less resources (e.g. training data).

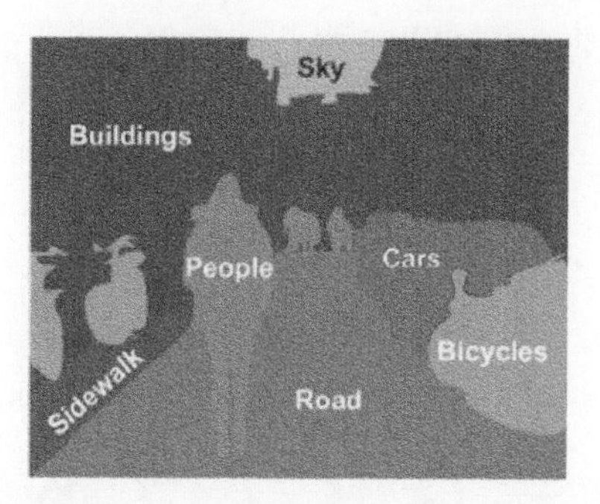

Figure 14.4 Semantic segmentation. *Source*: Adapted from Géron, 2019.

need some already-labelled images to have a proper training set. It is also out of the scope of this chapter to collect and annotate the images, which is the reason why we will mostly rely on pre-existing image databases (i.e., MINST or Fashion MINST) and pre-trained models (i.e., CNN architectures).

14.4.1 Basic Classification with Shallow Algorithms

In Chapter 8 we introduced you to the exciting world of machine learning and in Section 11.4 we introduced the *supervised* approach to classify texts. Most of the discussed models used so-called *shallow* algorithms such as Naïve Bayes or Support Vector Machines rather than the various large neural network models called *deep learning*. As we will see in the next section, deep neural networks are nowadays the best option for complex tasks in image classification. However, we will now explain how to conduct simple multi-class classification of images that contain numbers with a shallow algorithm.

Let us begin by training a model to recognize numbers using 70 000 small images of digits handwritten from the Modified National Institute of Standards and Technology (MNIST) dataset (Lecun et al., 1998). This popular training corpus contains gray-scale examples of numbers written by American students and workers and it is usually employed to test machine learning models (60 000 for training and 10 000 for testing). The image sizes are 28 × 28, which generates 784 features for each image, with pixels values from white to black represented by a 0–255 scales. In Figure 14.5 you can observe the first 10 handwritten numbers used in both training and test set.

You can download the MNIST images from its project web page[5], but many libraries also offer this dataset. In Example 14.9 we use the `read_mnist` function from the *dslabs* package (Data Science Labs) in R and the `fetch_openml` function from the *sklearn* package (`datasets` module) in Python to read and load a `mnist` object into our workspace. We then create the four necessary objects (`X_train`, `X_test`, `y_train`, `y_test`) to generate a ML model and print the first numbers in training and test sets and check they coincide with those in 14.5.

Once we are ready to model the numbers we choose one of the shallow algorithms explained in Section 8.3 to deploy a binary or multi-class image classification task. In the case of binary, we should select a number of reference (for instance "3") and then create the model of that number against all the others (to answer questions such as "What's the probability of this digit of being number 3?"). On the other hand, if we choose multi-class classification our model can predict any of the ten numbers (0, 1, 2, 3, 4, 5, 6, 7, 8, 9) included in our examples.

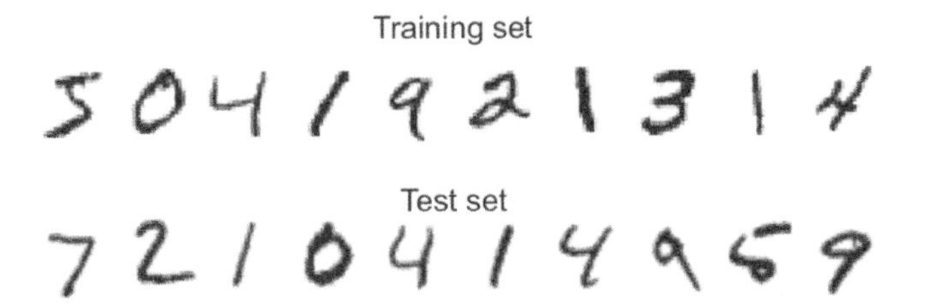

Figure 14.5 First 10 handwritten digits from the training and test set of the MNIST.

[5] http://yann.lecun.com/exdb/mnist/

Python Code

```python
mnist = fetch_openml("mnist_784", version=1)
X, y = mnist["data"], mnist["target"]
y = y.astype(np.uint8)
X_train, X_test = X[:60000], X[60000:]
y_train, y_test = y[:60000], y[60000:]
print("Shape = ", X.shape)
print("Numbers in training set= ", y_train[0:10])
print("Numbers in test set= ", y_test[0:10])
```

R Code

```r
1  mnist = read_mnist()
2
3  X_train = mnist$train$images
4  y_train = factor(mnist$train$labels)
5  X_test = mnist$test$images
6  y_test = factor(mnist$test$labels)
7
8  print("Shape = ")
9  dim(mnist$train$images)
10 print("Numbers in training set = ")
11 print(factor(y_train[1:10]), max.levels = 0)
12 print("Numbers in test set = ")
13 print(factor(y_test[1:10]), max.levels = 0)
```

Python Output

```
Shape = (70000, 784)
Numbers in training set= [5 0 4 1 9 2 1 3 1 4]
Numbers in test set= [7 2 1 0 4 1 4 9 5 9]
```

R Output

```
Shape =
[1] 60000 784
Numbers in training set =
[1] 5 0 4 1 9 2 1 3 1 4
Numbers in test set =
[1] 7 2 1 0 4 1 4 9 5 9
```

Example 14.9 Loading MNIST dataset and preparing training and test sets.

Now, we used the basic concepts of the Random Forest algorithm (see section 8.3.4) to create and fit a model with 100 trees (`forest_clf`). In Example 14.10 we use again the *randomForest* package in R and *sklearn* package in Python to estimate a model for the ten classes using the corpus of 60 000 images (classes were similarly balanced, ~ 9–11% each). As we do in the examples, you can check the predictions for the first ten images of the test set (`X_test`), which correctly correspond to the right digits, and also check the (`predictions`) for the whole test set and then get some metrics of the model. The accuracy is over 0.97 which means the classification task is performed very well.

This approach based on shallow algorithms seems to work pretty well for simple images, but has a lot of limitations for more complex images such as figures or real pictures. After all, the more complex the image and the more abstract the concept, the less likely it is that one can expect a direct relationship between a pixel color and the classification. In the next section we introduce the use of deep learning in image classification which is nowadays a more accurate approach for complex tasks.

14.4.2 Deep Learning for Image Analysis

Even though they require heavy computations, Deep Neural Networks (DNN) are nowadays the best way to conduct image classification because their performance is normally higher than shallow algorithms. The reason is that we broaden the learning process using intermediate hidden layers, so each of these layers can learn different patterns or aspects of the image with different levels of abstraction: e.g., from detecting lines or contours in the first layers to catching higher feature representation of an image (such as the color of skin, the shapes of the eyes or the noses) in the next layers. In Section 8.3.5 and Section 8.4 we introduced the general concepts of a DNN (such as

Python Code

```
forest_clf = RandomForestClassifier(
    n_estimators=100, random_state=42)
forest_clf.fit(X_train, y_train)
print(forest_clf)
print("Predict the first 10 numbers of our set:",
      forest_clf.predict(X_test[:10]))
predictions = forest_clf.predict(X_test)

print("Overall Accuracy: ", accuracy_score(
     y_test, predictions))
```

R Code

```
1  #Multiclass classification with RandomForest
2  rf_clf = randomForest(X_train, y_train, ntree=100)
3  rf_clf
4  predict(rf_clf, X_test[1:10,])
5  predictions = predict(rf_clf, X_test)
6  cm = confusionMatrix(predictions, y_test)
7  print(cm$overall["Accuracy"])
8
9
10
```

Python Output

```
RandomForestClassifier(bootstrap=True,
      class_weight=None,
criterion='gini',
            max_depth=None, max_features='auto',
                  max_leaf_nodes=None,
            min_impurity_decrease=0.0,
                  min_impurity_split=None,
            min_samples_leaf=1, min_samples_split=2,
            min_weight_fraction_leaf=0.0, n_
                  estimators =100,
            n_jobs=None, oob_score=False,
                  random_state =42, verbose=0,
            warm_start=False)
Predict the first 10 numbers of our set: [7 2
      1 0 4 1 4 9 5 9]
Accuracy: 0.9705
```

R Output

```
Call:
randomForest(x = X_train, y = y_train, ntree = 100)
               Type of random forest: classification
                     Number of trees: 100
No. of variables tried at each split: 28

        OOB estimate of error rate: 3.45%
Confusion matrix:
     0    1    2    3    4    5    6    7    8    9  class.error
0 5846    1    9    5    6    6   14    1   32    3  0.01300017
1    1 6637   37   14   11    5    9   10   12    6  0.01557401
2   27   12 5755   25   28    4   23   36   39    9  0.03407184
3    3    4   87 5817    5   71    6   48   58   32  0.05121514
4   10    7    9    3 5654    3   30   10   16  100  0.03218076
5   21    4    8    1    8 5195   42    5   39   28  0.04168973
6   25   11    4    1   12   39 5809    0   16    1  0.01841838
7    5   20   58    9   34    1    0 6061   11   66  0.03256185
8   12   38   36   49   20   53   26    6 5534   77  0.05417877
9   20    9   16   64   88   25    5   52   45 5625  0.05446293
1  2 3  4 5 6  7 8 9 10
7  2 1  0 4 1 4 9 6 9
Levels: 0 1 2 3 4 5 6 7 8 9
Accuracy
0.9691
```

Example 14.10 Modeling the handwritten digits with RandomForest and predicting some outcomes.

perceptrons, layers, hidden layers, back or forward propagation, and output functions), and now we will cover some common architectures for image analysis.

One of the simplest DNNs architectures is the Multilayer Perceptron (MLP) which contains one input layer, one or many hidden layers, and one output layer (all of them *fully* connected and with bias neurons except for the output layer). Originally in a MLP the signals propagate from the inputs to the outputs (in one direction), which we call a feedforward neural network (FNN), but using Gradient Decent as an optimizer we can apply *backpropagation* (automatically computing the gradients of the network's errors in two stages: one forward and one backward) and then obtain a more efficient training.

We can use MLPs for binary and multi-class classification. In the first case, we normally use a single output neuron with the *sigmoid* or *logistic* activation function (probability from 0 to 1) (see Section 8.3.2); and in the second case we will need one output neuron per class with the *softmax* activation function (probabilities from 0 to 1 for each class but they must add up to 1 if the classes are exclusive. This is the function used in multinomial logistic regression). To predict probabilities, in both cases we will need a *loss* function and the one that is normally recommended is the *cross entropy loss* or simply *log loss*.

The state-of-the-art library for neural networks in general and for computer vision in particular is *TensorFlow*[6] (originally created by Google and later publicly released) and the high-level Deep Learning API *Keras*, although you can find other good implementation packages such as *PyTorch* (created by Facebook), which has many straightforward functionalities and has also become popular in recent years (see for example the image classification tasks for social sciences conducted in *PyTorch* by Williams et al. (2020)). All these packages have current versions for both R and Python.

Now, let's train an MLP to build an image classifier to recognize fashion items using the Fashion MNIST dataset[7]. This dataset contains 70 000 (60 000 for training and 10 000 for test) gray scale examples (28×28) of ten different classes that include ankle boots, bags, coats, dresses, pullovers, sandals, shirts, sneakers, t-shirts/tops and trousers (Figure 14.6). If you compare this dataset with the MINST, you will find that figures of fashion items are more complex than handwritten digits, which normally generates a lower accuracy in supervised classification.

You can use *Keras* to load the Fashion MNIST. In Example 14.11 we load the complete dataset and create the necessary objects for modeling (`X_train_full`, `y_train_full`, `X_test`, `y_test`). In addition we rescaled all the input features from 0–255 to 0–1 by dividing them by 255 in order to apply Gradient Decent. Then, we obtained three sets with arrays: 60 000 in the training, and 10 000 in the test. We

Figure 14.6 Examples of Fashion MNIST items.

---

[6] We will deploy *TensorFlow 2* in our exercises.

[7] https://github.com/zalandoresearch/fashion-mnist

Python Code

```
fashion_mnist = keras.datasets.fashion_mnist
(X_train, y_train), (X_test, y_test) = \
    fashion_mnist.load_data()
class_names = ["T-shirt/top", "Trouser",
               "Pullover", "Dress", "Coat",
               "Sandal", "Shirt", "Sneaker",
               "Bag", "Ankle boot"]
X_train = X_train / 255.
X_test = X_test / 255.
print(X_train.shape, X_test.shape)
```

R Code

```
1  fashion_mnist <- dataset_fashion_mnist()
2  c(X_train, y_train) %<-% fashion_mnist$train
3  c(X_test, y_test) %<-% fashion_mnist$test
4  class_names = c("T-shirt/top","Trouser",
5       "Pullover","Dress", "Coat", "Sandal","Shirt",
6       "Sneaker", "Bag","Ankle boot")
7  X_train <- X_train / 255
8  y_test <- y_test / 255
9  print(dim(X_train))
10 print(dim(X_test))
```

Python Output

```
(60000, 28, 28) (10000, 28, 28)
```

R Output

```
[1] 60000 28 28
[1] 10000 28 28
```

Example 14.11 Loading Fashion MNIST dataset and preparing training test and validation sets.

could also generate here a validation set (e.g., `X_valid` and `y_valid`) with a given amount of records extracted from the training set (e.g., 5 000), but as you will later see *Keras* allows us to automatically generate the validation set as a proportion of the training set (e.g., 0.1, which would be 6 000 records in our example) when fitting the model (check the importance to work with a validation set to avoid over-fitting, explained in Section 8.5.2).

The next step is to design the architecture of our model. There are three ways to create the models in *Keras* (*sequential*, *functional*, or *subclassing*), but there are thousands of ways to configure a deep neural network. In the case of this MLP, we have to include first an input layer with the `input_shape` equal to the image dimension (28×28 for 784 neurons). At the top of the MLP you will need a output layer with 10 neurons (the number of possible outcomes in our multi-class classification task) and a *softmax* activation function for the final probabilities for each class.

In Example 14.12 we use the *sequential* model to design our MLP layer by layer including the above-mentioned input and output layers. In the middle, there are many options for the configuration of the *hidden* layers: number of layers, number of neurons, activation functions, etc. As we know that each hidden layer will help to model different patterns of the image, it would be fair to include at least two of them with different numbers of neurons (significantly reducing this number in the second one) and transmit its information using the *relu* activation function. What we actually do is create an object called `model` which saves the proposed architecture. We can use the method `summary` to obtain a clear representation of the created neural network and the number of parameters of the model (266 610 in this case!).

The next steps will be to `compile`, `fit`, and `evaluate` the model, similarly to what you have already done in previous exercises. In Example 14.13 we first include the parameters (loss, optimizer, and metrics) of the compilation step and fit the model, which might take some minutes (or even hours depending on your dataset, the architecture of you DNN and, of course, your computer).

When fitting the model you have to separate your training set into phases or *epochs*. A good rule of thumb to choose the optimal number of epochs is to stop a few

Python Code

```
model = keras.models.Sequential([
    keras.layers.Flatten(input_shape=[28, 28]),
    keras.layers.Dense(300, activation="relu"),
    keras.layers.Dense(100, activation="relu"),
    keras.layers.Dense(10, activation="softmax")
])
model.summary()
```

R Code

```
1  model = keras_model_sequential()
2  model %>%
3  layer_flatten(input_shape = c(28, 28)) %>%
4  layer_dense(units=300, activation="relu") %>%
5  layer_dense(units=100, activation="relu") %>%
6  layer_dense(units=10, activation="softmax")
7  model
```

Output

```
Model: "sequential"
_________________________________________________________________
Layer (type)           Output Shape         Param
=================================================================
flatten (Flatten)      (None, 784)          0
_________________________________________________________________
dense (Dense)          (None, 300)          235500
_________________________________________________________________
dense_1 (Dense)        (None, 100)          30100
_________________________________________________________________
dense_2 (Dense)        (None, 10)           1010
=================================================================
Total params: 266,610
Trainable params: 266,610
Non-trainable params: 0
```

Example 14.12 Creating the architecture of the MLP with *Keras*.

iterations after the test loss stops improving[8] (here we chose five epochs for the example). You will also have to set the proportion of the training set that will become the validation set (in this case 0.1). In addition, you can use the parameter `verbose` to choose whether to see the progress (1 for progress bar and 2 for one line per epoch) or not (0 for silent) of the training process. By using the method `evaluate` you can then obtain the final loss and accuracy, which in this case is 0.84 (but you can reach up 0.88 if you fit it with 25 epochs!).

Finally, you can use the model to predict the classes of any new image (using `predict_classes`). In Example 14.14 we used the model to predict the classes of the first six elements of the test set. If you go back to Figure 14.6 you can compare these predictions ("ankle boot", "pullover", "trouser", "trouser", "shirt", and "trouser") with the actual first six images of the test set, and see how accurate our model was.

Using the above-described concepts and code you may try to train a new MLP using color images of ten classes (airplane, automobile, bird, cat, deer, dog, frog, horse, ship, and truck) using the CIFAR-10 and CIFAR-100 datasets[9]!

---

[8] The train loss/accuracy will gradually be better and better. And the test loss/accuracy as well, in the beginning. But then, at some point train loss/acc improves but test loss/acc stops getting better. If we keep training the model for more epochs, we are just overfitting on the train set, which of course we do not want to. Specifically, we do not want to simply stop at the iteration where we got the best loss/acc for the test set, because then we are overfitting on the test set. Hence practitioners often let it run for a few more epochs after hitting the best loss/acc for the test set. Then, a final check on the validation set will really tell us how well we do out of sample.

[9] https://www.cs.toronto.edu/ kriz/cifar.html

Python Code

```python
model.compile(loss=
              "sparse_categorical_crossentropy",
              optimizer="sgd",
              metrics=["accuracy"])

history = model.fit(X_train, y_train, epochs=5,
                    verbose=2, validation_split=0.1)

print("Evaluation: ")
print(model.evaluate(X_test, y_test))
```

R Code

```r
model %>% compile(
  optimizer = "sgd",
  loss = "sparse_categorical_crossentropy",
  metrics = c("accuracy")
)
history = model %>% fit(X_train, y_train,
   validation_split=0.1, epochs=5, verbose= 2)
print(history$metrics)
score = model %>% evaluate(
    X_test, y_test, verbose = 0)
print("Evaluation")
print(score)
```

Python Output

```
Epoch 1/5
1688/1688 - 3s - loss: 0.7274 - accuracy: 0.7613
      - val_loss: 0.5562 - val_accuracy: 0.8040
Epoch 2/5
1688/1688 - 2s - loss: 0.4914 - accuracy: 0.8289
      - val_loss: 0.5015 - val_accuracy: 0.8198
Epoch 3/5
1688/1688 - 2s - loss: 0.4452 - accuracy: 0.8444
      - val_loss: 0.4435 - val_accuracy: 0.8375
Epoch 4/5
1688/1688 - 3s - loss: 0.4173 - accuracy: 0.8540
      - val_loss: 0.4054 - val_accuracy: 0.8543
Epoch 5/5
1688/1688 - 3s - loss: 0.3970 - accuracy: 0.8616
      - val_loss: 0.3991 - val_accuracy: 0.8570
Evaluation:
313/313 [==================] - 0s 1ms/step - loss:
       0.4241 - accuracy: 0.8486
[0.4241237938404083, 0.8485999703407288]
```

R Output

```
$loss
[1] 0.7255158 0.4923369 0.4453345 0.4178082
      0.3969852
$accuracy
[1] 0.7633519 0.8292222 0.8435556 0.8542778
      0.8622593
$val_loss
[1] 0.5811644 0.4684601 0.4389609 0.4139460
      0.4023611
$val_accuracy
[1] 0.7863333 0.8321667 0.8453333 0.8591667
      0.8561667
```

Example 14.13 Compiling fitting and evaluating the model for the MLP.

Python Code

```python
X_new = X_test[:6]
y_pred = np.argmax(model.predict(X_new), axis=-1)
class_pred = [class_names[i] for i in y_pred]
print(class_pred)
```

R Code

```r
img = X_test[1:6, , , drop = FALSE]
class_pred = model %>% predict_classes(img)
class_pred
```

Python Output

```
['Ankle boot', 'Pullover', 'Trouser', 'Trouser',
     'Shirt', 'Trouser']
```

R Output

```
[1] 9 2 1 1 6 1
```

Example 14.14 Predicting classes using the MLP.

14.4.3 Re-using an Open Source CNN

Training complex images such as photographs is normally a more sophisticated task if we compare it to the examples included in the last sections. On the one hand, it might not be a good idea to build a deep neural network from scratch as we did in Section 14.4.2 to train a MLP. This means that you can re-use some lower layers of other DNNs and deploy *transfer learning* to save time with less training data. On the other hand, we should also move from traditional MLPs to other kinds of DNNs such as Convolutional Neural Networks (CNNs) which are nowadays the state-of-the-art approach in computer vision. Moreover, to get good results we should also build or explore different CNN architectures that can produce more accurate predictions in image classification. In this section we will show how to re-use an open source CNN architecture and will suggest an example of how to fine-tune an existing CNN for a social science problem.

As explained in Section 8.4.1 a CNN is a specific type of DNN that has had great success in complex visual tasks (image classification, object detection or semantic segmentation) and voice recognition[10]. Instead of using *fully connected* layers like in a typical MLP, a CNN uses only *partially connected* layers inspired on how "real" neurons connect in the visual cortex: some neurons only react to stimuli located in a limited *receptive field*. In other words, in a CNN every neuron is connected to some neurons of the previous layer (and not to all of them), which significantly reduces the amount of information transmitted to the next layer and helps the DNN to detect complex patterns. Surprisingly, this reduction in the number of parameters and weights involved in the model works better for larger and more complex images, different from those shown in MNIST.

Building a CNN is quite similar to a MLP, except for the fact that you will have to work with *convolutional* and *pooling* layers. The convolutional layers include a *bias term* and are the most important blocks of a CNN because they establish the specific connections among the neurons. In simpler words: a given neuron of a high-level layer is connected only to a rectangular group of neurons (the receptive field) of the low-level layer and not to all of them[11]. For more technical details of the basis of a CNN you can can go to specific literature such as Géron (2019).

Instead of building a CNN from scratch, there are many pre-trained and open-source architectures that have been optimized for image classification. Besides a stack of convolutional and pooling layers, these architectures normally include some fully connected layers and a regular output layer for prediction (just like in MLPs). We can mention here some of these architectures: LeNet-5, AlexNet, GoogLeNet, VGGNet, ResNet, Xception or SENet[12]. All these CNNs have been previously tested in image

[10] CNNs have also a great performance in natural language processing.

[11] If the input layer (in the case of color images there are three sublayers, one per color channel) and the convolutional layers are of different sizes we can apply techniques such as *zero padding* (adding zeros around the inputs) or spacing out the receptive fields (each shift from one receptive field to the other will be a *stride*). In order to transmit the weights from the receptive fields to the neurons, the convolutional layer will automatically generate some *filters* to create *features maps*, which are the areas of the input that mostly activate those filters. Additionally, by creating subsamples of the inputs, the pooling layers will reduce the number of parameters, the computational effort of the network and the risk of overfitting. The pooling layers *aggregates* the inputs using a standard arithmetic function such as minimum, maximum or mean.

[12] The description of technical details of all of these architectures is out of the scope of this book, but besides the specific scientific literature of each architecture, some packages such as *keras* usually include basic documentation.

classification with promising results, but you still have to look at the internal composition of each of them and their metrics to choose the most appropriate for you. You can implement and train most of them from scratch either in *keras* or *PyTorch*, or you can just use them directly or even fine-tune the pre-trained model in order to save time.

Let's use the pre-trained model of a Residual Network (ResNet) with 50 layers, also known as *ResNet50*, to show you how to deploy a multi-class classifier over pictures. The ResNet architecture (also with 34, 101 and 152 layers) is based on residual learning and uses *skip connections*, which means that the input layer not only feeds the next layer but this signal is also added to the output of another high-level layer. This allows you to have a much deeper network and in the case of ResNet152 it has achieved a top-five error rate of 3.6%. As we do in Example 14.15, you can easily import into your workspace a ResNet50 architecture and include the pre-trained weights of a model trained with ImageNet (uncomment the second line of the code to visualize the complete model!).

ImageNet is a corpus of labelled images based on the WordNet hierarchy. ResNet uses a subset of ImageNet with 1000 examples for each of the 1000 classes for a total corpus of roughly 1 350 000 pictures (1 200 000 for training, 100 000 for test, and 50 000 for validation).

In Example 14.16 we crop a part of our second example picture of refugees arriving at the European coast (`myimg2_RGB`) in order to get just the sea landscape. With the created `model_resnet50` we then ask for up to three predictions of the class of the photograph in Example 14.17.

As you can see in the Python and R outputs[13], the best guess of the model is a *sandbar*, which is very close to the real picture that contains sea water, mountains and sky.

Python Code

```python
model_rn50 = resnet50.ResNet50(weights="imagenet")
```

R Code

```r
model_resnet50 = application_resnet50(
    weights="imagenet")
#model_resnet50
```

Example 14.15 Loading a visualizing the ResNet50 architecture.

Python Code

```python
def plot_color_image(image):
    plt.imshow(image, interpolation="nearest")
    plt.axis("off")
picture1 = np.array(myimg2_RGB)/255
picture2 = np.array(myimg2_RGB)/255
images = np.array([picture1, picture2])
see = [0, 0, 0.3, 0.3]
refugees = [0.1, 0.35, 0.8, 0.95]
tf_images = tf.image.crop_and_resize(images,
            [see, refugees], [0, 1], [224, 224])
plot_color_image(tf_images[0])
plt.show()
```

R Code

```r
picture1 = image_crop(myimg2_RGB, "224x224+50+50")
plot(picture1)
picture1 = as.integer(picture1[[1]])
#drop the extra channel for comparision
picture1 = picture1[,,-4]
picture1 = array_reshape(picture1,
                         c(1, dim(picture1)))
picture1 = imagenet_preprocess_input(picture1)
```

[13] Outputs in Python and R might differ a little bit since the cropping of the new images were similar but not identical.

Output: Python

Output: R

Example 14.16 Cropping an image to get a picture of a sea landscape.

Python Code

```
inputs = resnet50.preprocess_input(tf_images*255)
Y_proba = model_rn50.predict(inputs)
preds = resnet50.decode_predictions(Y_proba,top=3)
preds[0]
```

R Code

```
preds1 = model_resnet50 %>% predict(picture1)
imagenet_decode_predictions(preds1, top = 3)[[1]]
```

Python Output

```
[('n09421951',' sandbar', 0.0835789),
('n09428293',' seashore', 0.061473366),
('n09246464',' cliff', 0.05028373)]
```

R Output

```
class_name class_description score
1 n09421951 sandbar     0.07926153
2 n04347754 submarine   0.04810236
3 n02066245 grey_whale  0.04798749
```

Example 14.17 Predicting the class of the first image.

However, it seems that the model is confusing sand with sea. Other results in the Python model are *seashore* and *cliff*, which are also very close to real sea landscape. Nevertheless, in the case of the R prediction the model detects a *submarine* and a *gray whale*, which revels that predictions are not 100% accurate yet.

If we do the same with another part of that original picture and focus only on the group of refugees in a lifeboat arriving at the European coast, we will get a different result! In Example 14.18 we crop again (`myimg2_RGB`) and get a new framed picture. Then in Example 14.19 we re-run the prediction task using the model *ResNet50* trained with ImageNet and get a correct result: both predictions coincide to see a *lifeboat*, which is a good tag for the image we want to classify. Again, other lower-level predictions can seem accurate (*speedboat*) and totally inaccurate (*volcano*, *gray whale* or *amphibian*).

These examples show you how to use an open-source and pre-trained CNN that has 1000 classes and has been trained on images that we do not have control of. However, you may want to build your own classifier with your own training data, but using part of an existing architecture. This is called fine-tuning and you can follow a good example in social science in Williams et al. (2020) in which the authors reuse RestNet18 to

Python Code

```python
plot_color_image(tf_images[1])
plt.show()
```

R Code

```r
picture2 = image_crop(myimg2_RGB, "224x224+1000")
plot(picture2)
picture2 = as.integer(picture2[[1]])
#drop the extra channel for comparision
picture2 = picture2[,,-4]
picture2 = array_reshape(picture2,
                         c(1, dim(picture2)))
picture2 = imagenet_preprocess_input(picture2)
```

Output: Python

Output: R

Example 14.18 Cropping an image to get a picture of refugees in a lifeboat.

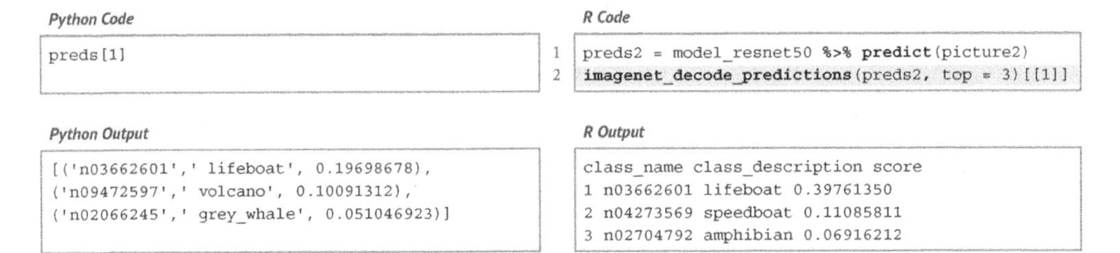

Python Code

```python
preds[1]
```

R Code

```r
preds2 = model_resnet50 %>% predict(picture2)
imagenet_decode_predictions(preds2, top = 3)[[1]]
```

Python Output

```
[('n03662601',' lifeboat', 0.19698678),
('n09472597',' volcano', 0.10091312),
('n02066245',' grey_whale', 0.051046923)]
```

R Output

```
class_name class_description score
1 n03662601 lifeboat 0.39761350
2 n04273569 speedboat 0.11085811
3 n02704792 amphibian 0.06916212
```

Example 14.19 Predicting the class of the second image.

build binary and multi-class classifiers adding their own data examples over the pre-trained CNN[14].

So far we have covered the main techniques, methods, and services to analyze multimedia data, specifically images. It is up to you to choose which library or service to use, and you will find most of them in R and Python, using the basic concepts explained in this chapter. If you are interested in deepening your understanding of multimedia analysis, we encourage you explore this emerging and exciting field of expertise given the enormous importance it will no doubt have in the near future.

[14] The examples are provided in Python with the package *PyTorch*, which is quite friendly if you are already familiar to *Keras*.

15

Scaling Up and Distributing

Abstract

Throughout this book, we have been working with examples that consist of code to conduct one specific analysis of data sets of modest size. But at some point, you may want to scale up. You may want others to be able to apply your code to their data; and you may want to be able to also use your own analyses on larger and more complex datasets. Or you may need to run analyses that your own computer cannot deal with. This chapter deals with such steps and points you to some techniques that become increasingly useful the larger your projects get.

Keywords databases, cloud computing, containerization, source code, version control

---

- Be able to scale up your analyses
- Know when to use databases
- Know when to use cloud computing
- Know about distributing source code and containers.

In this chapter, we provide a brief overview of techniques for scaling up computational analyses. In particular, we introduce SQL and noSQL databases, cloud computing platforms, version control systems, and Docker containers.

15.1 Storing Data in SQL and noSQL Databases

15.1.1 When to Use a Database

In this book, we have so far stored our data in files. In fact, before covering the wide range of methods for computational analysis, we discussed some basics of file handling (Chapter 5). Probably, you did not experience any major trouble here (apart from occasional struggles with non-standard encodings, or confusion about the delimiters in a csv file). On the other hand, the examples we used were still modest in size: usually, you were dealing with a handful of csv files; except for huge image classification datasets, the maximum you had to deal with were the 50 000 text files from the IMDB movie review dataset.

Computational Analysis of Communication: A Practical Introduction to the Analysis of Texts, Networks, and Images with Code Examples in Python and R, First Edition. Wouter van Atteveldt, Damian Trilling & Carlos Arcila Calderón.

In particular, when loading your data into a data frame, you copied all the data from your disk into memory[1]. But what if you want to scale up our analyses a bit (see Trilling and Jonkman, 2018)? Maybe you want to build up a larger data collection, maybe even share it with multiple team members, search and filter our data, or collect it over a larger timespan? An example may illustrate the problems that can arise.

Imagine you do some web scraping (Chapter 12) that goes beyond a few thousand texts. Maybe you want to visit relevant news sites on a regular basis (say, once an hour) and retrieve everything that's new. How do you store your data then? You could append everything to a huge csv file, but this file would quickly grow so large that you cannot load it into memory any more. Besides, you may run the risk of corrupting the file if something goes wrong in one of your attempts to extend the file. Or you could also write each article to a new, separate file. That's maybe more failsafe, but you would need to design a good way to organize the data. In particular, devising a method to search and find relevant files would be a whole project in itself.

Luckily, you can outsource all these problems to a database that you can install on your own computer or possibly on a server (in that case, make sure that it is properly secured!). In the example, the scraper, which is running once an hour, just sends the scraped data to the database instead of to a file, and the database will take care of storing it. Once you want to retrieve a subset of your articles for analysis, you can send a query to the database and read from it. Both Python and R offer integration for multiple commonly used databases. It is even possible to directly get the results of such a database query in the form of a data frame.

We can distinguish two main categories of databases that are most relevant to us (see also Günther et al., 2018): relational databases (or SQL-databases) and noSQL-databases. Strictly speaking, SQL ("structured query language") is a query language for databases, but it is so widespread that it is used almost synonymously for relational databases. Even though they have already been around for 50 years (Codd, 1970), relational databases are still very powerful and very widely used. They consist of multiple tables that are linked by shared columns (keys). For instance, you could imagine a table with the orders placed in a webshop that has a column `customer-id`, and a different table with addresses, billing information, and names for each `customer-id`. Using filter and join operations (like in Chapter 6, but then on the database directly), one can then easily retrieve information on where the order has to be shipped. A big advantage of such a relational database is that, if a customer places 100 orders, we do not need to store their address 100 times, but only once, which is not only more efficient in terms of storage, but also prevents inconsistencies in the data.

In contrast to SQL databases, noSQL databases are not based on tables, but use concepts such as "documents" or key-value pairs, very much like Python dictionaries or JSON files. These types of databases are particularly interesting when your data are less well-structured. If not all of your cases have the same variables, or if the content is not well-defined (let's say, you don't know exactly in what format the date of publication on a news site will be written), or if the data structure may change over time, then it is hard or impossible to come up with a good table structure for an SQL

[1] In fact, this is sometimes a reason to avoid data frames: for instance, it is possible to use a generator that reads data line-by-line from a file and yields them to *scikit-learn*. In this way, only *one* row of data is in your memory at the same time (see Section 3.3).

database. Therefore, in many "big data" contexts, noSQL databases are used, as they – depending on your configuration – will happily accept almost any kind of content you dump in them. This comes, of course, at the expense of giving up advantages of SQL databases, such as the avoidance of inconsistencies. But often, you may want to store your data first and clean up later, rather than risking that data collection fails because you enforced a too strict structure. Also, there are many noSQL databases that are very fast in searching full text – something that SQL databases, in general, are not optimized for.

Despite all of these differences, both SQL and noSQL databases can play the same role in the computational analysis of communication. They both help you to focus on data collection and data analysis without needing to devise an ingenious way to store your data. They both allow for much more efficient searching and filtering than you could design on your own. All of this becomes especially interesting when your dataset grows too large to fit in memory, but also when your data are continuously changed, for instance because new data are added while scraping.

15.1.2 Choosing the Right Database

Choosing the right database is not always easy, and has many consequences for the way you may conduct your analyses. As Günther et al. (2018) explain, this is not a purely technical choice, but impacts your social-scientific workflow. Do you want to enforce a specific structure from the very start, or do you want to collect everything first and clean up later? What is your trade-off between avoiding any inconsistency and risking throwing away too much raw information?

Acknowledging that there are often many different valid choices, and at the risk of oversimplifying matters, we will try to give some guidance in which databases to choose by offering some guiding questions.

How is your data structured? Ask yourself: can I organize my data in a set of relational tables? For instance, think of television viewing data: there may be a table that gives information on when the television set was switched on and which channel was watched and by which user id. A second table can be used to associate personal characteristics such as age and gender with the user id. And a third table may be used to map the time stamps to details about a specific program aired at the time. If your data looks like this, ask yourself: can I determine the columns and the data types for each column in advance? If so, then a SQL database such as *MySQL*, *PostgreSQL*, or *MariaDB* is probably what you are looking for. If, on the other hand, you cannot determine such a structure *a priori*, if you believe that the structure of your information will change over time, or if it is very messy, then you may need a more flexible noSQL approach, for instance using *MongoDB* or *ElasticSearch*.

How important is full-text searching for you? SQL databases can handle numeric datatypes as well as text datatypes, but they are usually not optimized for the latter. They handle short strings (such as usernames, addresses, and so on) just fine, but if you are interested in full-text searching, they are not the right tool for the job. This is in particular true if you want to be able to do fuzzy searches where, for instance, documents containing the plural of a word that you searched for as singular are also found. Databases of, for instance, news articles, tweets, transcripts of speeches, or other documents are much better accessed in a database such as *ElasticSearch*.

How flexible does it need to be? In relational databases, it is relatively hard to change the structure afterwards. In contrast, a noSQL database has no problem whatsoever with adding a new document that contains keys that did not exist before. There is no assumption that all documents contain the same keys. Therefore, if it is hard to tell in advance which "columns" or "keys" may represent your data best, you should stay clear of SQL databases. In particular, if you think of gradually extending your data and use it on a long timeline for re-use, potentially even by multiple teams, the flexibility of a noSQL database may be a game changer.

15.1.3 A Brief Example Using SQLite

Installing a database server such as *mysql*, *mariadb* (an open-source fork of mysql), *MongoDB*, or *Elasticsearch* is not really difficult (in fact, it may already be come pre-packaged with your operating system), but the exact configuration and setup may differ widely depending on your computer and your needs. Most importantly, especially if you store sensitive data in your database, you will need to think about authentication, roles, etc. — all beyond the scope of this book.

Luckily, there is a compromise between storing your data in the files that you need to manage yourself and setting up a database server, locally or remotely. The library *SQlite* offers a self-contained database engine – essentially, it allows you to store a whole database in one file and interact with it using the SQL query language. Both R and Python offer multiple ways of directly interacting with sqlite files (Example 15.1). This gives you access to some great functionality straight away: after all, you can issue (almost) any SQL command now, including (and maybe most importantly) commands for filtering, joining, and aggregating data. Or you could consider immediately writing each datapoint you get from an API or a webscraper (Chapter 12) without risking losing any data if connections time out or scraping fails halfway.

Of course, *SQlite* cannot give you the same performance as a "real" mysql (or similar) installation could offer. Therefore, if your project grows bigger, or if you have a lot of read- or write-operations per second, then you may have to switch at some point. But as you can see in Example 15.1, Python and R do not really care about the back end: all you need to do is to change the connection `conn` such that it points to your new database instead of the sqlite file.

15.2 Using Cloud Computing

Throughout this book, we assumed that all tasks can actually be performed on your own computer. And often, that is indeed the best thing to do: you normally want to maintain a local copy of your data anyway, and it may be the safest bet for ethical and legal reasons – when working with sensitive data, you need to know what you are doing before transferring them somewhere else.

Python Code

```python
import pandas as pd
import sqlite3

# Load a dataframe
url = "https://cssbook.net/d/gun-polls.csv"
d = pd.read_csv(url)

# connecting to a SQLite database
conn = sqlite3.connect("mydb.db")
# store the df as table "gunpolls" in the database
d.to_sql("gunpolls", con=conn)

# run a query on the SQLite database
sql = """SELECT support, pollster
        FROM gunpolls LIMIT 5;"""
d2 = pd.read_sql_query(sql, conn)
# close connection
conn.close()
d2
```

R Code

```r
library(tidyverse)
library(RSQLite)

# Load a dataframe
url = "https://cssbook.net/d/gun-polls.csv"
d = read_csv(url)

# connecting to a SQLite database
mydb = dbConnect(RSQLite::SQLite(), "mydb.sqlite")
# store the df as table "gunpolls" in the database
#dbWriteTable(mydb, "gunpolls", d)

# run a query on the SQLite database
sql = "SELECT support, pollster
        FROM gunpolls LIMIT 5;"
d2 = dbGetQuery(mydb, sql)
d2
# close connection
dbDisconnect(mydb)
```

Python Output

	Support	Pollster
0	72.0	CNN/SSRS
1	82.0	NPR/Ipsos
2	67.0	Rasmussen
3	84.0	Harris Interactive
4	78.0	Quinnipiac

Example 15.1 *SQLite* offers you database functionality without setting up a database server such as *mysql*.

However, once you scale up your project, problems may arise (see Trilling and Jonkman (2018)):

- Multiple people need to work on the same data
- Your dataset is too large to fit on your disk
- You do not have enough RAM or processing power
- Running a process simply takes too long (e.g., training a model for several days) or needs to be run in continuous intervals (e.g., scraping news articles once an hour) and you need your computer for other things.

This is the point where you need to start moving your project to some remote server instead. Broadly speaking, we can consider four scenarios:

1. A cloud service that just lets you run code. Here, you can just submit your code and have it run. You do not have full control, you cannot set up your own system, but you also do not have to do any administration.
2. A dedicated server. You (or your university) could buy a dedicated, physical server to run computational social science analyses. On the bright side, this gives you full control, but it is also not very flexible: after all, you make a larger investment once, and if it turns out that you need more (or less) resources, then it might be too late to change.

3. A virtual machine (VM) on a cloud computing platform. For most practical purposes, you can do the same as in the previous option, with the crucial difference that you rent the resources. If you need more, you just rent more; and when you are done, you just stop the machine.
4. A set of machines to run complex tasks using parallel computing. With large amounts of information (think about image or video data) and sophisticated modeling (such as deep learning) you may need to distribute the computation among several different computers at the same time.

An example for the first option is Google Colab. While it makes it easy to share and run notebooks, the free tier we used so far does not necessarily solve any of the scalability issues discussed. However, Google Colab also has a paid Pro version, in which additional hardware (such as GPUs, TPUs or extra memory) that you may not have on your own computer can be used. This makes it an attractive solution for enabling projects (e.g., involving resource-intensive neural networks) that otherwise would not be possible.

However, this is often not enough. For instance, you may want to run a database (Section 15.1) or define a so-called *cron* job, which runs a specific script (e.g., a web scraper) at defined intervals. Here, options 2 and 3 come into play – most realistically for most beginners, option 3.

There are different providers for setting up VMs in the cloud, the most well-known probably being Amazon Web Services (AWS) and Microsoft Azure. Some universities or (national) research infrastructure providers provide high-performance computing in the cloud as well. While the specific way to set up a virtual machine of your own on such an infrastructure varies, the processes are roughly similar: you select the technical specifications such as the number of CPU cores and the amount of memory you need, attach some storage, and select a disk image with an operating system, virtually always some Linux distribution (Figure 15.1). After a couple of minutes, your machine is ready to use.

While setting up such a machine is easy, some knowledge is required for the responsible and safe administration of the machine, in particular to prevent unauthorized access.

Imagine you have a script `myscript.py` that takes a couple of days to run. You can then use the tool `scp` to copy it to your new virtual machine, log on to your virtual machine using `ssh`, and then – now on your virtual machine! – run the script using a tool such as `nohup` or `screen` that will start your script and will keep running it

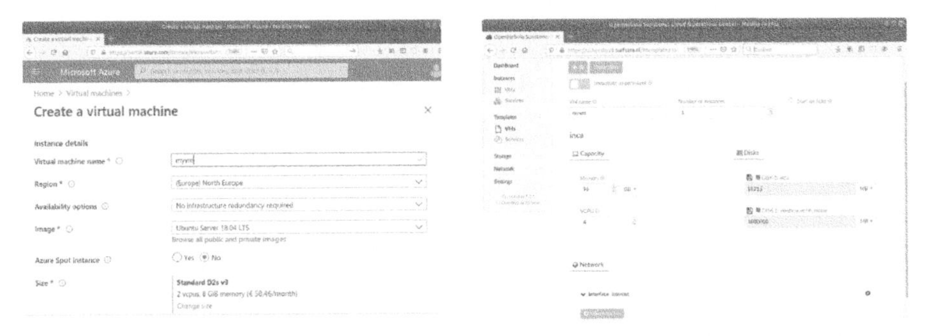

Figure 15.1 Creating a Virtual Machine on Microsoft Azure (left) and on a university cloud computing platform using OpenNebula (right). *Source:* Used with permission from Microsoft

(Figure 15.2). You can safely logout again, and your virtual machine in the cloud will keep on doing its work. The only thing you need to do is collect your results once your script is done, even if that's a couple of weeks later. Or you may want to add your script to the crontab (Google it!), which will automatically run it at set intervals.

You may want to have some extra luxury, though. Popular things to set up are databases (Section 15.1) and *JupyterHub*, which allows users such as your colleagues to connect through their web browser with your server and run their own Jupyter Notebooks on the machine. Do not forget to properly encrypt all connections, for instance using *letsencrypt*.

Finally, option 4 must be selected when the scale of your data and the complexity of the tasks cannot be deployed in a single server or virtual machine. For example, building a classification model by training a complex and deep convolutional neural network with millions of images and update this model constantly may require the use of different computers at the same time. Actually, in modern computers with multiple cores or processors you normally run parallel computing within a single machine. But when working at scale you will probably need to set a infrastructure of different computers such as that of a *grid* or a *cluster*.

Cloud services (e.g. AWS, Microsoft Azure, etc.) or scientific infrastructures (e.g. Supercomputers) offer the possibility to set these architectures remotely. For instance, in a computer cluster you can configure a group of virtual computers, where one will act as a main and the others as worker. With this logic the master can distribute the storage and analysis of data among the slaves and then resume the results: see for example the *MapReduce* or the *Resilient Distributed Dataset* (RDD) approaches used by the open-source softwares *Apache Hadoop* and *Apache Spark* respectively. For a specific example of parallel computing in computational analysis of communication you can take a look at the implementation of distributed supervised sentiment analysis, in which one of the authors of this book deployed supervised text classification in *Apache Spark* and connected this infrastructure with real-time analysis of tweets using *Apache Kafka* in order to perform streaming analytics (see Arcila-Calderón et al., 2019).

These architectures for parallel processing will significantly increase your computation capacity for big data problems but the initial implementation will consume time and (most of the time) money, which is the reason why you must think in advance if there is a simpler solution (such as a single but powerful machine) before implementing a more complex infrastructure in your analysis.

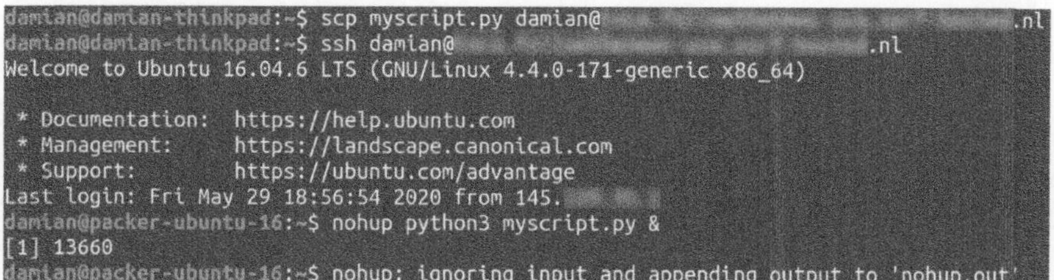

```
damian@damian-thinkpad:~$ scp myscript.py damian@                    .nl
damian@damian-thinkpad:~$ ssh damian@                    .nl
Welcome to Ubuntu 16.04.6 LTS (GNU/Linux 4.4.0-171-generic x86_64)

 * Documentation:  https://help.ubuntu.com
 * Management:     https://landscape.canonical.com
 * Support:        https://ubuntu.com/advantage
Last login: Fri May 29 18:56:54 2020 from 145.
damian@packer-ubuntu-16:~$ nohup python3 myscript.py &
[1] 13660
damian@packer-ubuntu-16:~$ nohup: ignoring input and appending output to 'nohup.out'
```

Figure 15.2 Running a script on a virtual machine. Note that the first two commands are issued on the local machine ("damian-thinkpad") and the next command on the remote machine ("packer-ubuntu-16"). *Source:* Used with permission from Microsoft

15.3 Publishing Your Source

Already in Section 4.3, we briefly introduced the idea of version control protocols such as *git*, and the most well-known online git repository *GitHub*. There are others, such as *Bitbucket* and the question of which one you use is not really of importance for our argument here. Already for small projects, it is a good idea to use version control so that you can always go back to earlier versions, but as soon as you start working with multiple people on one project, it becomes indispensable.

In particular, it is possible to work on multiple *branches*, different versions of the code that can later be merged again. In this way, it is possible to develop new features without interfering with the main version of the code. There are plenty of git tutorials available online, and we highly recommended using git from the beginning of a specific project on – be it your bachelor, master or doctoral thesis, a paper, or a tool that you want to create.

In the computational analysis of communication, it is becoming more and more the norm to publish all your source code together with an article, even though it is important to keep in mind ethical and legal restrictions (Van Atteveldt et al., 2019). Using a version control platform like *GitHub* from the beginning makes this easy: when publishing your paper, the only thing you have to do is to set access of your repository to "public" (in case it was private before), add a `README.md` file (in case you have not done so earlier), and preferably, get a persistent identifier, a `doi` for your code (see https://guides.*GitHub*.com/activities/citable-code/). And don't forget to add a license to your code, such as MIT, GPL, or Apache. All of these have specific implications on what others can or cannot do with your code (e.g., whether it can be used for commercial purposes or whether derivatives need to be published under the same license as well). Whatever you choose here, it is important that you make a choice, as otherwise, it may not be (legally) possible to use your code at all. If your code pertains to a specific paper, then we suggest you organize your repository as a so-called "research compendium", integrating both your code and your data. Van Atteveldt et al. (2020a) provide a template and tools for easily creating one [2].

In virtually all instances, your code will rely on libraries written by others, which are available free of charge. Therefore, it only seems fair to "give back" and make sure that any code that you wrote and that can be useful to others, is also available to them.

Just like in the case of a research compendium for a specific paper, publishing source code for more generic re-use also begins with a github repository. In fact, both R (with *devtools*) and Python (via *pip*) can install packages directly from github. In order to make sure that your code can be installed as a package, you need to follow specific instructions on how to name files, how to structure your directory, and so on (see https://packaging.python.org/tutorials/packaging-projects/ and http://r-pkgs.had.co.nz/).

Regardless of these specific technical instructions, you can make sure from the outset, though, that your code is easily re-usable. The checklist below can help making your code publishable from the outset.

[2] See url https://compendium.ccs.amsterdam

- Do not hard-code values. Rather than using "myoutputfile.csv" or 50 within your script, create constants like OUTPUTFILE="myoutputfile" and NUMBER_OF_TOPICS=50 at the beginning of your script and use these variables instead of the values later on. Even better, let the user provide these arguments as command line arguments or via a configuration file.
- Use functions. Rather than writing large scripts that are executed from the first line to the last in that order, structure the code in different functions that fulfill one specific task each, and can hence be reused. If you find yourself copy-pasting code, then most likely, you can write a function instead.
- Document your code. Use docstrings (Python) or comments (R) to make clear what each function does.

15.4 Distributing Your Software as Container

When publishing your software, you can think of multiple user groups. Some may be interested in building on and further developing your code. Some may not care about your code at all and just want your software to run. And many others will be somewhere in between.

Only publishing your source code (Section 15.3) may be a burden for those who want your code to "just run" once your code becomes more complex and has more dependencies. Imagine a scenario where your software requires a specific version of Python or R and/or some very specific (or maybe incompatible) libraries that you do not want to force the user to install.

And maybe your prospective user does not even know any R or Python.

For such cases, so-called containers are the solution, with as most prominent platform *Docker*. You can envision a container as a minimalistic virtual machine that includes everything to run your software. To the outside, none of that is visible – just a network port to connect to, or a command line to interact with, depending on your choices.

Software that is containerized using Docker is distributed as a so-called *Docker image*. You can build such an image yourself, but it can also be distributed by pushing it to a so-called registry, such as the *Docker Hub*. If you publish your software this way, the end user has to do nothing other than installing Docker and running the command `docker run nameofyourimage` – it will even be downloaded automatically if necessary. There are also GUI versions of Docker available, which lowers the threshold for some end user groups even more.

Let's illustrate the typical workflow with a toy example. Imagine you wrote the following script, myscript.py:

```
import numpy as np
from random import randint

a = randint(0,10)
print(f"exp({a}) = {np.exp(a)}")
```

You think that this is an awesome program (after all, it calculates *e* to the power of a random integer!), and others should be able to use it. And you don't want to bother them with setting up Python, installing numpy, and then running the script. In fact, they do not even need to *know* that it's a Python program. You could have written it as well in R, or any other language – for the user, that will make no difference at all.

What would a Docker image that runs this code need to contain? Not much: first some basic operating system (usually, a tiny Linux distribution), Python, numpy, and the script itself.

To create such a Docker image, you create a file named `Dockerfile` in the same directory as your script with the following content:

```
FROM python:3
ADD myscript.py /
RUN pip install numpy
CMD [ "python", "./myscript.py" ]
```

The first line tells Docker to build your new image by starting from an existing image that already contains an operating system and Python3. You could also start from scratch here, but this makes your life much easier. The next line adds your script to the image, and then we run `pip install numpy` within the image. The last line just specifies which command with which parameters needs to be executed when the image is run – in our case `python ./myscript.py`.

To create the image, you run `docker build -t dockertest .` (naming the image "dockertest"). After that, you can run it using `docker run dockertest` – and, if you want to, publish it.

Easy, right?

But when does it make sense to use Docker? Not in our toy example, of course. While the original code is only a couple of bytes, it now got bloated to hundreds of megabytes. But there are plenty of scenarios where this makes a lot of sense.

- To "abstract away" the inner workings of your code. Rather than giving potentially complicated instructions how to run your code, which dependencies to install, etc., you can just provide users with the Docker image, in which everything is already taken care of.
- To ensure that users get the same results. Though it doesn't form a huge problem on a daily basis for most computational scientists, different versions of different libraries on different systems may occasionally produce slightly different results. The container ensures that the code is run using the same software setup.
- To avoid interfering with existing installations. Already our toy example had a dependency, *numpy*, but often, dependencies can be more complex and a program we write may need very specific libraries, or even some other software beyond Python or R libraries. Distributing the source code alone means forcing the user to also install these; and there are many good reasons why people may be reluctant to do so. It may be incompatible with other software on their computer, there may be security concerns, or it just may be too much work. But if it runs inside of the Docker container, many of these problems disappear.

In short, the Docker image is rarely the *only* way in which you distribute your source code. But already adding a Dockerfile to your github repository so that users can build a Docker container can offer another and maybe better way of running your software to your audience.

16

Where to Go Next

Abstract:

This chapter summarizes the main learning goals of the book, and outlines possible next steps. Special attention is payed to an ethical application of computational methods, as well as to the importance of open and transparent science.

Keywords summary, open science, ethics

- Reflect on the learning goals of the book
- Point out avenues for future study
- Highlight ethical considerations for applying the techniques covered in the book
- Relate the techniques covered in the book to Open Science practices

This concluding chapter provides a broad overview of what was covered so far, and what interesting avenues there are to explore next. It gives pointers to resources to learn more about topics such as programming, statistical modeling or deep learning. It also discusses considerations regarding ethics and open science.

16.1 How Far Have We Come?

In this book, we introduced you to the computational analysis of communication. In Chapter 1, we tried to convince you that the computational analysis of communication is a worthwhile endeavor – and we also highlighted that there is much more to the subject than this book can cover. So here we are now. Maybe you skipped some chapters, maybe you did some additional reading or followed some online tutorials, and maybe you completed your first small project that involved some of techniques we covered. Time to recap.

You now have some knowledge of programming. We hope that this has opened new doors for you, and allows you to use a wealth of libraries, tutorials, and tools that may make your life easier, your research more productive, and your analyses better.

You have learned how to handle new types of data. Not only traditional tabular datasets, but also textual data, semi-structured data, and to some extent network data and images.

Computational Analysis of Communication: A Practical Introduction to the Analysis of Texts, Networks, and Images with Code Examples in Python and R, First Edition. Wouter van Atteveldt, Damian Trilling & Carlos Arcila Calderón.

You can apply machine-learning frameworks. You know about both unsupervised and supervised approaches, and can decide how they can be useful for finding answers to your research questions.

Finally, you have got at least a first impression of some cool techniques like neural networks and services such as databases, containers, and cloud computing. We hope that being aware of them will help you to make an informed decision whether they may be good tools to dive into for your upcoming projects.

16.2 Where To Go Next?

But what should you learn next?

Most importantly, we cannot stress enough that it should be the research question that is the guide, not the method. You shouldn't use the newest neural network module just because it's cool, when counting occurrences of a simple regular expression does the trick. But this also applies the other way around: if a new method performs much better than an old one, you should learn it! For too long, for instance, people have relied on simple bag-of-words sentiment analyses with off-the-shelf dictionaries, simply because they were easy to use – despite better alternatives being available.

Having said that, we will nevertheless try to give some general recommendations for what to learn next.

Become better at programming. In this book, we tried to find a compromise between teaching the programming concepts necessary to apply our methods on the one hand, and not getting overly technical on the other hand. After all, for many social scientists, programming is a means to an end, not a goal in itself. But as you progress, a deeper understanding of some programming concepts will make it easier for you to tailor everything according to your needs, and will – again – open new doors. There are countless books and online tutorials on "Programming in [Language of your choice]". In fact, in this "bilingual" book we have shown you how to program with R and Python (the most used languages by data scientists), but there are other programming languages that might also deserve your attention (e.g. Java, Scala, Julia, etc.) if you become a computational scientist.

Learn how to write libraries. A very specific yet widely applicable skill we'd encourage you to learn is writing your own packages ("modules" or "libraries"). One of the nice things about computational analyses is that they are very much compatible with an Open Science approach. Sharing what you have done is much easier if everything you did is already documented in some code that you can share. But you can go one step further: of course it is nice if people can exactly reproduce your analysis, but wouldn't it be even nicer if they could also use your code to run analyses using their own data? If you thought about a great way to compute some statistic, why not make it easy for others to do the same? Consider writing (and documenting!) your code in a general way and then publishing it on CRAN or pypi so others can easily install and use it.

Get inspiration for new types of studies. Try to think a bit out of the box and beyond classical surveys, experiments, and content analyses to design new studies. Books like *Bit by Bit* (Salganik, 2019) may help you with this. You can also take a look at other scientific disciplines such as computational biology that has reinvented its methods, questions and hypotheses. Keep in mind that computational methods have an impact on the theoretical and empirical discussions of communication processes, which in turn will call for novel types of studies. The emerging scientific fields such as Computational Communication Science, Computational Social Sciences and Digital Humanities show how theory and methods can develop hand in hand.

Get a deeper understanding of deep learning. For many tasks in the computational analysis of communication, classical machine learning approaches (like regression or support vector machines) work just fine. In fact, there is no need to always jump on the latest band wagon of the newest technique. If a simple logistic regression achieves an F1-score of 88.1, and the most fancy neural network achievs an 88.5 – would it be worth the extra effort and the loss of explainability? It depends on your use case, but probably not. Nevertheless, by now, we can be fairly certain that neural networks and deep learning are here to stay. We could only give a limited introduction in this book, but state-of-the-art analysis of text and especially visual material cannot do without it any more. Even though you may not train such models yourself all the time, but may use, for instance, pre-trained word embeddings or use packages like *spacy* that have been trained using neural networks, it seems worthwhile to understand these techniques better. Also here, a lot of online tutorials exist for frameworks such as `keras` or `tensorflow`, but also thorough books that provide a sound understanding of the underlying models (Goldberg, 2017; Géron 2019).

Learn more about statistical models. Not everything in the computational analysis of communication is machine learning. We used the analogy of the mouse trap (where we only care about the performance, not the underlying mechanism) versus better beforehand understanding, and argued that often, we may use machine learning as a "mouse trap" to enrich our data – even if we are ultimately interested in explaining some other process. For instance, we may want to use machine learning as one step in a workflow to predict the topic of social media messages, and then use a conventional statistical approach to understand which factors explain how often the message has been shared. Such data, though, often have different characteristics than data that you may encounter in surveys or experiments. In this case, for instance, the number of shares is a so-called count variable: it can take only positive integers, and thus has a lower bound (0) but no upper bound. That's very different than normally distributed data and requires regression models such as negative binomial regression. That's not difficult to do, but worth reading up on. Similarly, multilevel modelling will often be appropriate for the data you work with. Being familiar with this and other techniques (such as mediation and moderation analysis, or even structural equation modeling) will allow you to make better choices. On a different note, you may want to familiarize yourself with Bayesian statistics – a framework that is very different from the so-called frequentist approach that you probably know from your statistics courses.

And, last but not least: have fun! At least for us, that is one of the most important parts: don't forget to enjoy the skills you gained, and create projects that you enjoy!

16.3 Open, Transparent, and Ethical Computational Science

We started this book by reflecting on what we are actually doing when conducting computational analyses of communication. One of the things we highlighted in Chapter 1 was our use of open-source tools, in particular Python and R and the wealth of open-source libraries that extend them. Hopefully, you have also realized not only how much your work could therefore build on the work of others, but also how many of the resources you used were created as a community effort.

Now that you acquired the knowledge it takes to conduct computational research on communication, it is time to reflect on how to give back to the community, and how to contribute to an open research environment. At the same time, it is not as simple as "just putting everything online" – after all, researchers often work with sensitive data. We therefore conclude this book with a short discussion on open, transparent, and ethical computational science.

Transparent and Open Science. In the wake of the so-called reproducibility crisis, the call for transparent and open science has become louder and louder in the last years. The public, funders, and journals increasingly ask for access to data and analysis scripts that underly published research. Of course, publishing your data and code is not a panacea for all problems, but it is a step towards better science from at least two perspectives (Van Atteveldt et al., 2019): first, it allows others to reproduce your work, enhancing its credibility (and the credibility of the field as a whole). Second, it allows others to build on your work without reinventing the wheel.

So, how can you contribute to this? Most importantly, as we advised in Section 4.3: use a version control system and share your code on a site like github.com. We also discussed code-sharing possibilities in Section 15.3. Finally, you can find a template for organizing your code and data so that your research is easy to reproduce at https://github.com/ccs-amsterdam/compendium.

The privacy–transparency trade-off. While the sharing of code is not particularly controversial, the sharing of data sometimes is. In particular, you may deal with data that contain personally identifiable information. On the one hand, you should share your data to make sure that your work can be reproduced – on the other hand, it would be ethically (and depending on your jurisdiction, potentially also legally) wrong to share personal data about individuals. As boyd and Crawford (2012) write: "Just because it is accessible does not make it ethical." Hence, the situation is not always black or white, and some techniques exist to find a balance between the two: you can remove (or hash) information such as usernames, you can aggregate your data, you can add artificial noise. Ideally, you should integrate legal, ethical, and technical considerations to make an informed decision on how to find a balance such that transparency is maximized while privacy risks are minimized. More and more literature explores different possibilities (e.g. Breuer et al., 2020).

Other Ethical Challenges in Computational Analyses. Lastly, there are also other ethical challenges that go beyond the use of privacy-sensitive data. Many tools we use give us great power, and with that comes great responsibility. For instance, as we highlighted in Section 12.4, every time we scrape a website, we cause some costs somewhere. They may be neglectable for a single http request, but they may add up. Similarly, calculations on some cloud service cause environmental costs. Before starting a large-scale project, we should therefore make a trade-off between the costs or damage we cause, and the (scientific) gain that we achieve.

In the end, though, we firmly believe that as computational scientists, we are well-equipped to contribute to the move towards more ethical, open, and transparent science. Let's do it!

Bibliography

Javier J Amores, Carlos Arcila Calderón, and Mikolaj Stanek. Visual frames of migrants and refugees in the main western european media. *Economics & Sociology*, *12*(3): 147–161, 2019.

Kenneth Benoit, Kohei Watanabe, Haiyan Wang, Paul Nulty, Adam Obeng, Stefan Müller, and Akitaka Matsuo. quanteda: An R package for the quantitative analysis of textual data. *Journal of Open Source Software*, *3*(30):774, 2018. doi: 10.21105/joss.00774. URL https://quanteda.io.

David Blei and John Lafferty. Correlated topic models. *Advances in Neural Information Processing Systems*, *18*:147, 2006a.

David M Blei and John D Lafferty. Dynamic topic models. In *Proceedings of the 23rd International Conference on Machine Learning*, pages 113–120, 2006b.

David M Blei, Andrew Y Ng, and Michael I Jordan. Latent dirichlet allocation. *Journal of Machine Learning Research*, *3*:993–1022, 2003.

Vincent D Blondel, Jean-Loup Guillaume, Renaud Lambiotte, and Etienne Lefebvre. Fast unfolding of communities in large networks. *Journal of Statistical Mechanics: Theory and Experiment*, *2008*(10):P10008, 2008.

Mark Boukes, Bob van de Velde, Theo Araujo, and Rens Vliegenthart. What's the tone? easy doesn't do it: Analyzing performance and agreement between off-the-whelf sentiment analysis tools. *Communication Methods and Measures*, *online first* , 2019. doi: 10.1080/19312458.2019.1671966.

Jelle W. Boumans and Damian Trilling. Taking stock of the toolkit: An overview of relevant autmated content analysis approaches and techniques for digital journalism scholars. *Digital Journalism*, *4*(1):8–23, 2016. doi: 10.1080/21670811.2015. 1096598.

Danah boyd and Kate Crawford. Critical questions for Big Data. *Information, Communication & Society*, *15*(5):662–679, 2012. doi: 10.1080/1369118X.2012.678878.

Leo Breiman. Statistical modeling: The two cultures. *Statistical Science*, *16*(3):199–215, 2001. doi: 10.1214/ss/1009213726.

Johannes Breuer, Libby Bishop, and Katharina Kinder-Kurlanda. The practical and ethical challenges in acquiring and sharing digital trace data: Negotiating public-private partnerships. *New Media & Society*, *22*(11):2058–2080, 2020. doi: 10.1177/ 1461444820924622.

Computational Analysis of Communication: A Practical Introduction to the Analysis of Texts, Networks, and Images with Code Examples in Python and R, First Edition. Wouter van Atteveldt, Damian Trilling & Carlos Arcila Calderón.

Axel Bruns. After the 'APIcalypse': social media platforms and their fight against critical scholarly research. *Information, Communication & Society*, *22*(11):1544–1566, 2019. doi: 10.1080/1369118X.2019.1637447.

Alan Bryman. *Social research methods*. Oxford University Press, New York, NY, 4th edition, 2012.

Björn Burscher, Daan Odijk, Rens Vliegenthart, Maarten de Rijke, and Claes H. de Vreese. Teaching the computer to code frames in news: Comparing two supervised machine learning approaches to frame analysis. *Communication Methods and Measures*, *8*(3):190–206, 2014. doi: 10.1080/19312458.2014.937527.

Alberto Cairo. *How charts lie*. WW Norton & Company, 2019.

Carlos Arcila Calderón, Félix Ortega Mohedano, Mateo Álvarez, and Miguel Vicente Mariño. Distributed supervised sentiment analysis of tweets: integrating machine learning and streaming analytics for big data challenges in communication and audience research. *Empiria: Revista De Metodologia De Ciencias Sociales*, (42):113–136, 2019.

Frédéric Cazals and Chinmay Karande. A note on the problem of reporting maximal cliques. *Theoretical Computer Science*, *407*(1-3):564–568, 2008.

Chung-hong Chan, Joseph Bajjalieh, Loretta Auvil, Hartmut Wessler, Scott Althaus, Kasper Welbers, Wouter van Atteveldt, and Marc Jungblut. Four best practices for measuring news sentiment using 'off-the-shelf' dictionaries: a large-scale p-hacking experiment. *Computational Communication Research*, 3(1): 1-27, 2021. doi:10.5117/CCR2021.1.001.CHAN

Jonathan Chang, Sean Gerrish, Chong Wang, Jordan L Boyd-Graber, and David M Blei. Reading tea leaves: How humans interpret topic models. In *Advances in Neural Information Processing Systems*, pages 288–296, 2009.

Lizi Chen. News-Processed-Dataset, 2017. URL https://figshare.com/articles/News-Processed-Dataset/5296357.

Nicholas A Christakis and James H Fowler. *Connected: The surprising power of our social networks and how they shape our lives*. Little, Brown Spark, 2009.

Claudio Cioffi-Revilla. *Introduction to Computational Social Science: Principles and Applications*. Springer, London, UK, 2014.

Aaron Clauset, Mark EJ Newman, and Cristopher Moore. Finding community structure in very large networks. *Physical review E*, *70*(6):066111, 2004.

Edgar F. Codd. A relational model of data for large shared data banks. *Communications of the ACM*, *13*(6):377–387, 1970. doi: 10.1145/362384.362685.

Michael J Crawley. *The R book*. Wiley, 2nd edition, 2012.

Tom De Smedt, W Daelemans, and Tom De Smedt. Pattern for Python. *The Journal of Machine Learning Research*, *13*:2063–2067, 2012. URL http://dl.acm.org/citation.cfm?id=2343710.

Bryce J Dietrich, Matthew Hayes, and Diana Z. O'Brien. Pitch perfect: Vocal pitch and the emotional intensity of congressional speech. *American Political Science Review*, *113*(4):941–962, 2019.

David Eppstein, Maarten Löffler, and Darren Strash. Listing all maximal cliques in sparse graphs in near-optimal time. In *International Symposium on Algorithms and Computation*, pages 403–414. Springer, 2010.

Deen Freelon. Computational research in the post-API age. *Political Communication*, *35*(4):665–668, 2018. doi: 10.1080/10584609.2018.1477506.

Aurélien Géron. *Hands-on machine learning with Scikit-Learn, Keras, and Tensor-Flow: Concepts, tools, and techniques to build intelligent systems*. O'Reilly Media, 2019.

Yoav Goldberg. *Neural Network Models for Natural Language Processing*. Morgan & Claypool, 2017.

S. Gonzalez-Bailon and G. Paltoglou. Signals of public opinion in online communication: A comparison of methods and data sources. *The ANNALS of the American Academy of Political and Social Science*, *659*(1):95–107, 2015. doi: 10.1177/00027 16215569192.

Sandra González-Bailón. *Decoding the social world: Data science and the unintended consequences of communication*. MIT, Cambridge, MA, 2017.

Thomas L Griffiths, Michael I Jordan, Joshua B Tenenbaum, and David M Blei. Hierarchical topic models and the nested chinese restaurant process. In *Advances in Neural Information Processing Systems*, pages 17–24, 2004.

J. Grimmer and B. M. Stewart. Text as data: The promise and pitfalls of automatic content analysis methods for political texts. *Political Analysis*, *21*(3):267–297, 2013. ISSN 1047–1987. doi: 10.1093/pan/mps028.

Elisabeth Günther, Damian Trilling, and Bob van de Velde. But how do we store it? Data architecture in the social-scientific research process. In C.M. Stuetzer, M. Welker, and M. Egger, editors, *Computational social science in the age of Big Data. Concepts, methodologies, tools, and applications*, pages 161–187. Herbert von Halem, 2018.

Kieran Healy. *Data visualization: a practical introduction*. Princeton University Press, 2018.

Yusaku Horiuchi, Tadashi Komatsu, and Fumio Nakaya. Should candidates smile to win elections? an application of automated face recognition technology. *Political Psychology*, *33*(6):925–933, 2012.

Benjamin D. Horne, William Dron, Sara Khedr, and Sibel Adali. Sampling the News Producers: A Large News and Feature Data Set for the Study of the Complex Media Landscape. In *12th International AAAI Conference on Web and Social Media (ICWSM)*, pages 518–527, 2018. URL http://arxiv.org/abs/1803.10124.

Clayton J Hutto and Eric Gilbert. Vader: A parsimonious rule-based model for sentiment analysis of social media text. In *Eighth International AAAI Conference on Weblogs and Social Media*, 2014.

Daniel Jurafsky and James H Martin. *Speech and language processing: An introduction to natural language processing, computational linguistics, and speech recognition (2nd ed.)*. Prentice Hall, 2009.

David Kahle and Hadley Wickham. ggmap: Spatial visualization with ggplot2. *The R Journal*, *5*(1):144–161, 2013.

John D Kelleher, Brian Mac Namee, and Aoife D'arcy. *Fundamentals of machine learning for predictive data analytics: algorithms, worked examples, and case studies*. MIT, 2015.

Andy Kirk. *Data visualisation: A handbook for data driven design*. SAGE, London, UK, 2016.

Rob Kitchin. Big Data, new epistemologies and paradigm shifts. *Big Data & Society*, *1*(1):1–12, 2014a. doi: 10.1177/2053951714528481.

Rob Kitchin. *The data revolution: Big data, open data, data infrastructures and their consequences*. SAGE, London, UK, 2014b.

Klaus Krippendorff. *Content analysis: An introduction to its methodology*. SAGE, Thousand Oaks, CA, 2nd edition, 2004.

Thomas K Landauer, Danielle S McNamara, Simon Dennis, and Walter Kintsch. *Handbook of latent semantic analysis*. Psychology Press, 2013.

Yann LeCun, Léon Bottou, Yoshua Bengio, and Patrick Haffner. Gradient-based learning applied to document recognition. *Proceedings of the IEEE, 86*(11):2278–2324, 1998.

Jimmy Lin. On building better mousetraps and understanding the human condition: Reflections on big data in the social sciences. *The ANNALS of the American Academy of Political and Social Science, 659*(1):33–47, 2015.

Andrew L. Maas, Raymond E. Daly, Peter T. Pham, Dan Huang, Andrew Y. Ng, and Christopher Potts. Learning word vectors for sentiment analysis. In *Proceedings of the 49th Annual Meeting of the Association for Computational Linguistics: Human Language Technologies*, pages 142–150, Portland, Oregon, USA, June 2011. URL http://www.aclweb.org/anthology/P11-1015.

Drew B Margolin. Computational contributions: a symbiotic approach to integrating big, observational data studies into the communication field. *Communication Methods and Measures, 13*(4):229–247, 2019.

Viktor Mayer-Schönberger and Kenneth Cukier. *Big Data: A revolution that will transform how we live, work, and think*. Houghton Mifflin Harcourt, New York, NY, 2013.

David Mimno, Hanna Wallach, Edmund Talley, Miriam Leenders, and Andrew Mc-Callum. Optimizing semantic coherence in topic models. In *Proceedings of the 2011 Conference on Empirical Methods in Natural Language Processing*, pages 262–272, 2011.

Jacob Levy Moreno. *Who shall survive? A new approach to the problem of human interrelations*. Nervous and mental disease publishing co, 1934.

Mark EJ Newman and Michelle Girvan. Finding and evaluating community structure in networks. *Physical Review E, 69*(2):026113, 2004.

Joel Nothman, Hanmin Qin, and Roman Yurchak. Stop word lists in free open-source software packages. In *Proceedings of Workshop for NLP Open Source Software (NLP-OSS)*, pages 7–12, 2018.

Yilang Peng. Same candidates, different faces: Uncovering media bias in visual portrayals of presidential candidates with computer vision. *Journal of Communication, 68*(5):920–941, 2018.

Thomas Piketty. *Capital in the Twenty-First Century*. Harvard University Press, Cambridge, MA, 2014.

Cornelius Puschmann. An end to the wild west of social media research: a response to Axel Bruns. *Information, Communication & Society, 22*(11):1582–1589, 2019. doi: 10.1080/1369118X.2019.1646300.

Peng Qi, Yuhao Zhang, Yuhui Zhang, Jason Bolton, and Christopher D. Manning. Stanza: A Python natural language processing toolkit for many human languages. In *Proceedings of the 58th Annual Meeting of the Association for Computational Linguistics: System Demonstrations*, 2020. URL https://nlp.stanford.edu/pubs/qi2020stanza.pdf.

Usha Nandini Raghavan, Réka Albert, and Soundar Kumara. Near linear time algorithm to detect community structures in large-scale networks. *Physical Review E, 76*(3):036106, 2007.

Andrew J. Reagan, Christopher M. Danforth, Brian Tivnan, Jake Ryland Williams, and Peter Sheridan Dodds. Sentiment analysis methods for understanding large-scale texts: a case for using continuum-scored words and word shift graphs. *EPJ Data Science, 6*(1), 2017. doi: 10.1140/epjds/s13688-017-0121-9.

Joseph Redmon and Ali Farhadi. Yolov3: An incremental improvement. arXiv, 2018.

Bernhard Rieder. Scrutinizing an algorithmic technique: the Bayes classifier as interested reading of reality. *Information Communication and Society, 20*(1):100–117, 2017. doi: 10.1080/1369118X.2016.1181195.

Daniel Riffe, Stephen Lacy, Brendan R. Watson, and Frederick Fico. *Analyzing Media Messages: Using Quantitative Content Analysis in Research*. Routledge, New York, NY, 4th edition, 2019.

Margaret E Roberts, Brandon M Stewart, Dustin Tingley, Christopher Lucas, Jet-son Leder-Luis, Shana Kushner Gadarian, Bethany Albertson, and David G Rand. Structural topic models for open-ended survey responses. *American Journal of Political Science*, *58*(4):1064–1082, 2014.

Matthew Salganik. *Bit by bit: Social research in the digital age*. Princeton University Press, 2019.

Michael Scharkow. Thematic content analysis using supervised machine learning: An empirical evaluation using German online news. *Quality & Quantity*, *47*(2): 761–773, 2011. ISSN 0033–5177. doi: 10.1007/s11135-011-9545-7.

Michael Scharkow. Content analysis, automatic. In Jörg Matthes, Christine S. Davis, and Robert F. Potter, editors, *The International Encyclopedia of Communication Research Methods*,. Wiley, Hoboken, NJ, 2017. doi: 10.1002/9781118901731.iecrm0043.

Milan Straka and Jana Straková. Tokenizing, Pos tagging, lemmatizing and parsing ud 2.0 with udpipe. In *Proceedings of the CoNLL 2017 Shared Task: Multilingual Parsing from Raw Text to Universal Dependencies*, pages 88–99, Vancouver, Canada, August 2017. Association for Computational Linguistics. URL http://www.aclweb.org/anthology/K/K17/K17-3009.pdf.

Mike Thelwall, Kevan Buckley, and Georgios Paltoglou. Sentiment strength detection for the social web. *Journal of the American Society for Information Science and Technology*, *63*(1):163–173, 2012. doi: 10.1002/asi.21662.

Damian Trilling. *Following the news: Patterns of online and offline news consumption*. PhD theses, University of Amsterdam, 2013. URL https://hdl.handle.net/11245/1.394551.

Damian Trilling. Big Data, analysis of. In *The International Encyclopedia of Communication Research Methods*, John Wiley & Sons, Inc., Hoboken, NJ, 2017. doi: 10.1002/9781118901731.iecrm0014.

Damian Trilling and Jeroen G.F. Jonkman. Scaling up content analysis. *Communication Methods and Measures*, *12*(2–3):158–174, 2018. doi: 10.1080/19312458.2018.1447655.

Dean Knox and Christopher Lucas. A dynamic model of speech for the social sciences. *American Political Science Review*, 115(2): 649–666, 2021

Edward R Tufte. *Beautiful evidence*, volume *1*. Graphics Press, Cheshire, CT, 2006.

John W Tukey. *Exploratory data analysis*, volume *2*. Reading, Mass., 1977.

Stéphan Tulkens, Lisa Hilte, Elise Lodewyckx, Ben Verhoeven, and Walter Daelemans. A dictionary-based approach to racism detection in Dutch social media. In *Proceedings of the Workshop on Text Analytics for Cybersecurity and Online Safety (TA-COS 2016)*, pages 11–17, 2016. URL http://www.clips.ua.ac.be/bibliography/a-dictionary-based-approach-to-racism-detection-in-dutch-social-media.

Wouter Van Atteveldt, Tamir Sheafer, Shaul R. Shenhav, and Yair Fogel-Dror. Clause analysis: Using syntactic information to automatically extract source, subject, and predicate from texts with an application to the 2008–2009 Gaza War. *Political Analysis*, *25* (2):207–222, 2017.

Wouter Van atteveldt, Joanna Strycharz, Damian Trilling, and Kasper Welbers. Toward open computational communication science: A practical road map for reusable data and code *International Journal of Communication*, *13*:3935–3954, 2019.

Wouter van Atteveldt, Anne Kroon, Felicia Loecherbach, Mickey Steijaert, Joanna Strycharz, Damian Trilling, Mariken Van der Velden, and Kasper Welbers. Standardized research compendiums: Making open and transparent science fun and easy. In *International Communication Association (ICA)*, Gold Coast, Australia (online due to Corona crisis), 5 2020a.

Wouter van Atteveldt, Mariken A.C.G. Van der Velden, and Mark Boukes. The validity of sentiment analysis: Comparing manual annotation, crowd-coding, dictionary approaches, and machine learning algorithms. *Computational Methods and Measures*, 2020b.

Jake VanderPlas. *Python data science handbook: Essential tools for working with data.* O'Reilly, 2016.

Susan A. M. Vermeer. A supervised machine learning method to classify Dutch-language news items, 11 2018. URL https://figshare.com/articles/A_supervised_machine_learning_method_to_classify_Dutch-language_news_items/7314896.

Susan A. M. Vermeer, Theo Araujo, Stefan F. Bernritter, and Guda van Noort. Seeing the wood for the trees: How machine learning can help firms in identifying relevant electronic word-of-mouth in social media. *International Journal of Research in Marketing*, *36*(3):492–508, 2019. doi: 10.1016/j.ijresmar.2019.01.010.

Soroush Vosoughi, Deb Roy, and Sinan Aral. The spread of true and false news online. *Science*, *359*(6380):1146–1151, 2018.

Annie Waldherr. Emergence of news waves: A social simulation approach. *Journal of Communication*, *64*(5):852–873, 2014. doi: 10.1111/jcom.12117.

Duncan J Watts. *Six degrees: The science of a connected age.* WW Norton & Company, 2004.

Martin Wettstein. Simulating hidden dynamics: Introducing Agent-Based Models as a tool for linkage analysis. *Computational Communication Research*, *2*(1):1–33, 2020. doi: 10.5117/CCR2020.1.001.WETT.

Nora Webb Williams, Andreu Casas, and John D Wilkerson. Images as data for social science research: An introduction to convolutional neural nets for image classification. *Elements in Quantitative and Computational Methods for the Social Sciences*, 2020.

Index

Note: Page numbers with *f* and *t* refer to figures and tables

a
abstraction, levels of 69, 132, 273
activation function 271*t*, 274, 276
aesthetic mapping 89
agglomerative algorithm 102
aggregation data 69–70
  adding summary values 71–72
  combining multiple operations 70–71
AlexNet 279
Amazon's Recognition 262, 262*f*
Amazon Web Services (AWS) 288
Anaconda 6–7
Apache Hadoop (software) 289
Apache Spark (software) 289
application programming interface (API) 48, 261–263, 262*f*
  access for researchers 217
  documentation of 214, 216
  endpoint 214, 228
  libraries and 261–263, 262*f*
  response 216
  restrictions 217
  web (*See* web APIs)
a priori 115, 185
  assumptions 186
  groups 101
arrays 28, 30
  n-dimensional 33, 34
  one-dimensional 30
artificial graphs 254
artificial intelligence 259
  community 130
assignment statement 35
associations 114–115
attributes 218, 221
  structure of 235
audio 259–260, 261*f*
  features, analysis of 260
  sentiment analysis 259
authentication:
  and APIs 228–229
  and webpages 229–230
automated recognition of facial traits 270
automatic analysis of text 184–185
  deciding on right method 185–187
  dictionary approaches to 189–191
  obtaining review dataset 187–188
  supervised text analysis (*See* supervised text analysis)
  unsupervised text analysis (*See* unsupervised text analysis)
automatic generation of music 259
automatic text analysis 186

b
backpropagation algorithm 130, 274
barplot 14–15
  of support for refugees 90
Bayesian statistics 295
betweenness centrality 247
  estimations of 249
bias term 279

Computational Analysis of Communication: A Practical Introduction to the Analysis of Texts, Networks, and Images with Code Examples in Python and R, First Edition. Wouter van Atteveldt, Damian Trilling & Carlos Arcila Calderón.

big data 2, 289
  contexts 285
bigrams 174, 196, 197
BitBucket 44, 290
Bit by Bit (Salganik) 294
bitmap 271*t*
Black Lives Matter movement 270
blank pixels 268
boilerplate topics 207
booleans 25–26, 56
bottom-up approach algorithm 102
branches 290
built-in functions 39
byte-order markers (BOM) 60

c

capital ownership 72–73
Cascading Style Sheets (CSS) 218–219
centrality measures 246–248
character classes 147
character vector objects 26
circuit 242
  visualization of 243
classical machine learning 122, 132, 295
  decision trees and random forests 127–128, 127*f*
  Naïve Bayes 122–124
  neural networks 129–130, 130*f*
  regression analysis 124–125, 125*f*
  support vector machines (SVM) 125–126
classical neural network 131
classification 117
classifier 198–199
  finding 194–197
  logistic regression 124–125
  Naïve Bayes 124–125, 194, 195
  Random Forest 128
  SVMs 126
cleaning process 68, 145
cliques 249–251
closeness:
  centrality 247
  estimations of 248–249
cloud computing 232, 262, 286–289, 294
cluster/clustering 114–115, 248–257, 289
  coefficient and diameter 245–246
  and community detection 248–257
  hierarchical clustering 102–106
  k-means 101–102
  optimal number of 103
  principal component analysis and singular value decomposition 106–112
  visualization of 102, 105
code/coding 49
  basic dimension of 50
  comments in 51
  documentation 47
  dynamic 51
  GitHub 52
codepages 59
code writing 43
  debugging strategies 48–49
  GitHub 49–54, 52*f*
  messages 46–48, 47*f*
  Notebooks 49–54, 53*f*, 54*f*
  re-using code 43–45
  understanding errors and getting help 46
coherence 209–211
collocations 174–176, 183
  identifying and applying 175
color 238
  model, standard 264–265
columns 286
community detection 248–257
computational analysis of communication 1–3, 290, 295–296
  computer vision concepts 270, 271*t*
  ethical challenges in 296
  methods for 283
  SQL and noSQL databases 285
computational hygiene 43, 49, 50
computational methods 260, 294–295
computer vision, standard dimensions in 266
conditional statements 37–38
confusion matrix 120
  visual representation of 121*f*
containers/containerization 296
  software as 291–292
continuous variables 88, 92, 94, 95, 102
convolutional layers 131, 133, 200, 279

convolutional networks 131, 133
convolutional neural networks (CNNs) 131–133, 132, 132*f*, 199, 260, 279–282
  applied to text analysis 132*f*
  architecture 279–280
  open-source and pre-trained 281
  pretrained 282
  for social science problem 279
cookies 229–230
corpus analysis 15–16
correlation analysis 78
count variable 295
crawling websites 223–224, 226
cron job 288
cross entropy loss 275
cross-tabulation 84
  of support of refugees and gender 88
cross-validation 138–140, 195, 196
CSS selectors 219–220, 229*t*
csv file 59*f*
  values in 60
cumulative explained variance 110–111
custom tokenizer 158

d

data:
  augmentation 268, 271*t*
  collection 2, 284, 285
  columns 73
  conversion 68
  driven techniques 2
  equal units of analysis 72–75
  inner and outer joins 75–76, 76*f*, 76*t*
  merging and visualizing 81
  messy 82
  nested data 76–78
  private and public 74
  processing 43
  scarcity problem 173
  on top incomes 80*f*
  types of 295
data analysis 2, 285
  about objects and data types 24–25
  combining multiple values 28–32
  data frames 34
  dictionaries 32–33
  matrices and *n*-dimensional arrays 33, 34
  simple control structures 35–36
    conditional statements 37–38
    functions and methods 39–42
    loops 36–37
  storing single values 25–26
  storing text 26–27
databases 294
  categories of 284
data frame 34, 56–57, 73, 215, 284
  contents of 58
  creating 34, 57
  encodings and dialects 59–62, 59*f*
  handling 58, 58*t*, 61–62
  from online sources 62–64
  to plot heatmap 97
  reading 58
    and saving, role of files 57–58
    and writing files into 61
  regular expressions on 152
  transforming data into 215
data mining 101
dataset 76*f*, 77–78, 79
  primary 76
  re-use existing 63
  tidy 79
  types of joins between 76*t*
data types 24, 29, 56
  converting 29
  objects of 31
  in Python and R 25*t*
data visualization 87–88, 99
  plotting (*See* plotting)
data wrangling 65, 71, 79
  calculating values 67–69
  filtering, selecting, and renaming 66–67
  grouping and aggregating 69–70
    adding summary values 71–72
    combining multiple operations 70–71
  merging data 72
    equal units of analysis 72–75
    inner and outer joins 75–76, 76*f*, 76*t*
    nested data 76–78
  reshaping data 78–79

restructuring messy data 79–82, 80*f*
debuggers 48
decimal separator 60
decision trees 127–128, 127*f*
deep learning 130–131, 199–202, 272, 295
approach 272
convolutional neural networks 131–133, 132*f*
extensive treatment of 130–131
for image analysis 273–278
in image classification 273
training and testing 202
deep neural networks (DNN) 131, 276
architectures 274
degree centralities 248, 250, 254–255
degree of node 247
dendogram 105, 106
dense layer 200
dense matrix 161
descriptive variables 101
dialects 59–62, 59*f*
dictionaries 32–33
analogy for 33
approaches 186, 190
grades 32–33
key-value pairs in 32
sentiments 32, 189
digital/digitalization 259
content 259
data 130–131
images 263, 265
training data 131
dimensionality reduction 100–101
hierarchical clustering 102–106
k-means clustering 101–102
directed graph 237
creating 237
visualization of 238
directed graphs 248
Dirichlet distribution 204, 204*f*
hyperparameters 205
distributions 90
divisive algorithm 102
Docker Hub 291
Docker image 291
docstrings 51
documents/documentation 50, 51, 284
clustering 203
feature matrix 173
geometric interpretation of 164
document-term matrix (DTM) 15–16, 156, 157, 160, 164, 170–171, 198, 203
analysis of 174
as "bag of words,"162
clean and preprocess 182–183
clustering techniques to 203
content of 160
as sparse matrix 159–162
trimming 170–171
weighting 171–172
downstream task 2
dplyr function 66
dslabs package 272
dynamic web pages 225–227

e

edge-betweenness clustering 251
edges 234, 235, 237
incident 245
proportion of 245
eigenvalues 108
eigenvector:
centrality 247
estimations of 248–249
ElasticSearch 285, 286
eli5, 199–200
else if clauses 37–38
embedding layer 200
embedding vector 176
emotional intensity 260
encodings 27, 59–62, 59*f*, 60, 142
ensemble model 128
epochs 276–277
error 46
code 47
debugging strategies 48–49
handling 49
messages 46–48, 47*f*
network's 274
in programming 46
estimation algorithm 207
ethical computational science 296
Eurobarometer 84–85, 92, 99
load data from 85

exploratory data analysis (EDA) 83–84, 99
clustering and dimensionality reduction 100–101
hierarchical clustering 102–106
k-means clustering 101–102
principal component analysis and singular value decomposition 106–112
goals of 98
simple 84–87
visualizing data 87–88 (*See* data visualization)

f
Facebook 228, 234, 237, 242, 245–246
facial emotions 270
false negatives (FN) 120
false positives (FP) 120
Fashion MNIST dataset 275, 275*f*
loading 276
feature engineering 118–119, 172–173
feedforward neural network (FNN) 274
file formats 58
file handling, data frame 61–62
Firefox 229
flattened images 268
floating-point numbers 25–26
floats 56
for-loops 36–37, 143
formatted strings 27
formed cluster 101
friends 28–32
functions 39–42
used repeatedly 41
writing 40

g
Gaussian kernels 126
GDPR regulations 231
generalization 51
generative model 203–204
generators 42
gensim (Python) 207
geographic networks 21–22
geopandas 98
geospatial data 98
ggplot2 function 95, 98
GGPlot syntax 89
Gibbs sampling 205
Girvan–Newman approach 251–252
clustering with 252
community detection with 252
GitHub 44, 49–54, 52*f*, 228, 290
online repository 52*f*
Google 228
Google Books 214–215
Google Colab 53, 288
Jupyter notebook in 54*f*
GoogLeNet 279
graph:
basics of working with 236
communities within 248–250
components of 238
creating 235
definition of 234
directed 237
distances in 243
edges and nodes of 237
lines 92, 97
structures, overview of 234
theory 234
type of 239
visualization of 126, 236
graphical processing units (GPUs) 131
Greedy Optimization algorithm 253–255
grid 289
gridsearch 138–140, 195–196
grouping, data 69–70
adding summary values 71–72
combining multiple operations 70–71

h
hidden layers 276
hierarchical clustering 102–106, 108
histograms 87
of age 91
homophily, principle of 249
HTML 221–222
dumping 226
electronic versions of newspapers in 142
retrieving and parsing 217–222
tags and character 149

human annotations 190–191, 192
hyperlinks 165, 218
hypothetical decision tree 127

i
if clauses 37–38
igraph package 19–20
imagemagik 263, 265, 268
ImageNet 280
images 259–261, 261*f*, 293
  analysis, deep learning for 273–278
  classification 261, 268, 270–271, 271*t*, 283
  cropping 282
  detect similarities between 269
  feature representation of 273–274
  function to crop 268
  to gray-scale 265
  manipulation 263
  orientation of 268
  predicting class 281, 282
  storing, representing, and converting 263–269, 264*f*
  three-dimensional matrix of 268
imagine 47, 288–289
IMDB database 187–188
incident edges 244
indentation 35–36
indexing 28, 29
induced subgraphs 239, 241
  for Democrats and Republicans 241–242
integers 25–27, 56
Integrated Development Environment (IDE) 6
Internet, adoption of 234
interpretability 209
inverse document frequency 171

j
JavaScript 227
JPG:
  file 265
  pictures 264
JSON files 62, 284
JupyterHub 289
Jupyter Notebook 231–232
  installing 9–11

k
keras 199–200, 276, 280
kernel function 126
keys 286
key-value pairs 284
k-means clustering 101–102, 104
knowledge 296
  of programming 295

l
labeled dataset 117
latent dirichlet allocation (LDA) 203–205, 204*f*
  analyzing and inspecting 208
  beyond 209–211
  generative model 203
  mixture model 204
  model, fitting 206–207
  task of 204
  topic model 206
Latent Semantic Analysis (LSA) 203
latent variables, hidden layer of 129
leaves 127
lemmatization 177–178, 183
LeNet-5 279
letsencrypt 289
libraries 261–263, 262*f*
line ending 60
linguistic preprocessing 177–182, 178*t*
LinkedIn 228
Linux 232
  distribution 288, 288*f*
list-of-lists technique 56
lists 28–32
  comprehension 143
  operations on 28
literate programming 52
logistic activation function 274
logistic regression 115, 124, 127, 194, 196
  classifiers 124–125
  equation 118
log loss 275
long data 78–79, 81
long short-term memory 131, 133
loops 36–37
loss function 271*t*
Louvain algorithm 253–254

m

machine learning (ML) 100–101, 115, 118*t*
  algorithms 129–130
  frameworks 294
  model 129*f*
  supervised 185
  unsupervised 185
MacOS 232
Magick 261
MapReduce approaches 289
MariaDB 285, 286
matplotlib function 93
matrices 33, 34
maximal cliques 250–251
  in undirected graph 251
max-pooling layer 200
mean distance 244
messy data 82
metadata, primary results and county-level 77
methods 39–42
microblogging service 19–20
Microsoft Azure 288, 288*f*
Modified National Institute of Standards and Technology (MNIST) dataset 272, 272*f*, 273
MongoDB 285, 286
multilayer perceptron (MLP) 274
  color images 277
  fitting and evaluating 278
  with keras 277
  predicting classes using 278
multimedia:
  analysis 282
  content 259
  files 259
multimedia data 258
  basic classification with shallow algorithms 272–273
  deep learning for image analysis 273–278
  image classification 270–271
  images, audio and video 259–260, 261*f*
  re-using an open source CNN 279–282
  storing, representing, and converting images 263–269, 264*f*
  using existing libraries and APIs 261–263, 262*f*
multiple decision trees 128
multiple key-value pairs 214
multiple user groups 291
music, automatic generation of 259
mutable objects, behavior of 31
MySQL 285, 287

n

nagisa package 158
Naïve Bayes classifier 118–119, 122–125, 131, 193, 195
Named Entity Recognition 2, 181–182
naming 50
natural language processing 2, 63
n-dimensional arrays 33, 34
neighbours, incident 245
network:
  of American politicians 243
  analysis 20, 234
  clusters in 254–255
  edges of 254
  errors 274
  estimating distances in 245
  fundamental components of 235
  fun with 19–22
  of Spanish politicians 256–257
  structure of 235
  visualization of 236, 241
network data 233, 293
  representing and visualizing networks 234–241
  social network analysis (SNA) (*See* social network analysis (SNA))
neural networks 129–130, 130*f*, 296
  classical 131
  with hidden layers 130, 131
  for sentiment analysis 129
neurons:
  combinations of 129
  connections between 129
n-grams 173–174, 183
  generating 173
  for whole corpus 174
nodes (components) 234, 235, 237
noise 167–170

crowding 164
nominal data 88
noSQL databases 283–285
choosing 285–286
notebooks 53
nouns 177–178

o
OAuth 216, 228–229
objects 24
detection 270–271, 272*f*
mutable 31
references 31
off-the-shelf sentiment analysis 197
one-dimensional arrays 30
online data:
authentication 228
and APIs 228–229
and webpages 229–230
ethical, legal, and practical considerations 230–232
web APIs (*See* web APIs)
online services 213–214
open data platform 213
open research environment 296
Open Science approach 294, 296
optical character recognition (OCR) 261, 262
optimization 271*t*
optional parameters 40
ordinary least square regression (OLS) 115–117, 124, 125, 205

p
package:
of reference 45
selection 45
versions 44
pandas function 4, 85, 86, 88
panel surveys 78
parallel processing, architectures for 289
parsing:
HTML page 217–222
link texts 222
part-of-speech (POS) tagging 177, 178*t*
paths 242–246, 265
Pearson correlation coefficient 94, 96
perceptron 129
perplexity 209
photograph of refugees 262, 262*f*
picture:
information 268
rotating 45 degrees 269
standard dimensions in 267
PIL 261
pixel 271*t*
numerical representation of 263–264, 264*f*
plagiarism detection 45
plotting:
frequencies and distributions 88–91
geospatial data 98
multiple lines 93
relationships 92–98
PNG pictures 264
pointers 24
pooling layers 279
PostgreSQL 285
precision 120, 124
calculating 121
and recall 133–137
trade-off 122
primary dataset 75, 76
principal component analysis (PCA) 101, 106–112, 115
of data frame 109
dimensionality and k-means 112
privacy-sensitive data 296
privacy-transparency trade-off 296
private capital data 73
procedural topics 205
programming 296
languages 43–44
proportion of variance 110
pseudocode 47
public capital data 73
publishing:
software 291–292
your source 290–291
punctuation 167–170
pyplot function 92
Python/R 3–4, 84–85, 157, 187, 198, 206, 261, 273, 282, 294–296
attributes in 107–108

checklist 226
code 49
data types in 25*t*
dictionaries 284
documentation 46
formats in 58, 234
implementations for 45
installing 5–11
lambda functions in 170
logging module in 48
for machine learning 192
measures in 246
networks 19–22, 235
objects 28
predicting class 281
quanteda package 161–162
splitting and joining strings, and extracting multiple matches 151–154
state-of-the-art libraries in 270–271
stopwords package in 166
string operations in 144, 144*t*
textual data 15–17
tokenizer 158
with tweets 14–15
using regular expressions in 150–151
with visualizing geographic information 17–19
PyTorch 275, 280

q
quanteda function 4, 157, 161–162, 169

r
radial basis function (RBF) 126
Random Forest 127–128, 127*f*
algorithm 273
classifier 128
handwritten digits with 274
random seed 16
reachability 242–246
real-time video 260, 261*f*
recall 120, 121, 124
calculating 121
precision and 133–137
trade-off 122
receiver operator characteristic (ROC) 134–135, 135*f*, 137
receptive field 279
reciprocity 246
recurrent networks 131
recurrent neural network 199
recycling code 44
regular expressions 142, 145–146
in Python and R 150–151
syntax 146–147, 148*t*, 149*t*
relational databases 284
relative document frequency 170–171
relu activation function 276
repeat (do-while) loops 36–37
re-run hierarchical clustering 107
research compendium 290
research questions 166
Resilient Distributed Dataset (RDD) approaches 289
ResNet 279–282
retrieving HTML page 217–222
review dataset 187–188
RGB (red, green and blue) 264–265
color model 266
mathematical representation 265
R/Python. *See* Python/R
RStudio 8, 8*f*, 231–232
history 9
installing 7–9
projects 8–9
rule-based approaches 187

s
sandbar 280–281
Scandinavian characters 59
scatterplot 78
of average support 95
with regression line 95–96
scikit-learn in Python 128, 165
seaborn function 93, 98
Selenium 227, 229
semantic segmentation 270
implementation of 270–271, 271*f*
semi-structured data 293
SENet 279–280
sentiment analysis tasks 197
sequential model to design 276

sets 31
shallow algorithms 272–273
shared columns 74, 284
sigmoid activation function 274
sigmoid function 124, 125*f*
simple control structures 35–36
  conditional statements 37–38
  functions and methods 39–42
  loops 36–37
simple decision tree 127, 127*f*
singular value decomposition (SVD) 101, 106–112, 203
skepticism of legal arguments 260
skip connections 280
sklearn function 120
sklearn package 272
social actions 234–235
social context 263
social group, individuals of 234
social media 52
  messages 26, 295
  network sites 234
  platforms 212, 230
social network analysis (SNA) 233, 241–242
  centrality measures 246–248
  clustering and community detection 248–257
  paths and reachability 242–246
social networks:
  interesting descriptors of 245
  profiles of 270
social researchers 261–262
social sciences:
  network approach in 234
  textbooks 2
softmax activation function 274–275
sound classification 259
sparse matrix representation 159
Speech and Language Processing (Jurafsky and Martin) 182
Speech-to-Text recognition 260–262, 262*f*
SPSS 34, 55, 57
Spyder 231–232
SQL databases 283–285
  choosing 285–286
SQlite 286
  database functionality 287
Stackover Flow 47
standard color model 264
state-of-the-art analysis 295
statistical models 295
statsmodels 4
stopwords 61, 166
  inspecting and customizing 167
storing:
  single values 25–26
  text 26–27
strings 26, 56
  and bytes 26–27
  of characters, text as 142–143
  formatted 27
  Levenshtein distance 45
structured query language (SQL) 284
studies, types of 294
subfigures, creating 94
subgroups:
  analysis of 249–250
  identification of 249
summarization process 132
summary values 71–72
supervised machine learning (SML) 2, 113–115, 185, 187
  classification and regression problems 114
  concepts and principles 117–122, 118*t*, 121*f*
  statistical modeling and prediction 115–117
  validation and best practices 133–140
supervised text analysis 191
  deep learning 199–202
  finding best classifier 194–197
  using model 198–199
  workflow 191–194
support vector machines (SVM) 125–126
  advantage of 126
  classifier 126

t

tab completion 39
tabular data 56
tag cloud 16, 169
tagging 183

tags 218, 221
Tarde's theory of social imitation 1–2
technical (pre)processing 183
TensorFlow 275
term-document matrix (TDM) 156
term frequency 171
test dataset, 118*t* 120
text:
  advanced representation 172–173
    collocations 174–176
    linguistic preprocessing 177–182, 178*t*
    n-grams 173–174
    word embeddings 176–177
  automatic analysis of (*See* automatic analysis of text)
  bag of words and the term-document matrix 156–157, 159–163
  cleaning approaches 145
  computational analysis of 185–186
  preprocessing 182–183
  storing 26–27
  weighting and selecting documents and terms 164–165, 170–172
    removing punctuation and noise 167–170
    removing stopwords 165–167
text, processing 141–142
  example patterns 147–150
  methods for dealing with 144–145, 144*t*
  regular expressions 145–146
    syntax 146–147, 148*t*, 149*t*
  single and multiple 143
  as string of characters 142–143
textual data 142, 293
  topics from 115
tfcdotidf weighting 183
theory-driven techniques 2
third-party packages, installing 12
threshold value 134–135
"tidy" dataset 79
tidyverse environment 4
time-series analysis 78
tokenization 157–159, 165
  differences between 157
  of Japanese verse 158
  in Python 158–159
topic intrusion 209
topic models 16, 203, 207, 210
  goal of 203
  validating and inspecting 208–209
training dataset 118*t*
transfer learning 271*t*
transitivity 245
transparent 296
Treebank tokenizer 158
true negatives (TN) 120
true positives (TP) 120
tuples 31
Twitter/tweets 17, 165, 228, 235, 237, 254
  about COVID 14, 18
  API 254
  barplot of 15
  TOS 230
txt files 59

u

UDPipe 181
  natural language processing toolkit 178, 181
undirected graphs 248
  maximal cliques in 251
unicode 60, 142
unigrams 196, 197
Universal Approximation theorem 130
unsupervised machine learning (UML) 100–101, 114–115, 185
unsupervised text analysis 203
  beyond LDA 209–211
  fitting LDA model 206–207
  latent dirichlet allocation (LDA) 203–205, 204*f*
  validating and inspecting topic models 208–209
URLs 149, 223
user agent 222

v

validation 208–209
  set 276
values 56
variables 24
  analysis 84–85
vector 28–32, 164

elements of 30
graphics 265, 271*t*
objects 28, 29
operations on 28
slicing 29
vectorizer 195, 198
versions 51–52
VGGNet 279–280
videos 259–261, 261*f*
computational methods to 260
of social interest 259
virtual machine 288, 288*f*
master and slaves 289
running script 289, 289*f*
visual content, numerical representation of 268
visualization:
of circuit 243
of cluster/clustering 102, 105
data (*See* data visualization)
of graph 236
of network 236, 241
of weighted graph 239, 240
voice recognition 259

w
warning message 48
web APIs 213–216
crawling websites 223–224
dynamic web pages 225–227
retrieving and parsing HTML page 217–222
webpage 52
web scraping 227, 230
web sites, operators of 230
weighted graph 237
visualization of 239, 240
while loops 36–37
wide data 78–79
within-cluster sum of squares (WSS) 102
word cloud 163
word embeddings 176–177
word intrusion 209

x
Xception 279–280
XPath 219–220, 219*t*
advantage of 220
parsing websites using 220
patterns 220

y
Youtube 259

z
zip code pattern 150

Printed and bound by CPI Group (UK) Ltd, Croydon, CR0 4YY
18/05/2022
03124731-0001